Current Topics in Microbiology and Immunology

Volume 407

More information about this series at http://www.springer.com/series/82

Eric Hunter · Klaus Bister
Editors

Viruses, Genes, and Cancer

Responsible series editor: Richard W. Compans

Springer

Editors
Eric Hunter
Emory Vaccine Center
Emory University
Atlanta, GA
USA

Klaus Bister
Institute of Biochemistry and Center for Molecular Biosciences
University of Innsbruck
Innsbruck
Austria

ISSN 0070-217X ISSN 2196-9965 (electronic)
Current Topics in Microbiology and Immunology
ISBN 978-3-319-61803-6 ISBN 978-3-319-61804-3 (eBook)
DOI 10.1007/978-3-319-61804-3

Printed on acid-free paper

This Springer imprint is published by Springer Nature
The registered company is Springer International Publishing AG
The registered company address is: Gewerbestrasse 11, 6330 Cham, Switzerland

Preface

In this special volume of *Current Topics in Microbiology and Immunology (CTMI)*, the scientific chapters are focused on current research developments in the fields to which Peter Vogt made preeminent contributions: viruses, genes, and cancer. Several chapters specifically highlight virus-host interactions, the role of infectious agents in human cancer, or HIV-host interactions relevant to pathogenesis and cure. Other chapters review the pivotal role of oncogenes and tumor suppressor genes as major cancer drivers—such as *MYC*, *RAF*, *PI3K*, or *TP53*—or explore the emerging role of microRNAs in tumorigenesis and cancer therapeutics.

Peter Vogt was born on March 10, 1932, in Broumov, a town with a large German-speaking population at that time, located in a region of the former state of Czechoslovakia that is now part of the Czech Republic. In 1950, Peter crossed the border from East to West Germany and moved to the City of Würzburg, where he received a B.S. in biology from the University of Würzburg in 1955. It was in this city that he also took classes with the painter Josef Versl. In 1955, he joined the Max-Planck-Institute for Virus Research (now: MPI for Developmental Biology) in Tübingen for graduate studies, and obtained a Ph.D. degree from the University of Tübingen in 1959. Peter then moved to the United States, to work as a Damon Runyon Cancer Research Fellow in the laboratory of Harry Rubin at the University of California in Berkeley from 1959 through 1962. Peter was Assistant and Associate Professor of Pathology at the University of Colorado School of Medicine at Denver from 1962 to 1967, and Associate Professor and Professor of Microbiology at the University of Washington School of Medicine in Seattle from 1967 to 1971. He then moved to Los Angeles to become Hastings Distinguished Professor of Microbiology and Chairman of the Department of Microbiology at the University of Southern California School of Medicine. Peter has been a Professor in the Department of Molecular and Experimental Medicine (now: Department of Molecular Medicine), The Scripps Research Institute, in La Jolla since 1993, serving as Executive Vice President for Scientific Affairs at this institution from 2012 through 2015. He is a member of several prestigious academies, including the US National Academy of Sciences and the German National Academy of Sciences Leopoldina. He has received numerous awards and prizes, including the Ernst Jung

Prize for Medicine, the Paul-Ehrlich and Ludwig-Darmstaedter Prize, the Pezcoller Foundation-AACR International Award for Cancer Research, and the IHV Lifetime Achievement Award for Scientific Contributions.

Throughout his scientific career, Peter has continued to make outstanding contributions to virology and cancer research. In his early work, he studied mechanisms of virus-host interactions, specifically avian retroviral cell entry, leading to a comprehensive definition of host range determinants. His focus then shifted to the genetics of retroviruses and the mechanisms of virus-induced cell transformation. This work culminated in the determination of the genetic map of Rous sarcoma virus (RSV), in the isolation of temperature-sensitive mutants of RSV, and in the first physical identification of a cancer gene, the oncogenic principle (v-*src*) of RSV, reported together with Peter Duesberg from the University of California at Berkeley in a seminal 1970 *PNAS* paper. The search for the origin of v-*src* led to the landmark discovery of the cellular origin of retroviral oncogenes by Harold Varmus and Mike Bishop from the University of California at San Francisco, in cooperation with Peter, recalled in the accompanying essay. Work in Peter's lab led to the identification of several retroviral oncogenes whose human cellular homologs (proto-oncogenes) are now recognized as major cancer driver genes, including *MYC*, *JUN*, and *PI3K*. Recent work in Peter's lab is focused on human cancer genetics, including the definition of tumor-specific mutations in cancer driver genes. Peter's research is also aimed at the isolation of inhibitors of oncogene protein products, such as MYC, eventually leading to the development of drugs suitable for pharmacological cancer treatment.

We are honored to act as volume editors for this very special issue of *CTMI* in appreciation of Peter Vogt, an eminent virologist and cancer researcher, and long-time editor of this series. Both of us had the privilege to work as post-doctoral fellows in Peter's lab in Los Angeles in the nineteen seventies, and we actually overlapped for some time. We vividly recall the exciting spirit, the collaborative atmosphere, and the scientific rigor of the lab at that time. Peter led by example and was at the microscope most mornings reviewing the results of experiments he had planned for that week. Lab meetings held at his home up on the Pasadena Hills were a time for stimulating scientific discussions, great food prepared by Peter, and an introduction to wonderful German wines! We also gladly remember the informal but tremendously stimulating West Coast Meetings of the Vogt, Bishop/Varmus, and Duesberg labs, held alternately in L.A. or San Francisco. Having worked with Peter was formative for our scientific careers, and we are very grateful. He has been a wonderful mentor and friend over the years. We also thank all authors of this volume for their great contributions.

Atlanta, USA Eric Hunter
Innsbruck, Austria Klaus Bister

Peter K. Vogt

A Brief Homage to Peter Vogt

This volume celebrates Peter Vogt's remarkable service as an editor of *Current Topics in Microbiology and Immunology*. An astounding feature of his tenure is its length. We have known Peter well for a very long time—over forty-five years—yet he was already half a decade into his editorial duties at *CTMI* by the time we were introduced to him in the early 1970s.

That introduction came about in the best possible way: through mutual interests in combining different kinds of information and methods to do experiments that would probably otherwise not be done. The two of us and our colleague Warren Levinson first wrote to Peter in 1971, in a rather formal letter on UCSF letterhead, to ask for biological materials that might allow us to understand some peculiar features of the reassociation kinetics of the in vitro products of reverse transcription by Rous sarcoma virus (RSV). About the same time, two of Peter's senior post-doctoral fellows, Robin Weiss and Robert Friis, had written to ask if we'd be willing to use molecular probes to seek endogenous RSV-related proviruses in DNA from unusual birds.

From these simple (and initially not particularly productive) beginnings, our research team in San Francisco developed a long-standing relationship with Peter and his trainees (who had just moved from Seattle to Los Angeles)—a relationship that proved to be essential to much of our own later success and was from the start both scientifically stimulating and socially enjoyable. At the time, it was also recognized as unusual. Now we can see that it was an important forerunner of the kinds of "team science" and "multi-disciplinary" projects that are widely embraced today.

What was at first an occasional meeting between Peter's laboratory at the University of Southern California and ours at UCSF expanded gradually over the next decade to include many other RNA tumor virologists and their students and fellows—at first, from UC Berkeley; later from Cal Tech, UC Irvine, the Salk Institute, UC San Diego, and occasionally the Fred Hutchison Cancer Center. Regionally constrained by the lack of today's digital technology and dependent on a willingness to travel regularly to California campuses (fostered by miniscule fares on the slightly louche and long-since deceased airline PSA), this loose consortium, now remembered as the West Coast Tumor Virus Cooperative, evolved over the ensuing decade to think together about virtually any aspect of RNA tumor virology. Focused initially on RSV, the topics eventually spanned a wide variety of retroviruses and exploited experimental tools that ranged from biological and genetic assays in cell culture to electron microscopy and molecular dissection of viral genomes. The discussions were unusual, even at that time, and particularly useful because the participants were generally willing, even eager, to share raw data not yet in manuscript form.

Peter provided the intellectual cornerstone of these meetings, even as the subjects and methods diversified, because he was the person most securely grounded in the foundations of our field. As emphasized elsewhere in this volume (in a brief biography by Klaus Bister and Eric Hunter), Peter grew up intellectually at a time—in part, during his training with Harry Rubin in Berkeley; in part, through his own labors in Seattle—when quantitative biological assays for replication and transformation by RSV and its cousins were being developed. These fundamental methods made the genetics and the molecular biology of tumor viruses not just possible but meaningful. Those of us who trained by studying other problems—and had an orientation that ran the risk of valuing the molecular over the biological—came to depend upon Peter's familiarity with the history of retroviruses and their biological properties in the design of our own work. Such lore is essential for experimental success, and it can be learned more easily from generous practitioners than from published work. For this reason, some of the moments we most vividly remember from the meetings of our research collective occurred during Peter's expositions on the genealogy, phenotypic effects, and genetics of RSV. His ability to draw simple messages from a decade or two of papers peppered with arcane terminology encouraged clear thinking and inspired all of us to conceive more meaningful experiments.

Meetings of the West Coast Tumor Virus Cooperative were occasions devoted to two principal purposes: to exchange new findings with other members and to discuss collaborative experiments. Our relationship with Peter and his group began with some immature ideas about joint projects, and the ideas deepened and proliferated over the ensuing decade. The best-remembered parts of our work together will inevitably be those that led to what is generally viewed as the most important outcome: the discovery of the first proto-oncogene, c-*src*. A fundamental feature of those experiments was Peter's provision of the ideal genetic reagents for them: clones of wild type RSV, fully competent for replication and transformation, from the Bratislava 77 strain and non-conditional, transformation-defective deletion mutants derived from those clones. These mutant/wild type pairs allowed our group to prepare molecular probes that proved to be nearly perfect representations of the viral transforming gene, v-*src*. Thus they were ideal for seeking the gene's cellular progenitor. Furthermore, in the era that preceded molecular cloning, they were essential for experimental success. There was never any doubt that Peter would be a coauthor of the publication that announced the discovery of c-*src*.

In the years before the discovery of c-*src*, Peter had also taught us about the value of working with birds other than chickens as hosts for replication of RSV—especially ducks and quail, from which useful cells and DNA could be abundantly prepared and assayed without any confusion with the RSV-related proviruses found endogenously in most chickens. Thinking about those avian species, and yet other birds and other animals, prompted our early studies of the conservation of the c-*src* gene, efforts that were critical to interpretation of our first findings at a time when molecular evolution was still in its infancy. Peter's work with cells from various species, including rodents, as hosts for RSV also allowed us to study important steps in the RSV life cycle—especially DNA synthesis during infection, proviral integration, several aspects of gene expression, and virus entry—revealing the influences of host cells on viral gene expression and virus production.

We look back with pleasure and take pride in the dozen or so papers that Peter co-authored with one or (more typically) both of us between 1972 and the mid-1980's, a collaborative effort that has been only rarely equaled for its persistence and productivity. Still, as is often true of collaborations that require investigators to overcome the barriers that separate different institutions and places, we eventually went in different directions scientifically.

But our affection and admiration for Peter have been lasting, in part because his laser-like focus on research belies a broader cast of mind, and our relationships have come to depend as much on cultural as on scientific commonalities of interest. He is an accomplished water colorist and pains-taking student of the visual arts, to the point of using opera glasses to examine the fine detail of brush work in paintings; one can only imagine how he views the current mania for accumulating smart-phone photographs of museum displays without so much as a glance at the real thing. As his career began to take him to far-flung places, he fell head-over-heels for the culture of Japan—its people (one of whom he eventually married), traditions, cuisine, arts, and crafts. His pursuit of this passion has been characterized by the depth of scholarship, meticulous attention to detail and appreciation of history that have distinguished his career as a scientist.

Peter's celebratory watercolors inscribed to us and dispatched to our homes on October 9th, 1989 (see illustrations) remain among the most generous and meaningful tributes that we received after our Nobel Prize was announced on that day. So it is gratifying for us to be able to reciprocate now by offering this belated (and less colorful!) tribute to Peter on the occasion of the golden anniversary of his dedication to scholarship as an editor of *CTMI*.

San Francisco, USA J. Michael Bishop
New York, USA Harold E. Varmus

Contents

Exchange of Genetic Sequences Between Viruses and Hosts

Robin A. Weiss

Abstract Although genetic transfer between viruses and vertebrate hosts occurs less frequently than gene flow between bacteriophages and prokaryotes, it is extensive and has affected the evolution of both parties. With retroviruses, the integration of proviral DNA into chromosomal DNA can result in the activation of adjacent host gene expression and in the transduction of host transcripts into retroviral genomes as oncogenes. Yet in contrast to lysogenic phage, there is little evidence that viral oncogenes persist in a chain of natural transmission or that retroviral transduction is a significant driver of the horizontal spread of host genes. Conversely, integration of proviruses into the host germ line has generated endogenous retroviral genomes (ERV) in all vertebrate genomes sequenced to date. Some of these genomes retain potential infectivity and upon reactivation may transmit to other host species. During mammalian evolution, sequences of retroviral origin have been repurposed to serve host functions, such as the viral envelope glycoproteins crucial to the development of the placenta. Beyond retroviruses, DNA viruses with complex genomes have acquired numerous genes of host origin which influence replication, pathogenesis and immune evasion, while host species have accumulated germline sequences of both DNA and RNA viruses. A codicil is added on lateral transmission of cancer cells between hosts and on migration of host mitochondria into cancer cells.

R.A. Weiss (✉)
Division of Infection and Immunity, University College London,
Gower Street, London WC1E 6BT, UK
e-mail: r.weiss@ucl.ac.uk

Current Topics in Microbiology and Immunology (2017) 407:1–30
DOI 10.1007/82_2017_21

Published Online: 28 May 2017

Contents

1 A Year of Virological Anniversaries

In addition to Peter Vogt's 50 years on the Editorial Board of this journal, 2017 is a year to acknowledge several remarkable anniversaries in virology and mobile genetic elements. Exactly 300 years ago, the Western world was introduced to immunization against smallpox by Lady Mary Wortley Montagu and 40 years ago, total eradication of this scourge was achieved (Weiss and Esparza 2015). One hundred years ago, Félix d'Hérelle isolated and propagated bacteriophage (d'Herelle 1917). Fifty years ago, Lynn Margulis (Sagan 1967) postulated that eukaryotic cells evolved by inclusion into the host cell of a number of previously independent microbes that became self-replicating endosymbionts which I shall discuss at the end of this article. This year also marks a personal 50th anniversary since I published my first virological paper (Weiss 1967) with a finding that was simultaneously observed by Peter Vogt (1967), and which led to the discovery of endogenous retroviruses. Three years later, I joined Peter Vogt's laboratory for 21 months which I regard as the most formative period in my career. Thus it is a great privilege for me to contribute to this volume honoring Peter's immense contributions.

1.1 Bacteriophage and Prokaryote Transduction

Bacteriophage lysis was first observed in 1915 by the British microbiologist (Twort 1915) as a filterable anti-bacterial factor. Twort's findings were not conclusive and the Nobel laureate, Jules Bordet, considered that the lytic effect was due to an enzyme rather than a transmissible agent (Bordet and Ciuca 1922; Summers 1999). However, while investigating dysentery among the allied troops, the French-Canadian microbiologist at Institut Pasteur, Félix d'Hérelle, observed lysis of *Shigella dysenteriae* by a replicating filterable agent (d'Herelle 1917). Whether d'Hérelle was aware of Twort's report is not known but he did not cite it. D'Hérelle was the first to establish multi-passage phage cultures and to quantitatively measure the extraordinary high titers of infection. He coined the terms *bacteriophage* for the agent and *plaque* for the clear areas of lysis in the bacterial films on the sloping agar plates of his cultures. He was convinced that bacteriophage could be used to cure sepsis and exaggerated his success in so doing (Summers 1999). Owing to the difficulties other investigators had in attempts to treat infection (not least because the host bacteria tended to become resistant), the use of bacteriophage to treat infections fell out of fashion in the West, although it continued to be promoted in the Soviet Union. Today, phage therapy is experiencing a renaissance (Cisek et al. 2017).

D'Hérelle's discovery 100 years ago eventually opened the field of molecular genetics following World War II, led by Max Delbrück and called the Phage Group (Cairns et al. 2007; Summers 1999). Lysogeny was not a particular focus of the Phage Group but was later investigated at the Institute Pasteur led by André Lwoff, whose discovery of transduction of host genes mediated by lysogenic phage (Lwoff 1953) not only became a useful tool in molecular genetics and gene regulation (Ptashne 2004), but together with plasmid transfer was also revealed to be a major means of exchange of host sequences among prokaryotes (Lane 2015). In the 40 years since the discovery of the *Archaea* (Woese and Fox 1977), the monophyletic origin of life forms on this planet have been confirmed for the evolutionary roots of prokaryotes, but the enormous extent of horizontal exchange confounds attempts to draw accurate phylogenetic trees with linear pedigrees within the major realms of *Archea* and *Bacteria* (Choi and Kim 2007; Martin 2011; Lane 2015).

Phage studies continue to be of great relevance to modern molecular biology. They serve as models for evolutionary theory and real-time experiments (Stern and Sorek 2011; Refardt et al. 2013). The discovery of DNA restriction enzymes came out of host range restriction of phage (Loenen et al. 2014), and the CRSPR-Cas9 mechanism of gene editing also has its origins in the natural control of invading phage by bacterial cells (Hartmann 2017).

1.2 The Debt of Tumor Virus Research to Phage Genetics

The linear pedigree of tumor virus research is much clearer than that of prokaryote evolution referred to above! At Caltech in the 1950s, Renato Dulbecco adapted the bacterial lawn culture practices of the Phage Group to the recently established technique of animal cell monolayers. He developed plaque assays for lytic viruses (Dulbecco 1952) and transformation assays for DNA tumor viruses (Vogt and Dulbecco 1960). His associate, Harry Rubin investigated Rous sarcoma virus (RSV) research and with his student, Howard Temin, developed a quantitative transformation assay for RSV (Temin and Rubin 1958). After Rubin moved to Berkeley, Peter Vogt joined his laboratory and detected a non-transforming, replication-competent Rous associated virus (RAV) in stocks of RSV (Rubin and Vogt 1962).

Temin (1960) observed variants of Rous transformed cells which had an elongated, fusiform morphology which bred true according to the variant of RSV used to infect them (Fig. 1) and were later shown to carry mutations within the C-terminus of the *src* gene (Rohrschneider and Reynolds 1985). This finding led Temin (1962) further noted a distinction between cell transformation and viral replication. He proposed that there was genetic information carried by the virus which persisted in the host cells and affected their phenotype, which he compared to lysogenic prophage (Sankaran 2014). In the same year, Jan Svoboda (who sadly died earlier this year) observed that rats non-productively inoculated at birth with RSV still contained the virus in adult tumors which lent support to the persistence of a latent genome (Svoboda 1960). One year later, however, Crawford and Crawford (1961) demonstrated that the genome of RSV was composed of RNA, not DNA. Temin went on to postulate his provirus hypothesis, proposing that RSV

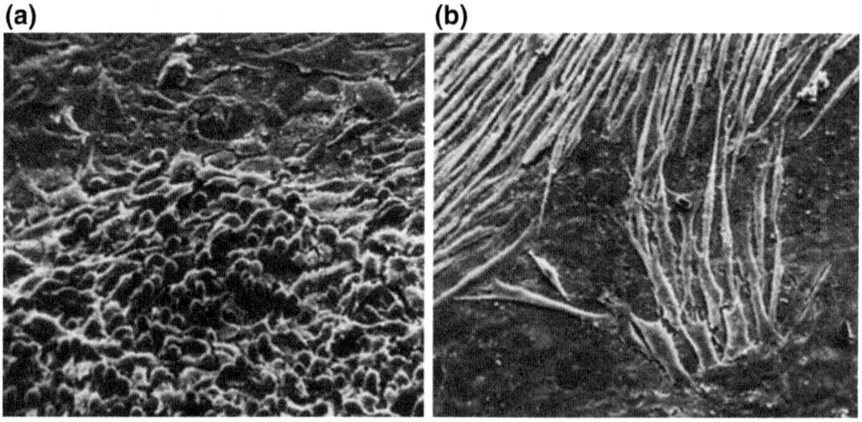

Fig. 1 Scanning electron micrographs of cells transformed by Rous sarcoma virus. **a** Wild-type round morphology cells; **b** Mutant fusiform cells (reproduced from Weiss RA, 1969 PhD Thesis, University of London)

forms a DNA copy of the genome in the infected cell which integrates into the host genome (Temin 1964). Although indirect evidence supported the provirus hypothesis, it remained speculative until the discovery of reverse transcriptase by Temin and Mizutani (1970) and Baltimore (1970).

1.3 Fifty-Five Years of Peter Vogt's Contributions to Retroviruses and Oncogenes

Although Peyton Rous discovered his eponymous virus over 100 years ago (Rous 1911) and the transforming function could be titrated on the chorioallantoic membrane of chick eggs (Keogh 1938), it was not until the quantitative virus assay in vitro was established by Temin and Rubin (1958) that RNA tumor research really blossomed (Weiss and Vogt 2011). Peter's first paper on RSV showed that the Bryan high titer strain of RSV virus used at that time in the USA actually comprised two components, a virus that transformed chick embryo fibroblasts and also an avian leukosis virus (Rubin and Vogt 1962). This finding led directly to the discovery that this strain of RSV was defective for replication, missing the gene *env* encoding the envelope glycoproteins (Hanafusa et al. 1963). These glycoproteins were provided by the accompanying leukosis virus or 'helper' virus (Hanafusa et al. 1963) which determined the specificity of antibody neutralization (Ishizaki and Vogt 1966) and of host range through the use of different cell surface receptors (Vogt and Ishizaki 1965). However, strains of RSV used in European laboratories were non-defective, possessing the *src* gene in addition to replication genes. The B77, Carr-Zilber, Prague, and Schmidt-Ruppin strains were derived from RSV that had been used to induce non-productive tumors in rats and rescued by inoculation into chickens or co-cultivation with permissive chicken cells (Svoboda 1966). The genome structures (Vogt and Hu 1977) were later depicted as:

Replication-competent leukosis virus:	LTR-*gag-pol-env*-LTR
Replication-defective RSV:	LTR-*gag-pol-src*-LTR
Replication-competent RSV:	LTR-*gag-pol-env-src*-LTR

From these seminal studies, the distinction between replication genes and oncogenes came to light (Toyoshima and Vogt 1969; Duesberg and Vogt 1970; Martin 1970) culminating in the demonstration of the host origin of oncogenes (Stehelin et al. 1976), and the identification of the Src protein (Brugge and Erikson 1977) as the first example of a tyrosine kinase (Hunter and Sefton 1980). Other retroviral oncogenes similarly have a cellular origin. This fascinating story has been reviewed by Steven Martin (2004), Peter Vogt (2012), Klaus Bister (2015), and the recent memoir on tumor viruses and cancer cell biology by Harold Varmus (2017).

In the same year that Peter Vogt was appointed to the Editorial Board of CTMI, I published my first virological paper on the release of infectious RSV particles possessing novel envelope properties in the apparent absence of a helper virus

(Weiss 1967). To my surprise but gratification, a similar paper was published by Peter a month later Vogt (1967). I was still a doctoral student of Michael Abercombie who discovered contact inhibition but was not a virologist, and my mentor in Abercrombie's laboratory, Warren Levinson, had returned to San Francisco. I felt rather nervous that my findings seemed to contradict what was known about the defectiveness of RSV and Peter's paper helped me to gain confidence in my own findings. Moreover, Peter's was the more elegant study because he showed that the virus which he called RSV(0) more readily infected Japanese quail cells rather than chick cells. It represented the first demonstration of what later became known for murine leukemia viruses as xenotropism (Levy 1978). Our 1967 papers provided the first step in evidence that led to the elucidation of endogenous retroviruses, a story I have told elsewhere (Weiss 2006).

2 Acquisition of Host Genes by Viruses

2.1 Oncogenesis by Simple Retroviruses and Transduction of Oncogenes

Leukemogenic retroviruses lacking oncogenes integrate at near-random sites mainly in 'open' regions of host chromosomal DNA. Among the millions of cells infected, some proviruses integrate next to host proto-oncogenes and activate their expression through promoter and enhancer sequences in the retroviral long terminal repeat. The first to be defined was c-*myc* in avian lymphoid leukosis (Hayward et al. 1981). The transduction of cellular oncogenes begins in this way (Vogt 2012; Varmus 2017).

With avian lymphoid leukosis, activation of c-*myc* appears to be sufficient for leukemogenesis of B cells in the bursa of Fabricius. However, with gamma-retroviruses such as murine and feline leukemia viruses, the emergence of transformed leukemic cells requires more complex recombination events between infectiously transmitted retroviruses and related endogenous genomes before the recombinant retrovirus activates cellular oncogenes (Rosenberg and Jolicoer 1997). For example, in cats, feline leukaemia virus subtype A (FeLV-A) is the major transmissible agent, but in most cats that develops leukemia, recombination with an endogenous retroviral genome encoding a subtype B *env* must occur (Roy-Burman 1995). Multiple interactions occur between exogenous and endogenous LTRs and *env* regions giving rise to variants with different pathogenic attributes (Neil et al. 1991; Bolin and Levy 2011; Stewart et al. 2011).

The various types of retroviral oncogenes show no structural similarities and they act at many sites in cell signaling and regulation, including the transcriptional regulators *myc* and *jun* studied by Peter Vogt (Stefan and Bister 2017). Thus the transduction of oncogenes by retroviruses does not depend on sequence homology but rather on read-through transcription and packaging of RNA transcripts into virus particles, followed by reverse transcription and integration into the next infected cell via normal replicative events (Vogt 2012). Apart from the European strains of RSV passed through mammals referred to above, oncogene-bearing retroviruses appear to be replication-defective and have been rescued through aberrant splicing. All that is required in the transcript is for the virion packaging signal the 5′ end of *gag* to be contiguous with the host sequence.

There appears to be little selective advantage to the virus to carry oncogenes, other than that tumor cells tend to be more permissive to retrovirus replication. In fact, there is scant evidence that *onc*-bearing viruses are naturally transmitted from host to host unless they come to the attention of pathologists and virologists who deliberately propagate them. The exception is the *cyclin* oncogene of replication-competent epsilon-retroviruses in fish, but it is probably not of host origin (Rovnak and Quackenbush 2010). Oncogene transduction appears to be restricted to the 'simple' retroviruses, alpha- and gamma-retroviruses. It remains a puzzle why cellular gene incorporation into and beta-, delta-, spuma- and lenti-viruses has not been found to date.

Host oncogenes have been transduced by retroviruses in many species including primates (Thielen et al. 1971). Outbred animals such as cats and chickens as well as inbred strains of mice give rise to *onc*-bearing viruses. Towards the end of my period in Peter Vogt's laboratory, I spent two weeks at a chicken abattoir harvesting solid tumors that the veterinary inspectors identified along the processing line. Approximately one tumor per 400 chickens was found, a high number considering that they were broiler fowl less than six months old. My time with Peter came to an end before I was able to establish what proportion of the tumor-bearing chickens were infected with avian leukosis and whether the tumors had acquired transduced oncogenes. Indeed, there is little quantitative evidence to this day on oncogene transduction in any animal species although Miles and Robinson (1985) detected frequent transduction of the c-*erb*B oncogene in chicks with erythroblastosis following experimental inoculation of RAV-1. Further novel oncogenes came to light in avian sarcomas, not least Peter's discovery of *jun* (Cavalieri et al. 1985; Maki et al. 1987), a key player in transcriptional regulation (Vogt 2002). Overall, the study of retroviral oncogenes had an immense impact because so many oncogenes found in human cancer were first identified in animal retroviruses (Varmus 2017).

2.2 *Why Is Retroviral Transduction not a Major Driver of Virus or Host Evolution?*

It is clear that retroviruses are equipped both to transduce host genes and to infect the germ lines of the same and foreign species. It is therefore germane to ask why retroviruses have not become generalized transducing agents by horizontal transfer to new hosts. Since alpha- and gamma-retroviruses readily acquire diverse oncogenes, they must surely pick up other gene transcripts even though they will not be detected through the clonal expansion of tumor cells. If lysogenic bacteriophages are a driving force in prokaryotic evolution, why don't retroviruses serve a similar function in vertebrate evolution?

The first consideration is that almost all oncogene-bearing retroviruses of birds and mammals are replication-defective and therefore need helper viruses to continue successive rounds of replication. However, the defective retroviruses should be able to propagate in vivo and spread to new hosts if the concentration of helper virus is sufficient. After all, infection by hepatitis delta virus and adeno-associated viruses rely on helper viruses, and plant partitiviruses with split genomes packaged in different particles are successful pathogens. Even if defective *onc*-bearing retroviruses do not spread infectiously through a host population, it would need only one round of integration into a germ cell for hereditary transduction to be accomplished. Moreover, if there were strong selective forces benefitting the host for transduction, replication-competent transducing viruses would probably emerge, like replication-competent strains of RSV.

A second point is that if transduction occurred only within the same host species, there would be little selective pressure for horizontal transmission because it would not present an advantageous alternative genetic exchange to sexual reproduction. But the widespread phenomenon of xenotropism and infection of unrelated germ-lines discussed in Sect. 3.3 below shows that retroviruses could potentially be highly successful in effecting lateral transfer of host sequences.

A third, a more serious limitation, is that retroviral transduction involves reverse transcription of spliced or partially spliced RNA transcripts. Therefore sequences introduced into a new host would represent pseudogenes, as mediated by retrotransposons. There may be difficulty in controlling expression of the transduced gene if it lacked regulatory sequences and micro-RNA sequences normally present in the intron (Roy and Gilbert 2006). However, introns can be gained during evolution (Yenerall and Zhou 2012) so that this hurdle is not insurmountable.

It is possible that we have not looked in sufficient depth for evidence of horizontal gene transfer by retroviruses. If it does occur between distantly related hosts one would expect whole genome sequence analysis to reveal genes or pseudogenes with greater homology to the species whence the retrovirus came than to pre-existing homologs in its adopted host.

2.3 Incorporation of Host Genetic Sequences into Complex Viruses

When retroviral oncogenes were first shown to be derived from host genes (Stehelin et al. 1976), the gene content of viruses with large, complex genomes was not known. With the advent of whole genome sequencing and bioinformatics, it became possible to interrogate the genomes of large DNA viruses for sequences that have been hi-jacked from ancient and recent hosts and repurposed (exapted) to fine tune the virus's life style, whether it be replication, latency or immune evasion.

Krupovic and Koonin (2017) argue that such basic components of viruses as capsid proteins originally evolved from host cells, but here I wish to examine acquisitions from the host that were imported into virus genomes long after viruses became established as replicating entities.

Herpesviruses, Poxviruses and Polydnaviruses of insects carry numerous genes originally derived from their hosts. Throughout virus-host evolution, viruses possessing a substantial packaging capacity have incorporated host genes and modified them to serve viral functions (Haig 2001). Although poxviruses replicate in the cytoplasm, that has not prevented them adopting multiple host genes (Bugert and Darai 2000; Odom et al. 2009; Austin et al. 2010). Trafficking of DNA between nucleus and cytoplasm is mediated by Importin 7 (Danoya et al. 2013), which may aid genetic exchange between virus and host. Some of the viral genes have introns and probably represent the horizontal transfer of host DNA rather than RNA transcripts that have been reverse transcribed.

The extent of host gene 'invasion' of large DNA viruses can be seen by comparing the differences between viruses of a single group that separated millions of years ago. For example, the two human gamma-herpesviruses, Epstein-Barr virus and Kaposi's sarcoma virus, have homologies among their core replication genes but extensive differences in the genes acquired from the host. Comparative genome analysis across the various human herpesviruses reveals multiple host contributions during virus evolution (Holzerlandt et al. 2002) as depicted in Fig. 2. Immunomodulatory genes are essential to the natural history of viruses in vivo, but tend to be lost during extensive passage in vitro, as seen for standard laboratory strains of Cytomegalovirus (Cunningham et al. 2010) and in the evolution of modified Vaccinia Ankara from wild type Vaccinia virus (Volz and Sutter 2017). Several of the acquired genes of oncogenic herpesviruses may play a role in oncogenesis, such as the chemokine homologs (v-*mip*) that we studied in Kaposi's sarcoma herpesvirus (Boshoff et al. 1997).

MicroRNA sequences are present in some viruses (Guo and Steitz 2014) and their precursors presumably were derived from cellular sequences. Viruses also alter host miRNA functions to their own advantage. For instance, Epstein-Barr virus EBNA3A and EBNA3C proteins affect regulation of c-*myc* in B-cells by inducing miR-221/miR-222 expression in transformed B-cells (Bazot et al. 2015).

The remarkable evolution of polydnaviruses in parasitic wasps further exemplifies the two-way genetic exchange between virus and host (Herniou et al. 2013). These large DNA viruses become endogenous in their hosts (Strand and Burke 2014) and probably act as endosymbionts (Herniou et al. 2013). The genetic exchange between amoebae and their intracellular viral and bacterial parasites (or endosymbionts) is also extraordinary. The 'giant' Mimiviruses of amoebae contain numerous genes of host origin as well as being themselves prey to smaller viruses, the so-called virophages (Colson et al. 2017). Perhaps there is something special about amoebae because Legionella species (which are intracellular parasites of amoebae as well as occasionally colonizing alveolar macrophages in humans) also acquire hundreds of host genes (Gomez-Valero et al. 2013; Burstein et al. 2016). Conversely, some hosts can acquire whole or partial viral and bacterial genomes,

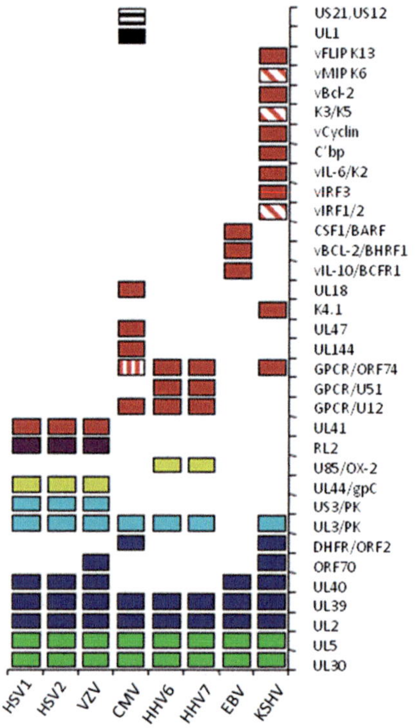

Fig. 2 Genes in human herpesviruses with human homologs. Some genes are common to all types of human herpesvirus whereas others have been acquired individually. Bars are color coded according to functional class: *green*, DNA replication; *dark blue*, nucleotide repair/metabolism; *light blue*, enzyme; *purple*, gene expression regulation; *yellow*, glycoprotein; *red*, host-virus interaction; and *black*, unknown. Diagonal lines within a box indicate two gene copies (per viral genome); *vertical lines*, three copies; and *horizontal lines*, up to 10 copies (adapted with permission from Holzerlandt et al. 2002)

such as the transfer of endosymbiont Wolbachia to insect and nematode nuclear genomes (Dunning Hotopp et al. 2007).

3 Acquisition of Viral Genes by Hosts

3.1 Viral Genomes in Host DNA: Retroviruses, DNA Viruses and RNA Viruses

Whole genome sequencing has led us to appreciate that every vertebrate species examined contains multiple endogenous retroviral genomes (ERV) (Stoye 2012; Hayward et al. 2015; Magiorkinis et al. 2015). As displayed in Fig. 3, approximately 8% of human DNA is derived from germ-line infection by retroviruses

(Griffiths 2001) and a larger proportion of our genome is represented by retrotransposons such as LINE elements (Cordaux and Batzer 2009; Rebollo et al. 2012). Rather than being relics of an earlier RNA world converted into genes, one may view ERV as 'fossils' of highly evolved, sophisticated viruses because they have the hallmarks of having been acquired by infection (Belshaw et al. 2004).

With the detection of endogenous spumavirus genomes in multiple hosts (Ruboyianes and Worobey 2016) and the discovery of an endogenous single copy delta-retrovirus genome in bats (Farkašová et al. 2017), it is now apparent that all seven genera of retroviruses have endogenous counterparts. The defective delta-retrovirus is distantly related to human T-cell leukemia virus and bovine leukosis virus and probably diverged over 25 million years ago, while the foamy viruses entered the germ-line even earlier.

Some retroviruses are promiscuous in integration so long as DNA is on hand. In addition to insertion into the host chromosomal genome, avian reticuloendotheliosis virus (REV) also integrates into the genomes of DNA viruses co-infecting the same chickens and turkeys, namely Marek's disease herpesvirus (Isfort et al. 1992) and fowlpox virus (Hertig et al. 1997). REV is a gamma-retrovirus of mammalian origin with related viruses endogenous in mongoose species. It has invaded birds and avian DNA viruses in the recent evolutionary past (Etienne and Emerman 2013). Niewiadomska and Gifford (2013) argue REV was introduced iatrogenically from mammals to chickens and turkeys through administration of live attenuated viral vaccines. Although 25 years have passed since the initial report of REV integration in a DNA viral genome (Isfort et al. 1992), I am not aware of integration of other retroviruses into DNA viral genomes, so this may be a special property of REV.

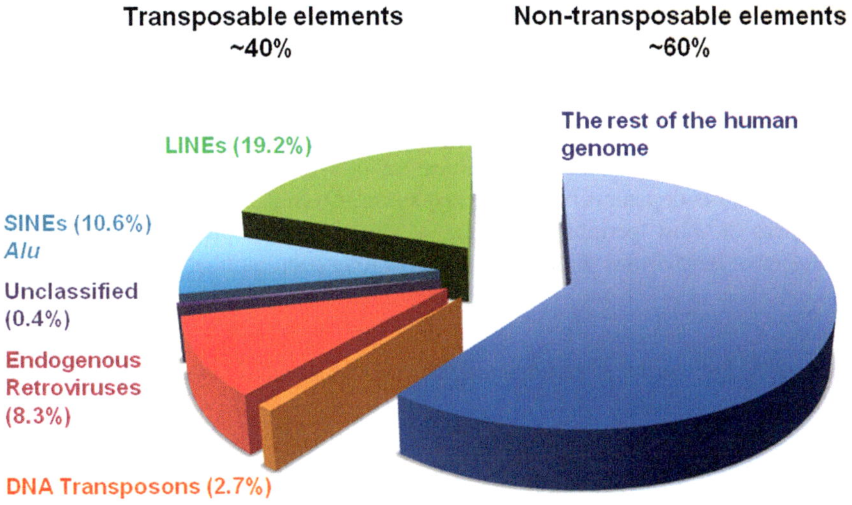

Fig. 3 Proportion of transposable elements in the human genome. *Lines* and *Sines* are retrotransposons (adapted with permission from Cordaux and Batzer 2009)

Retroviruses are not the only viruses to invade the host genome. It has become evident that some DNA herpesvrisues and hepadnaviruses endogenize in host genomes. Mendelian inheritance of the beta-herpesvirus, human herpes virus 6 (HHV-6) was first reported by Daibata et al. (1999). Chromosomally integrated HHV-6 is found in about 1% of the human Caucasian population (Pellet et al. 2012; Hill et al. 2016) and debate continues whether these endogenous genomes are linked to disease. An endogenous gamma-herpesvirus has similarly been detected in the prosimian tarsier (Aswad and Katzourakis 2014). Avian hepadnaviruses such as duck hepatitis virus are rapidly evolving DNA viruses with a reverse transcription step in their life cycle. Gilbert and Feschotte (2010) detected an endogenous hepadnavirus in the zebra finch which appears to have colonized the passerine germline nearly 20 million years ago. Beyond viruses, there is evidence of bacterial genomes becoming incorporated into the genomes of multicellular eukaryotes (Dunning Hotopp et al. 2007).

The presence of chromosomally integrated DNA that is complementary to gene sequences of RNA viruses was first reported by Zhdanov (1975) for measles virus and then forgotten for 20 years until recalled (Weiss and Kellam 1997) when Klenerman et al. (1997) detected cDNA of hepatitis C virus, soon followed by the discovery of bornavirus and filovirus sequences (Belyi et al. 2010; Horie et al. 2010; Taylor et al. 2010). The reverse transcription and integration is linked to LINE retroposon elements (Belyi et al. 2010). In an experimental study of vesicular stomatitis virus infection in vitro, the synthesis and integration of cDNA was dependent on expression of LINE-1 (Shimizu et al. 2014).

Germline inserts of RNA virus genome fragments were first detected for bornaviruses (Belyi et al. 2010; Horie et al. 2010) and filoviruses related to Marburg and Ebola viruses (Taylor et al. 2010). The ability to search whole animal and human genomes for homologs of viruses has revealed complementary DNA sequences related to RNA viruses in the germline (Ferschotte and Gilbert 2012). Bornavirus cDNA is widely dispersed in animal genomes (Horie et al. 2013). Whether they play a functional role for the host is not known, but endogenous bornavirus sequences are expressed in human tissues and they may inhibit exogenous infection (Honda and Tomonaga 2016), like the immunity to superinfection exerted by some endogenous retroviruses referred to below. It remains questionable whether bornavirus is currently infectious for humans as clinical reports of detection may result from expression of endogenous bornavirus sequences. Overall, the accumulation of viral sequences in host germline DNA is a much broader phenomenon than the endogenous retroviruses with which we have long been familiar.

3.2 *Evolutionary Dynamics of Exogenous and Endogenous Retroviruses*

It is customary to view parasite-host relations as an 'evolutionary arms race', the term coined by van Valen (1973) when he introduced the predator-prey 'Red Queen' hypothesis: "It takes all the running you can do to keep in the same place," the Red Queen tells Alice in *Through the Looking Glass*, "If you want to get somewhere else, you must run at least twice as fast as that". Rapidly replicating pathogens can outpace their hosts in immune escape and the development of drug resistance.

There is, however, a stark difference between the replication rate of exogenous viruses and that of their endogenous counterparts, as noted by Lee et al. (2013) for retroviruses and by Gilbert and Feschotte (2010) for avian hepadnavirus. Exogenous retroviruses undergo millions of replication cycles, including the reverse transcription step which is not subject to repair of errors, whereas endogenous retroviruses are by definition part of the host and replicate at the pace of the host germline, at less than 1000th the rate of virus replication (Aiewsakun and Katzourakis 2015). Retroviruses parked in the genome of one host may act as a hidden reservoir for infection to another.

Potentially pathogenic ERV and retrotransposons are controlled by host intracellular restriction factors (Sanz-Ramos and Stoye 2013; Goodier 2016). In addition, the Env glycoprotein of ERV can itself act as a restriction factor against exogenous infection by blocking receptors (Malfavon-Borja and Feschotte 2015), a phenomenon that we first reported for avian ERV (Payne et al. 1971). However, such restrictions often do not apply if the virus gains access to an unrelated host species and it is clear that replication-competent ERVs can make large leaps to infect distant host species. For instance, the baboon ERV that horizontally crossed hosts to become an ERV (RD114) in cats (Benveniste and Todaro 1974) retains its ability to replicate and to re-infect cats in vivo (Shimode et al. 2015). This phenomenon of xenotropism led us to examine the potential infection hazard of porcine ERV in pig-to-human xenotransplantation (Patience et al. 1997; Le Tissier et al. 1997) although fortunately human infection has not been reported to date.

Another interesting example of cross-species infection by replication-competent ERVs are the gamma-retroviruses related to gibbon ape leukemia virus (GALV). Isolates of GALV were made among captive lar gibbons in Thailand (Kawakami et al. 1972, 1978) although this exogenous retrovirus has not been recorded in wild gibbons. ERV sequences related to GALV were detected in the DNA of several species of South-East Asian rodents (Benveniste et al. 1977), including one in Australia and Indonesia (Simmons et al. 2014; Alfano et al. 2016). The koala retrovirus (KoRV) is also closely related to GALV (Hanger et al. 2000). It was first observed in 1988 (Canfield et al. 1988) but analysis of preserved taxidermy specimens indicates that KoRV has been present in koalas for at least 120 years (Avila-Arcos et al. 2013) and possibly much longer (Ishida et al. 2015). KoRV is associated with leukemia (Canfield et al. 1988; Tarlinton et al. 2006). The current leukemia epidemic appears to be related to an envelope variant (KoRV-B or KoRV-J) co-existing with the original form (KoRV-A) (Shojima et al. 2013; Xu et al. 2013). KoRV-A utilizes the Pit-1 cell surface receptor like GALV and FeLV-B, whereas KoRV-J utilises a thiamine transport receptor like FeLV-A (Shojima et al. 2013).

It thus appears that an ERV of rodents has crossed host species at least twice, to become an exogenous pathogenic retrovirus in both gibbons and koalas. Moreover,

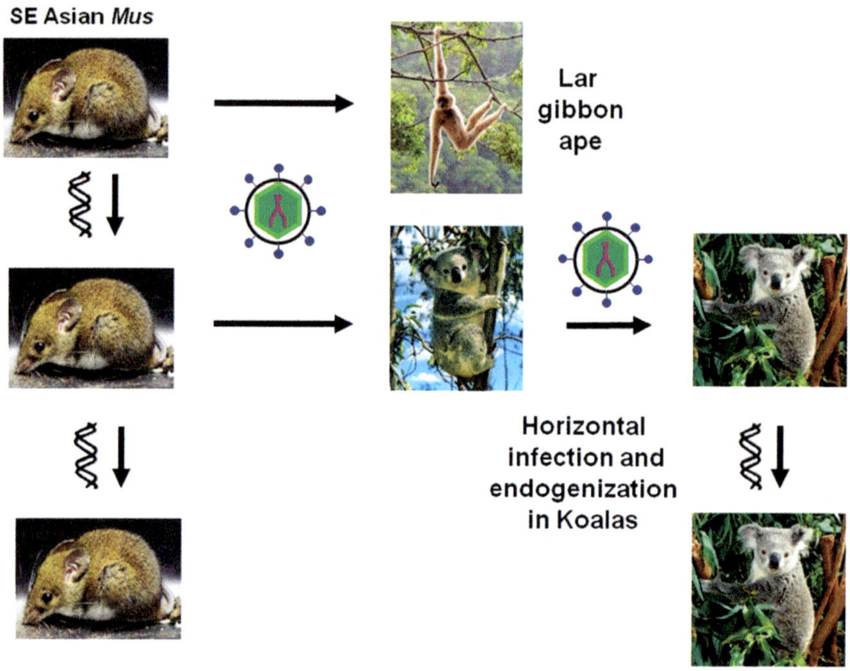

Fig. 4 Endogenous retrovirus in one host species can act as a 'reservoir' for millions of years and emerge to invade unrelated hosts. Gamma-retroviruses resident as mendelian genomes in South-East Asian rodents have transferred horizontally to lar gibbons (GALV) and to Koalas (KoRV). KoRV is becoming endogenous in its new host. Based on findings reported by Benveniste et al. (1977), Kawakami et al. (1978) and Tarlington et al. (2006)

KoRV is in the process of becoming endogenized in the koala germ line (Tarlinton et al. 2006; Ishida et al. 2015). Since many rodent species of the family *Muridae* carry related ERVs (Alfano et al. 2016), we may be confident that the precursor of GALV evolved in rodents and subsequently spread horizontally to the koala and the ape (Fig. 4).

3.3 Pathogenic and Beneficial Attributes of Endogenous Retroviruses

ERVs and other retrotransposons are open to natural selection operating on the host which presents a two-edged sword. On one side, integration and expression of certain ERV is associated with disease; on the other, the presence of transposable elements may be of overall benefit to the host population (Rebollo et al. 2012; Babaian and Mager 2016). There has been much discussion about both deleterious and beneficial effects of ERVs (Moyes et al. 2007; Stoye 2012) and new ERV insertions may disrupt essential host genes (Chuong et al. 2017). Recently integrated ERVs can be oncogenic, contributing to leukemia in mice and cats and to mammary cancer in mice (Rosenberg and Jolicoeur 1997; Ross 2010).

Whether human ERV sequences (HERV) play a causative role in human cancers is less definitive (Magiorkinis et al. 2015; Babaian and Mager 2016; Kassiotis and Stoye 2017). Beta-retrovirus HERV-K (HML2) genomes have been implicated in testicular tumors, melanoma and breast cancer. Certain HERV-K loci are more highly expressed in tumor tissue, but that does not provide conclusive evidence for causality; tumor cells impose fewer restrictions on HERV-K so that its expression may be a consequence rather than a cause of the malignancy.

Human endogenous retroviruses may be beneficial to the host through non-coding sequences providing new sources of gene regulation and by coding sequences providing useful proteins. For example, HERV-K is expressed in pre-implantation embryos and is linked with pluripotency but becomes transcriptionally silenced upon differentiation (Fuchs et al. 2013; Grow et al. 2015). HERV-H genomes and LINE elements are also tightly linked to transcriptional regulation in pluripotent cells and exert an influence on early development (Robbez-Masson and Rowe 2015; Schlesinger and Goff 2015). Modulation of expression of host enzymes by ERVs has also been turned to use by the host (Rebollo et al. 2012). For instance, the expression of human salivary amylase in the parotid gland is controlled by a novel HERV-E insertion in the primate lineage which became amplified in hominids (Ting et al. 1992). In the HERV-E LTR, the promoter and tissue-specific enhancers activate amylase expression which is otherwise restricted to the pancreas. It may have helped our forebears adapt from a diet mainly of fruit to one containing starch.

3.4 Role of Endogenous Retroviral Envelopes in the Placenta

The most striking example of ERVs becoming a benefit to their hosts is the role of Env in effecting cell fusion to form the syncytiotrophoblast of the mammalian placenta. The notion that HERV envelope might be involved in placental development interested me from the early 1990s, but was notably carried forward by the laboratories of Thierry Heidmann, John M McCoy and François-Loïc Cosset.

In the human placenta, we noted high expression of a defective HERV with an open reading frame for *env*, called ERV-3 (Boyd et al. 1993). ERV-3 expression was tightly linked to the syncytiotrophoblast (Fig. 3) and we postulated that a functional retroviral Env glycoprotein would be able to induce the cell-to-cell fusion (Venables et al. 1995). Indeed, Lin et al. (1999) demonstrated that transfection of ERV-3 into BeWo choriocarcinoma cells (a malignant version of cytotrophoblast) induced cell fusion and differentiation into syncytiotrophoblast. However, it was found that some humans lack the ERV-3 genome (de Parseval and

Heidmann 1998) although they must have been gestated with a healthy placenta. Gorillas also lack ERV-3 (Hervé et al. 2004). But our hypothesis of ERV driving placental differentiation survived albeit involving different ERVs. Blond et al. (2000) and Mi et al. (2000) showed that HERV-W Env also induced the syncytiotrophoblast fusion. Lavillette et al. (2002) showed that HERV-W interacts with several amino-acid transporters already known to act as beta-retrovirus receptors to initiate fusion. Mi et al. (2000) coined a new term for the HERV-W envelope glycoprotein: syncytin (Figs. 3 and 5).

The syncytin story became grew more convoluted with the discovery of a second human Env glycoprotein, syncytin-2, encoded by HERV-FRD (Blaise et al. 2003). Low expression of syncytin 1 and 2 is correlated with poor placental development and pre-eclampsia (Vargas et al. 2011). Moreover, different orders of placental mammal employ the Env glycoproteins of quite distinct ERVs to induce cell fusion of the trophoblast (Lavialle et al. 2013) and even in a proto-placenta in the marsupial opossum (Cornelis et al. 2015).

If the evolution of the placenta was a monophyletic event, why have placental mammals repeatedly entrained different ERVs to effect trophoblast differentiation into a syncytium? Imakawa et al. (2015) postulate a 'baton pass' hypothesis, in which multiple successive ERV variants take over cell-fusion roles, resulting in variations in placental structures and enhanced reproductive success in placental mammals. They speculate that ERVs replaced more ancient mediators of trophoblast fusion. Imakawa et al. (2015) support this view with the observation that several of the ERVs encoding syncytins have become endogenous in their respective host genomes only within the past 12–80 million years, more recently than the evolution of the mammalian placenta. In humans, redundancy in retroviral-driven trophoblast fusion explains why ERV-3, HERV-W and HERV-FRD can each induce fusion of the cytotrophoblast to form the syncytiotrophoblast.

HERV Env may have a dual role in the placenta: to be locally immunosuppressive at the maternal-fetal interface (the syncytiotrophoblast) in order to protect the fetus from maternal rejection, as well as to induce the trophoblast fusion (Denner 2016). The immunosuppressive domain in the transmembrane envelope protein is present in ERV-3 and HERV-W *env* (Boyd et al. 1993; Lavialle et al. 2013). Syncytins have maintained their virus entry capacity. For instance, HERV-W Env functionally pseudotypes retroviral vectors (Blond et al. 2000). HERV-W is also expressed in multinuclear macrophages such as osteoclasts (Søe et al. 2009), but there is no evidence that other syncytial tissues (e.g., striated muscle) utilize ERV syncytins to effect cell fusion. Nonetheless, placental mammals have given a virtuoso performance playing on a keyboard of retrovirus envelope-cell surface receptor engagement to repurpose ERVs to provide fusion proteins beneficial to the host.

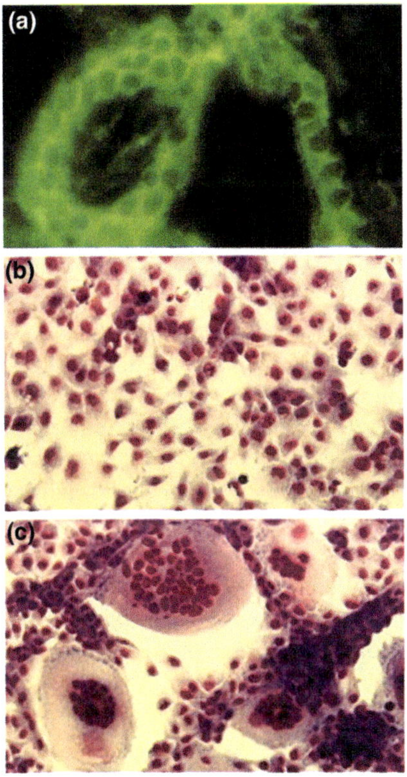

Fig. 5 Endogenous Env glycproteins function in the human placenta as 'syncytins'. **a** Indirect immunofluorescence of ERV-3 envelope glycoprotein in the syncytiotrophoblast of a full term human placenta (reproduced with permission from Venables et al. 1995); **b** Cos cells transfected with reverse orientation HERV-W syncytin 1 and **c** syncytium formation in Cos cells transfected by correctly oriented HERV-W syncytin 1 (reproduced with permission from Mi et al. 2000)

4 Endosymbiont Organelles in Eukaryotic Cells and Their Horizontal Transfer in Cancer

4.1 Evolution of Complex Cells and Their Organelles

The year in which Peter Vogt joined the Editorial Board of CTMI also marks the birth of the hypothesis that eukaryotic cells originated from the combination of prokaryotes to form complex cells containing organelles. This notion was postulated by Lynn Margulis, publishing under her first married name Sagan (1967); like many novel concepts was initially greeted with much skepticism (Margulis 2009). Lynn Margulis herself enjoyed stoking up controversies and she was attracted to the notion that HIV does not cause AIDS, holding that AIDS was nothing more than syphilis masquerading under a new name (Margulis et al. 2009).

Today, we accept that mitochondria, chloroplasts, cilia and the nuclear membrane had distinct origins and coalesced together to form eukaryotic cells (Lane 2014; Martin et al. 2015) although how this happened is open to debate (Baum and Baum 2014). During the long period of endosymbiosis, genes have switched location; for example, many genes involved in oxidative respiration appear to have transferred to the nuclear genome as the mtDNA episome became reduced (Allen 2003; Gershoni et al. 2009). If each kind of subcellular organelle in eukaryotes is monophyletic to what extent are they transferred laterally between species? In animals, the genomes of mitochondria have co-evolved with the nuclear genome because both types of genetic sequence have concordant phylogenies. Yet there is evidence of widespread lateral mobility of mitochondria and plastids in higher plants (Bergthorsson et al. 2003; Warren et al. 2016). This difference may be explained by mitochondrial variation driving evolution of sexes but also the germline-soma distinction that exists in metazoan animals but not in higher plants (Radzvilavicius et al. 2016).

Mitochondrial transfer between animal cells is illustrated by research into transmissible tumor cells in which I became involved a few years ago. My interest was aroused by a note published by Hayes et al. (1983) just before the discovery of HIV speculating—wrongly it turned out when a causative virus was discovered (Chang et al. 1994)—that canine transmissible venereal tumor (CTVT) might be a model for Kaposis's sarcoma in AIDS.

4.2 *Horizontal Spread of Cancer Cells*

Certain transmissible tumors are spread from one individual to another not by an oncogenic virus but by the migration of the tumor cell itself. Common marker chromosomes indicated that the transmissible agent of CTVT is the cancer cell itself (Cohen 1985) and it was shown that a LINE-1 retrotransposon insertion near c-*myc* is unique to the tumor (Katzir et al. 1985; Amariglio et al. 1991). Using forensic DNA markers we confirmed the cellular transmission of CTVT, and demonstrated that the tumor represents a single clone which has colonized dogs worldwide (Murgia et al. 2006).

CTVT was first described in 1876 (Novinski 1876) and it played an important role in the early years of cancer research because it was the only tumor that could be experimentally transplanted from one animal to another before the developmental of inbred lines. We estimated that CTVT emerged in an ancient dog breed ~11,000 years ago (Murchison et al. 2014). Thus CTVT represents a naturally occurring cancer cell clone some 2000 times older than HeLa cells. It has undergone thousands of somatic mutations, deletions, amplifications and chromosome rearrangements since it first emerged (Murchison et al. 2014) and it is a useful tool to examine how diverse a tumor cell can be while retaining its proliferative properties. For a cellular parasite that no longer requires many host functions (e.g., a sense of smell), it is noteworthy that CTVT has not shown a massive reduction of the size compared to the canine genome. Since household genes are interspersed with specialty genes, any deletions of unnecessary genes would have to occur on a case by case basis.

The Devil facial tumor disease (DFDT) is also transmitted horizontally as a tumor cell (Pearse and Swift 2006; Murchison et al. 2012) in the Tasmanian Devil (*Sarcophilus harrisii*), an endangered marsupial species which has relatively low genetic diversity (Siddle and Kaufman 2015). There are two independent DFDT clones circulating in Devils (Pye et al. 2016) both of recent provenance. David Metzger in Stephen Goff's laboratory has shown that some clam species are infested with clonal tumor cells (Metzger et al. 2015), including one in which the host species differs from that in which the tumor first arose (Metzger et al. 2016). The modes of transmission of these tumors differ: in dogs the tumor is mainly spread sexually, in Devils through biting, and among clams by filter feeding water containing tumor cells.

The emergence of transmissible tumor cells remains a rare phenomenon and we do not fully understand how the tumors evade the host immune response. The lack of major histocompatability antigens or their down regulation would promote the chance of tumor emergence (Murgia et al. 2006; Siddle and Kaufman 2015). Looking into the literature, however, I found only one example of 'naturally' transmissible tumors among inbred strains of laboratory rodents: a leukemia in Syrian hamsters which remarkably could be transmitted by mosquitoes (Banfield et al. 1965). In humans, there are several examples of horizontal tumor transmission from donors to immunosuppressed transplant recipients (Nalesnik et al. 2011). Leukemia has also been transmitted in utero between fetuses sharing a placenta (Greaves et al. 2003).

4.3 *Colonization of Cancer Cells by Host Mitochondria*

When we attempted to determine the date the most recent common ancestor of the tumor cell clone we obtained contradictory data for nuclear microsatellite DNA and mitochondrial DNA, as the latter appeared to have diversified for a longer time. We tentatively suggested that the two major clades of mtDNA in CTVT might have distinct origins (Murgia et al. 2006), and this hypothesis was verified by Rebbeck et al. (2011). It is now apparent that host mtDNA was acquired at least five times during the passage of the tumor clone through countless canine hosts (Strakova et al. 2016). Mitochondrial function may be a driver in tumor progression and recombination between mtDNA genomes is found in some CTVT tumors (Strakova et al. 2016). If mutations in the mtDNA accrue during serial passage of the tumor, acquisition of mtDNA from the host may contribute to the fitness of the tumor to persist. Thus a tumor that has spread worldwide as a somatic cell parasite has itself been colonized by host mitochondria via lateral transfer.

It would be interesting to determine whether similar colonization of tumors by host mitochondria occurs in the other transmissible tumor cells that have recently come to light. In particular, the tumor that has colonized a different species of clam than that from which it arose (Metzger et al. 2016) would be suitable to look for evidence of cross-species transfer of mitochondria and other organelles.

Mitochondria have lost many of their genes to the nuclear genome (Lane 2010), perhaps mediated by Importin 7 (Danoya et al. 2013). Thus there are likely to be constraints on cytoplasmic organelles containing DNA to co-evolve with the nuclear genome, owing to interplay of non-coding RNA elements (Vendramin et al. 2017) and to protein complexes with components encoded by both genomes. Therefore lateral transfer of mitochondria to distantly related species is unlikely to occur. Horizontal spread of mitochondria does occur within individuals, however, and has been detected in cancer. In an experimental murine system using tumor cells devoid of mitochondria, tumor progression was activated upon transfer of mitochondria from host stroma (Tan et al. 2015). Human tumor cells may also acquire host mitochondria from endothelial cells (Pasquier et al. 2013). Tumor progression and relapse may sometimes involve colonization of tumor cells by 'fitter' mitochondria imported from the host.

5 Concluding Remarks

The horizontal exchange of genetic sequences is widespread in all forms of life. Viruses embed themselves in host DNA, and host genes are transferred to viruses. Endogenous retroviruses can lie 'dormant' in the germ line for millions of years and re-emerge as replication-competent viruses infecting distantly related species. Host cells can emerge as transmissible malignant clones, survive longer than any other vertebrate somatic cells, and be colonized in turn by mitochondria from new hosts many transplant generations later. There is far more fluidity of genomes and cells within the eukaryotic world than we imagined 50 years ago.

Acknowledgements I am indebted to my mentors early in my career, Warren Levinson, Jan Svoboda and Peter K Vogt for providing a stimulating intellectual environment combined with a rigorous approach to experimental research. I am grateful to Klaus Bister, Ariberto Fassati and Paul Kellam for constructive discussions.

References

Aiewsakun P, Katzourakis A (2015) Endogenous viruses: connecting recent and ancient viral evolution. Virology 479–480:26–37. doi:10.1016/j.virol.2015.02.011

Alfano N, Michaux J, Morand S, Aplin K, Tsangaras K, Löber U, Fabre PH, Fitriana Y, Semiadi G, Ishida Y, Helgen KM, Roca AL, Eiden MV, Greenwood AD (2016) Endogenous gibbon ape leukemia virus identified in a rodent (*Melomys burtoni* subsp.) from Wallacea (Indonesia). J Virol 90:8169–8180. doi:10.1128/JVI.00723-16

Allen JF (2003) Why chloroplasts and mitochondria contain genomes. Comp Funct Genomics 4:31–36. doi:10.1002/cfg.245

Amariglio EN, Hakim I, Brok-Simoni F, Grossman Z, Katzir N, Harmelin A, Ramot B, Rechavi G (1991) Identity of rearranged LINE/c-MYC junction sequences specific for the canine transmissible venereal tumor. Proc Natl Acad Sci U S A 88:8136–8139

Aswad A, Katzourakis A (2014) The first endogenous herpesvirus, identified in the tarsier genome, and novel sequences from primate rhadinoviruses and lymphocryptoviruses. PLoS Genet 10: e1004332

Austin L, Hughes AL, Irausquin S, Friedman R (2010) The evolutionary biology of poxviruses. Infect Genet Evol 10:50. doi:10.1016/j.meegid.2009.10.001

Ávila-Arcos MC, Ho SY, Ishida Y, Nikolaidis N, Tsangaras K, Hönig K, Medina R, Rasmussen M, Fordyce SL, Calvignac-Spencer S, Willerslev E, Gilbert MT, Helgen KM, Roca AL, Greenwood AD (2013) One hundred twenty years of koala retrovirus evolution determined from museum skins. Mol Biol Evol 30:299–304

Babaian A, Mager DL (2016) Endogenous retroviral promoter exaptation in human cancer. Mob DNA 7:24. doi:10.1186/s13100-016-0080-x

Baltimore D (1970) RNA-dependent DNA polymerase of RNA tumour viruses. Nature 226: 1209–1211

Banfield WG, Woke PA, Mackay CM, Cooper HL (1965) Mosquito transmission of a reticulum cell sarcoma of hamsters. Science 148:1239–1240

Baum DA, Baum B (2014) An inside-out origin for the eukaryotic cell. BMC Biol 12: 76. doi: 10. 1186/s12915-014-0076-2

Bazot Q, Paschos K, Skalska L, Kalchschmidt JS, Parker GA, Allday MJ (2015) Epstein-Barr virus proteins EBNA3A and EBNA3C together induce expression of the oncogenic microRNA cluster miR-221/miR-222 and ablate expression of its target p57KIP2. PLoS Pathog 11:e1005031. doi:10.1371/journal.ppat.1005031

Belshaw R, Pereira V, Katzourakis A, Talbot G, Paces J, Burt A, Tristem M (2004) Long-term reinfection of the human genome by endogenous retroviruses. Proc Natl Acad Sci U S A 101:4894–4899

Belyi V, Levine AJ, Skalka AM (2010) Unexpected inheritance: multiple integrations of ancient bornavirus and ebolavirus/marburgvirus sequences in vertebrate genomes. PLoS Pathog 6: e1001030

Benveniste RE, Todaro GJ (1974) Evolution of C-type viral genes: inheritance of acquired viral genes. Nature 252:456–459

Benveniste RE, Callahan R, Sherr CJ, Chapman V, Todaro GJ (1977) Two distinct endogenous type C viruses isolated from the Asian rodent *Mus cervicolor*: conservation of virogene sequences in related rodent species. J Virol 21:849–862

Bergthorsson U, Adams KL, Thomason B, Palmer JD (2003) Widespread horizontal transfer of mitochondrial genes in flowering plants. Nature 424:197–201

Bister K (2015) Discovery of oncogenes: the advent of molecular cancer research. Proc Natl Acad Sci U S A. 112:15259–15260. doi:10.1073/pnas.1521145112

Blaise S, de Parseval N, Bénit Land Heidmann T (2003) Genome wide screening for fusogenic human endogenous retrovirus envelopes identifies syncytin 2, a gene conserved on primate evolution. Proc Natl Acad Sci U S A. 100:13013–13018

Blond JL, Lavillette D, Cheynet V, Bouton O, Oriol G, Chapel-Fernandes S, Mandrand B, Mallet F, Cosset FL (2000) An envelope glycoprotein of the human endogenous retrovirus HERV-W expressed in the human placenta and fuses cells expressing the D-type mammalian retrovirus receptor. J Virol 74:3321–3329

Bolin LL, Levy LS (2011) Viral determinants of FeLV infection and pathogenesis: lessons learned from analysis of a natural cohort. Viruses 3:1681–1698. doi:10.3390/v3091681

Bordet J, Ciuca MC (1922) Concerning the theories of the so-called 'bacteriophage'. Br Med J 2:296

Boshoff C, Endo Y, Collins PD, Takeuchi Y, Reeves JD, Schweickart VL, Siani MA, Sasaki T, Williams TJ, Gray PW, Moore PS, Chang Y, Weiss RA (1997) Angiogenic and HIV-inhibitory functions of KSHV-encoded chemokines. Science 278:290–294

Boyd MT, Bax CMR, Bax BE, Bloxam DL, Weiss RA (1993) The human endogenous retrovirus ERV-3 is upregulated in differentiating placental trophoblast cells. Virology 196:905–909
Brugge JS, Erikson RL (1977) Identification of a transformation-specific antigen induced by by an avian sarcoma virus. Nature 269:346–348
Bugert JJ, Darai G (2000) Poxvirus homologues of cellular genes. Virus Genes 21:111–133
Burstein D, Amaro F, Zusman T, Lifshitz Z, Cohen O, Gilbert JA, Pupko T, Shuman HA, Segal G (2016) Genomic analysis of 38 Legionella species identifies large and diverse effector repertoires 48: 167–175. doi:10.1038/ng.3481
Cairns J, Stent GS, Watson JD (2007) Phage and the origins of molecular biology. Cold Spring Harbor Laboratory Press, New York
Canfield PJ, Sabine JM, Love DN (1988) Virus particles associated with leukaemia in a koala. Aust Vet J 65:327–328
Cavalieri F, Ruscio T, Tinoco R, Benedict S, Davis C, Vogt PK (1985) Isolation of three new avian sarcoma viruses: ASV 9, ASV 17, and ASV 25. Virology 143:680–683
Chang Y, Cesarman E, Pessin MS, Lee F, Culpepper J, Knowles DM, Moore PS (1994) Identification of herpesvirus-like DNA sequences in AIDS-associated Kaposi's sarcoma. Science 265: 1865–1869
Choi IG, Kim SH (2007) Global extent of horizontal gene transfer. Proc Natl Acad Sci U S A 104:4489–4494. doi:10.1073/pnas.0611557104
Chuong EB, Elde NC, Feschotte C (2017) Regulatory activities of transposable elements: from conflicts to benefits. Nat Rev Genetics 18:71–86. doi:10.1038/nrg.2016.139
Cisek AA, Dabrowska I, Gregorczyk KP, Wyzewski Z (2017) Phage therapy in bacterial infections treatment: one hundred years after the discovery of bacteriophages. Curr Microbiol 74:277–283. doi:10.1007/s00284-016-1166-x
Cohen D (1985) The canine transmissible venereal tumor: a unique result of tumor progression. Adv Cancer Res 43:75–112
Colson P, La Scola B, Levasseur A, Caetano-Anollés G, Raoult D (2017) Mimivirus: leading the way in the discovery of giant viruses of amoebae. Nat Rev Microbiol 15:243–254. doi:10.1038/nrmicro.2016.197
Cornelis G, Vernochet C, Carradec Q, Souquere S, Mulot B, Catzeflis Nilsson MA, Menzies BR, Renfree MB, Pierron G, Zeller U, Heidmann O, Dupressoir A, Heidmann T (2015) Retroviral envelope gene captures and syncytin exaptation for placentation in marsupials. Proc Natl Acad Sci U S A. 112:E487–E496. doi:10.1073/pnas.1417000112
Cordaux R, Batzer MA (2009) The impact of retrotransposons on human genome evolution. Nat Rev Genet 10:691–703. doi:10.1038/nrg2640
Crawford LV, Crawford EM (1961) The properties of Rous sarcoma virus purified by density gradient centrifugation. Virology 13:227–232
Cunningham C, Gatherer D, Baluchova K, Dargan DJ, Thomson M, Griffiths PD, Wilkinson GWG, Schulz TF, Davison AJ (2010) Sequences of complete human cytomegalovirus genomes from infected cell cultures and clinical specimens. J Gen Virol 91:605–615. doi:10.1099/vir.0.015891-0
Daibata M, Taguchi T, Nemoto Y, Taguchi H, Miyoshi I (1999) Inheritance of chromosomally integrated human herpesvirus 6 DNA. Blood 94:1545–1549
Danoya A, Wang T, Keshavarz-Moore E, Fassati A, Chain BM (2013) Importin-7 mediates nuclear trafficking of DNA in mammalian cells. Traffic 14:165–175
Denner J (2016) Expression and function of endogenous retroviruses in the placenta. APMIS 124:31–43. doi:10.1111/apm.12474
d'Herelle F (1917) An invisible microbe that is antagonistic to the dysentery Bacillus. Compt Rend Acad Sci Paris 165:373–375
de Parseval N, Heidmann T (1998) Physiological knockout of the envelope gene of the single-copy ERV-3 human endogenous retrovirus in a fraction of the Caucasian population. J Virol 72:3442–3445
Duesberg PH, Vogt PK (1970) Differences between the ribonucleic acids of transforming and nontransforming avian tumor viruses. Proc Natl Acad Sci U S A 67:1673–1680

Dulbecco R (1952) Production of plaques in monolayer tissue cultures by single particles of an animal virus. Proc Natl Acad Sci U S A 38:747–752

Dunning Hotopp JC, Clark ME, Oliveira DC, Foster JM, Fischer P, Muñoz Torres MC, Giebel JD, Kumar N, Ishmael N, Wang S, Ingram J, Nene RV, Shepard J, Tomkins J, Richards S, Spiro DJ, Ghedin E, Slatko BE, Tettelin H, Werren JH (2007) Widespread lateral gene transfer from intracellular bacteria to multicellular eukaryotes. Science 317:1753–1756

Etienne L, Emerman M (2013) The mongoose, the pheasant, the pox, and the retrovirus. PLoS Biol 11:e1001641. doi:10.1371/journal.pbio.1001641

Farkašová H, Hron T, Pačes J, Hulva P, Benda P, Gifford RJ, Elleder D (2017) Discovery of an endogenous Deltaretrovirus in the genome of long-fingered bats (Chiroptera: Miniopteridae). Proc Natl Acad Sci U S A. 2017 Mar 9. pii: 201621224. doi:10.1073/pnas.1621224114

Feschotte C, Gilbert C (2012) Endogenous viruses: insights into viral evolution and impact on host biology. Nat Rev Genet 13:283–296. doi:10.1038/nrg3199

Fuchs NV, Loewer S, Daley GQ, Izsvák Z, Löwer J, Löwer R (2013) Human endogenous retrovirus K (HML-2) RNA and protein expression is a marker for human embryonic and induced pluripotent stem cells. Retrovirology 10:115. doi:10.1186/1742-4690-10-115

Gershoni M, Templeton AR, Mishmar D (2009) Mitochondrial bioenergetics as a major motive force of speciation. BioEssays 31:642–650. doi:10.1002/bies.200800139

Gilbert C, Feschotte C (2010) Genomic fossils calibrate the long-term evolution of hepadnaviruses. PLoS Biol 8:e1000495

Gomez-Valero L, Buchrieser C. (2013) Genome dynamics in Legionella: the basis of versatility and adaptation to intracellular replication. Cold Spring Harb Perspect Med 3: pii: a009993. doi:10.1101/cshperspect.a009993

Goodier JL (2016) Restricting retrotransposons: a review. Mobile. DNA 7:16. doi:10.1186/s13100-016-0070

Greaves MF, Maia AT, Wiemel JL, Ford AM (2003) Leukemia in twins: lessons in natural history. Blood 102:2321–2333

Griffiths D (2001) Endogenous retroviruses in the human genome sequence. Genome Biol 2: reviews 1017.1-reviews 1017.5

Grow EJ, Flynn RA, Chavez SL, Bayless NL, Wossidlo M, Wesche D, Martin L, Ware C, Blish CA, Chang HY, Reijo Pe RA, Wysocka J (2015) Intrinsic retroviral reactivation in human preimplantation embryos and pluripotent cells. Nature 522: 221–225. doi: 10.1038/nature14308

Guo YE, Steitz JA (2014) Virus meets host microRNA: the destroyer, the booster, the hijacker. Mol Cell Biol 34:3780–3787

Haig DM (2001) Subversion and piracy: DNA viruses and immune evasion. Res Vet Sci 70:205–219

Hanafusa H, Hanafusa T, Rubin H (1963) The defectiveness of Rous sarcoma virus. Proc Natl Acad Sci U S A 49:572–580

Hanger JJ, Bromham LD, McKee JJ, O'Brien TM, Robinson WF (2000) The nucleotide sequence of koala (*Phascolarctos cinereus*) retrovirus: a novel type C endogenous virus related to gibbon ape leukemia virus. J Virol 74:4264–4272

Hartmann G (2017) Nucleic Acid immunity. Adv Immunol 133:121–169. doi:10.1016/bs.ai.2016.11.001

Hayes HM, Biggar RJ, Pickle LW, Hoover R, Toft JD (1983) Canine transmissible venereal tumor: a model for Kaposi's sarcoma? Am J Epidemiol 117:108–109

Hayward A, Cornwallis CK, Jern P (2015) Pan-vertebrate comparative genomics unmasks retrovirus macroevolution. Proc Natl Acad Sci U S A 112:464–469

Hayward WS, Neel BG, Astrin SM (1981) Activation of a cellular *onc* gene by promoter insertion ALV-induced lymphoid leukosis. Nature 290:475–480

Herniou EA, Huguet E, Thézé J, Bézier A, Periquet G, Drezen JM (2013) When parasitic wasps hijacked viruses: genomic and functional evolution of polydnaviruses. Philos Trans R Soc Lond B 368:20130051. doi:10.1098/rstb.2013.0051

Hertig C, Coupar BE, Gould AR, Boyle DB (1997) Field and vaccine strains of fowlpox virus carry integrated sequences from the avian retrovirus, reticuloendotheliosis virus. Virolgy 235:367–376
Hervé CA, Forrest G, Löwer R, Griffiths DJ, Venables PJ (2004) Conservation and loss of the ERV3 open reading frame in primates. Genomics 83:940–943
Hill JA, Ruth HallSedlak R, Magaret A, Huang ML, Zerr DM, Jerome KR, Boeckh M (2016) Efficient identification of inherited chromosomally integrated human herpesvirus 6 using specimen pooling. J Clin Virol 77:71–76
Holzerlandt R, Orengo C, Kellam P, Albà MM (2002) Identification of new herpesvirus gene homologs in the human genome. Genome Res 12:1739–1748. doi:10.1101/gr.334302
Honda T, Tomonaga K (2016) Endogenous non-retroviral RNA virus elements evidence a novel type of antiviral immunity. Mob Genet Elements 6:e1165785. doi:10.1080/2159256X.2016.1165785
Horie M, Honda T, Suzuki Y, Kobayashi Y, Daito T, Oshida T, Ikuta K, Jern P, Gojobori T, Coffin JM, Tomonaga K (2010) Endogenous non-retroviral RNA virus elements in mammalian genomes. Nature 463:84–87. doi:10.1038/nature08695
Horie M, Kobayashi Y, Suzuki Y, Tomonaga K (2013) Comprehensive analysis of endogenous bornavirus-like elements in eukaryote genomes. Phil Trans R Soc B 368:20120499. doi:10.1098/rstb.2012.0499
Hunter T, Sefton BM (1980) Transforming gene product of Rous sarcoma virus phoshorylates tyrosine. Proc Natl Acad Sci U S A 77:1311–1315
Imakawa K, Nakagawa S, Miyazawa T (2015) Baton pass hypothesis: successive incorporation of unconserved endogenous retroviral genes for placentation during mammalian evolution. Genes Cells 10:771–778
Isfort RJ, Jones D, Kost R, Witter R, Kung H-J (1992) Retrovirus insertion into herpesvirus in vitro and in vivo. Proc Natl Acad Sci U S A 89:991–995
Ishida Y, Zhao K, Greenwood AD, Roca AL (2015) Proliferation of endogenous retroviruses in the early stages of a host germ line invasion. Mol Biol Evol 32:109–120
Ishizaki R, Vogt PK (1966) Immunological relationships among envelope antigens of avian tumor viruses. Virology 30:375–387
Kassiotis G, Stoye JP (2017) Making a virtue of necessity: the pleiotropic role of human endogenous retroviruses in cancer. Phil Trans R Soc B (in press)
Katzir N, Rechavi G, Cohen JB, Unger T, Simoni F, Segal S, Cohen D, Givol D (1985) "Retroposon" insertion into the cellular oncogene c-myc in canine transmissible venereal tumor. Proc Natl Acad Sci U S A 82:1054–1058. doi:10.1073/pnas.82.4.1054
Kawakami TG, Huff SD, Buckley PM, Dungworth DL, Synder SP, Gilden RV (1972) C-type virus associated with gibbon lymphosarcoma. Nat New Biol 235:170–171
Kawakami TG, Sun l, McDowell TS (1978) Natural transmission of gibbon leukemia virus. J Natl Cancer Inst 61:1113–1115
Keogh EV (1938) Ectodermal lesions produced by the virus of Rous sarcoma. Brit J Exp Pathol 19:1–9
Klenerman P, Hengartner H, Zinkernagel RM (1997) A non-retroviral RNA virus persists in DNA form. Nature 390: 298–301
Krupovic M, Koonin EV (2017) Multiple origins of viral capsid proteins from cellular ancestors. Proc Natl Acad Sci U S A 114:E2401–E2410. doi:10.1073/pnas.1621061114
Lane N (2010) Life ascending: the ten great inventions of evolution. Profile Books, London
Lane N (2014) Bioenergetic constraints on the evolution of complex life. Cold Spring Harb Perspect Biol 6:a015982. doi:10.1101/cshperspect.a015982
Lane N (2015) The vital question: why is life the way it is?. Profile Books, London
Lavillette D, Marin M, Ruggieri A, Mallet F, Cosset FL, Kabat D (2002) The envelope glycoprotein of human endogenous retrovirus type W uses a divergent family of amino acid transporters/cell surface receptors. J Virol 76:6442–6652

Lavialle C, Cornelis G, Dupressoir A, Esnault C, Heidmann O, Vernochet C, Heidmann T (2013) Paleovirology of 'syncytins', retroviral env genes exapted for a role in placentation. Philos Trans R Soc Lond B 368:20120507. doi:10.1098/rstb.2012.0507

Lee A, Nolan A, Watson J, Tristem M (2013) Identification of an ancient endogenous retrovirus, predating the divergence of the placental mammals. Philos Trans R Soc Lond B Biol Sci 368:20120503

Le Tissier P, Stoye JP, Takeuchi Y, Patience C, Weiss RA (1997) Two sets of human-tropic pig retrovirus. Nature 389:681–682

Levy JA (1978) Xenotropic type C viruses. Curr Top Microbiol Immunol 79:111–213

Lin L, B. Xua B, Rote NS (1999) Expression of endogenous retrovirus ERV-3 induces differentiation in BeWo, a choriocarcinoma model of human placental trophoblast. Placenta 20:109–118

Loenen WAM, Dryden DTF, Raleigh EA, Geoffrey GG, Murray NE (2014) Highlights of the DNA cutters: a short history of the restriction enzymes. Nucleic Acids Res 42(1):3–19. doi:10.1093/nar/gkt990

Lwoff A (1953) Lysogeny. Bacteriol Rev 17:269–337

Magiorkinis G, Belshaw R, Katzourakis A (2015) 'There and back again': revisiting the pathophysiological roles of human endogenous retroviruses in the post-genomic era. Phil Trans R Soc B 368:20120504

Maki Y, Bos TJ, Davis C, Starbuck M, Vogt PK (1987) Avian sarcoma virus 17 carries the *jun* oncogene. Proc Natl Acad Sci U S A 84:2848–2852

Malfavon-Borja R, Feschotte C (2015) Fighting fire with fire: Endogenous retrovirus envelopes as restriction factors. J Virol 89:4047–4050

Margulis L, Maniotis A, MacAllister J, Scythes BO, Hall J et al (2009) Spirochete round bodies, syphilis, Lyme disease & AIDS: resurgence of "the great imitator"? Symbiosis 47:51–58

Margulis L (2009) Genome acquisition in horizontal gene transfer: symbiogenesis and macromolecular sequence analysis. Methods Mol Biol 532:181–191. doi:10.1007/978-1-60327-853-9_10

Martin GS (1970) Rous sarcoma virus: a function required for the maintenance of the transformed state. Nature 227:1021–1023

Martin GS (2004) The road to *Src*. Oncogene 23:7910–7917

Martin WF (2011) Early evolution without a tree of life. Biol Direct 6:36

Martin WF, Garg S, Zimorski V (2015) Endosymbiotic theories for eukaryote origin. Phil Trans R Soc B 370:20140330. doi:10.1098/rstb.2014.0330

Metzger MJ Reinisch C, Sherry J, Goff SP (2015) Horizontal transmission of clonal cancer cells causes leukemia in soft-shell clams. Cell 161: 255–263

Metzger MJ, Villalba A, Carballal MJ, Iglesias D, Sherry J, Reinisch C, Muttray AF, Baldwin SA, Goff SP (2016) Widespread transmission of independent cancer lineages within multiple bivalve species. Nature 534:705–709

Mi S, Lee X, Li X, Veldman GM, Finnerty H, Racie L, LaVallie E, Tang XY, Edouard P, Howes S, Keith JC, McCoy JM (2000) Syncytin is a captive retroviral envelope protein involved in human placental morphogenesis. Nature 403:785–789

Miles BD, Robinson HL (1985) High-frequency transduction of c-erbB in avian leukosis virus-induced erythroblastosis. J Virol 54:295–303

Moyes D, Griffiths DJ, Venable PJ (2007) Insertional polymorphisms: a new lease of life for endogenous retroviruses in human disease. Trends Genet 23:326–333

Murchison EP, Schulz-Trieglaff OB, Ning Z, Alexandrov LB, Bauer MJ, Fu B, Hims M, Ding Z, Ivakhno S, Stewart C, Ng BL, Wong W, Aken B, White S, Alsop A, Becq J, Bignell GR, Cheetham RK, Cheng W, Connor TR, Cox AJ, Feng ZP, Gu Y, Grocock RJ, Harris SR, Khrebtukova I, Kingsbury Z, Kowarsky M, Kreiss A, Luo S, Marshall J, McBride DJ, Murray L, Pearse AM, Raine K, Rasolonjatovo I, Shaw R, Tedder P, Tregidgo C, Vilella AJ, Wedge DC, Woods GM, Gormley N, Humphray S, Schroth G, Smith G, Hall K, Searle SM, Carter NP, Papenfuss AT, Futreal PA, Campbell PJ, Yang F, Bentley DR, Evers DJ,

Stratton MR (2012) Genome sequencing and analysis of the Tasmanian devil and its transmissible cancer. Cell 148:780–791

Murchison EP, Wedge DC, Alexandrov LB, Fu B, Martincorena I, Ning Z, Tubio JM, Werner EI, Allen J, De Nardi AB, Donelan EM, Marino G, Fassati A, Campbell PJ, Yang F, Burt A, Weiss RA, Stratton MR (2014) Transmissible dog cancer genome reveals the origin and history of an ancient cell lineage. Science 343:437–440

Murgia C, Pritchard KJ, Kim SK, Fassati A, Weiss RA (2006) Clonal origin and evolution of a transmissible cancer. Cell 126:477–487

Nalesnik MA, Woodle ES, DiMaio JM, Vasudev B, Teperman LW, Covington S, Taranto S, Gockerman JP, Shapiro R, Sharma V, Swinnen LJ, Yoshida A, Ison MG (2011) Donor-transmitted malignancies in organ transplantation: assessment of clinical risk. Amer J Transplant 11:1140–1147. doi:10.1111/j.1600-6143.2011.03565.x

Neil JC, Fulton R, Rigby M, Stewart M (1991) Feline leukemia virus: generation of pathogenic and oncogenic variants. Curr Top Microbiol Immunol 171:67–93

Niewiadomska AM, Gifford RJ (2013) The extraordinary evolutionary history of the reticuloendotheliosis viruses. PLoS Biol 11:e1001642

Novinski MA (1876) Zur Frage uber die Impfung der Krebsigen Geschwulste. Zentralbl Med Wissensch 14:790–791

Odom MR, Hendrickson RC, Lefkowitz EJ (2009) Poxvirus protein evolution: family wide assessment of possible horizontal gene transfer events. Virus Res 144:233–249. doi:10.1016/j.virusres.2009.05.006

Pasquier J, Guerrouahen BS, Al Thawadi H, Ghiabi P, Maleki M, Abu-Kaoud N, Jacob A, Mirshahi M, Galas L, Rafii S, Le Foll F, Rafii A (2013) Preferential transfer of mitochondria from endothelial to cancer cells through tunneling nanotubes modulates chemoresistance. J Transl Med 11:94. doi:10.1186/1479-5876-11-94

Patience C, Takeuchi Y, Weiss RA (1997) Infection of human cells by an endogenous retrovirus of pigs. Nat Medicine 3:282–286

Payne LN, Pani PK, Weiss RA (1971) A dominant epistatic gene which inhibits cellular susceptibility to RSV(RAV-O). J Gen Virol 13:455–462

PearseAM, Swift K (2006) Allograft theory: transmission of devil facial-tumour disease. Nature 439: 549

Pellett PE, Ablashi DV, Ambros PF, Agut H, Caserta MT, Descamps V, Flamand L, Gautheret-Dejean A, Hall CB, Kamble RT, Kuehl U, Lassner D, Lautenschlager I, Loomis KS, Luppi M, Lusso P, Medveczky PG, Montoya JG, Mori Y, Ogata M, Pritchett JC, Rogez S, Seto E, Ward KN, Yoshikawa T, Razonable RR Raymund R (2012) Chromosomally integrated human herpesvirus 6: questions and answers. Rev Med Virol 22:144–155

Ptashne M (2004) A genetic switch: phage lambda revisited, 3rd edn. Cold Harbor Spring Laboratory Press, New York. ISBN 978-0-87969-716-7

Pye RJ, Pemberton D, Tovar C, Tubio JM, Dun KA, Fox S, Darby J, Hayes D, Knowles GW, Kreiss A, Siddle HV, Swift K, Lyons AB, Murchison EP, Woods GM (2016) A second transmissible cancer in Tasmanian devils. Proc Natl Acad Sci U S A. 113(2):374–379. doi:10.1073/pnas.1519691113

Radzvilavicius AL, Hadjivasiliou Z, Pomiankowski A, Lane N (2016) Selection for mitochondrial quality drives evolution of the germline. PLoS Biol 14:e2000410. doi:10.1371/journal.pbio.2000410

Rebbeck CA, Leroi AM, Burt A (2011) Mitochondrial capture by a transmissible cancer. Science 331: 303

Rebollo R, Romanish MT, Mager DL (2012) Transposable elements: an abundant and natural source of regulatory sequences for host genes. Annu Rev Genet 46:21–42

Refardt D, Bergmiller T, Kümmerli R (2013) Altruism can evolve when relatedness is low: evidence from bacteria committing suicide upon phage infection. Proc R Soc B 280:20123035

Robbez-Masson L, Rowe HM (2015) Retrotransposons shape species-specific embryonic stem cell networks. Retrovirology 12:45. doi:10.1186/s12977-015-0173-5

Rohrschneider L, Reynolds S (1985) Regulation of cellular morphology by the Rous sarcoma virus src gene: analysis of fusiform mutants. Mol Cell Biol 5:3097–3107

Rosenberg N, Jolicoeur P (1997) Retroviral pathogenesis. In: Coffin JM, Hughes SH, Varmus HE (eds) Retroviruses. Cold Spring Harbor Laboratory Press, pp 475–585

Ross SR (2010) Mouse mammary tumor virus molecular biology and oncogenesis. Viruses 2:2000–2012

Rous P (1911) A sarcoma of the fowl transmissible by an agent separable from the tumor cells. J Exp Med 13: 397–411. doi:10.1084/jem.13.4.397

Rovnak J, Quackenbush SL (2010) Walleye dermal sarcoma virus: molecular biology and oncogenesis. Viruses 2:1984–1999

Roy SW, Gilbert W (2006). The evolution of spliceosomal introns: patterns, puzzles and progress. Nat Rev Genetics. 7: 211–221. doi:10.1038/nrg1807

Roy-Burman P (1995) Endogenous env elements: partners in generation of pathogenic feline leukemia viruses. Virus Genes 11:147–161

Rubin H, Vogt PK (1962) An avian leukosis virus associated with stocks of Rous sarcoma virus. Virology 17:184–194

Ruboyianes R, Worobey M (2016) Foamy-like endogenous retroviruses are extensive and abundant in teleosts. Virus Evol 2: vew032. doi: 10.1093/ve/vew03

Sagan L (1967) On the origin of mitosing cells. J Theoret Biol 14: 225–274. doi:10.1016/0022 5193(67)90079-3

Sankaran N (2014) When viruses were not in style: parallels in the histories of chicken sarcoma viruses and bacteriophages. Stud Hist Philos Biol Biomed Sci 48:189–199

Sanz-Ramos R, Stoye JP (2013) Capsid-binding retrovirus restriction factors: discovery, restriction specificity and implications for the development of novel therapeutics. J Gen Virol 94:2587–2598. doi:10.1099/vir.0.058180-0

Shimizu A, Nakatani Y, Nakamura T, Jinno-Oue A, Ishikawa O, Boeke JD, Takeuchi Y, Hoshino H (2014) Characterisation of cytoplasmic DNA complementary to non-retroviral RNA viruses in human cells. Sci Rep 4:5074. doi:10.1038/srep05074

Schlesinger S, Goff SP (2015) Retroviral transcriptional regulation and embryonic stem cells: war and peace. Mol Cell Biol 35:770–777

Shimode S, Nakagawa S, Miyazawa T (2015) Multiple invasions of an infectious retrovirus in cat genomes. Sci Rep 5:8164. doi:10.1038/srep08164

Shojima T, Yoshikawa R, Hoshino S, Shimode S, Nakagawa S, Ohata T, Nakaoka R, Miyazawa T (2013) Identification of a novel subgroup of koala retrovirus from koalas in Japanese zoos. J Virol 87:9943–9948

Siddle HV, Kaufman J (2015) Immunology of transmissible tumours. Immunology 144:11–20

Simmons G, Clarke D, McKe J, Young P, Meers J (2014) Discovery of a novel retrovirus sequence in an Australian native rodent (*Melomys burtoni*): a putative link between gibbon ape leukemia virus and koala retrovirus. PLoS ONE 9:e106954

Søe K, Andersen TL, Hobolt-Pedersen AS, Bjerregaard B, Larsson LI, Delaissé JM (2009) Involvement of human endogenous retroviral syncytin-1 in human osteoclast fusion. J Mol Biol 392:301–318

Stefan E, Bister K (2017) MYC and RAF: key effectors in cellular signaling and major drivers in human cancer. Curr Topics Microbiol Immunol (this Volume)

Stehelin D, Varmus HE, Bishop JM, Vogt PK (1976) DNA related to the transforming gene(s) of avian sarcoma viruses is present in normal avian DNA. Nature 260:170–173

Stern A, Sorek R (2011) The phage-host arms-race: shaping the evolution of microbes. BioEssays 33:43–51. doi:10.1002/bies.201000071

Stewart H, Jarrett O, Hosie MJ, Willett BJ (2011) Are endogenous feline leukemia viruses really endogenous? Vet Immunol Immunopathol 143:325–331

Stoye JP (2012) Studies of endogenous retroviruses reveal a continuing evolutionary saga. Nat Rev Microbiol 10:395–406

Strakova A, Ní Leathlobhair M, Wang GD, Yin TT, Airikkala-Otter I, Allen JL, Allum KM, Bansse-Issa L, Bisson JL, Castillo Domracheva A, de Castro KF, Corrigan AM,

Cran HR, Crawford JT, Cutter SM, Delgadillo Keenan L, Donelan EM, Faramade IA, Flores Reynoso E, Fotopoulou E, Fruean SN, Gallardo-Arrieta F, Glebova O, Häfelin Manrique RF, Henriques JJ, Ignatenko N, Koenig D, Lanza-Perea M, Lobetti R, Lopez Quintana AM, Losfelt T, Marino G, Martincorena I, Martínez Castañeda S, Martínez-López MF, Meyer M, Nakanwagi B, De Nardi AB, Neunzig W, Nixon SJ, Onsare MM, Ortega-Pacheco A, Peleteiro MC, Pye RJ, Reece JF, Rojas Gutierrez J, Sadia H, Schmeling SK, Shamanova O, Ssuna RK, Steenland-Smit AE, Svitich A, Thoya Ngoka I, Viţălaru BA, de Vos AP, de Vos JP, Walkinton O, Wedge DC, Wehrle-Martinez AS, van der Wel MG, Widdowson SA, Murchison EP (2016) Mitochondrial genetic diversity, selection and recombination in a canine transmissible cancer. eLife 5: e14552. doi:10.7554/eLife.14552

Strand MR, Burke GR (2014) Polydnaviruses: nature's genetic engineers. Annu Rev Virol 1:335–354. doi:10.1146/annurev-virology-031413-085451

Summers WC (1999) Felix d'Herelle and the origins of molecular biology. Yale University Press, New Haven

Svoboda J (1960) Presence of chicken tumour virus in the sarcoma of adult rat inoculated after birth with Rous sarcoma virus. Nature 186:980–981

Svoboda J (1966) Basic aspects of the interaction of oncogenic viruses with heterologous cells. Int Rev Exp Pathol 5:25–66

Tan AS, Baty JW, Dong LF, Bezawork-Geleta A, Endaya B, Goodwin J Bajzikova M, Kovarova J, Peterka M, Yan B, Pesdar EA, Sobol M, Filimonenko A, Stuart S, Vondrusova M, Kluckova K, Sachaphibulkij K, Rohlena J, Hozak P, Truksa J, Eccles D, Haupt LM, Griffiths LR, Neuzil J, Berridge MV (2015) Mitochondrial genome acquisition restores respiratory function and tumorigenic potential of cancer cells without mitochondrial DNA. Cell Metab 21:81–94

Tarlinton RE, Meers J, Young PR (2006) Retroviral invasion of the koala genome. Nature 442: 79–81

Taylor DJ, Leach RW, Bruenn J (2010) Filoviruses are ancient and integrated into mammalian genomes. BMC Evol Biol 10:193. doi: 10.1186/1471-2148-10-193

Temin HM (1960) The control of cellular morphology on embryonic cells infected with Rous sarcoma virus in vitro. Virology 10:182–197

Temin HM (1962) Separation of morphological conversion and virus production in Rous sarcoma virus infection. Cold Spring Harb Symp Quant Biol 27:407–414

Temin HM (1964) Nature of the provirus of Rous sarcoma. Natl Cancer Inst Monog 17:557–570

Temin HM, Mizutani S (1970) RNA-dependent DNA polymerase in virions of Rous sarcoma virus. Nature 226:1211–1213

Temin HM, Rubin H (1958) Characteristics of an assay for Rous sarcoma virus and Rous sarcoma cells in tissue culture. Virology 6:669–688

Thielen GH, Goud D, Fowler M, Dungworth DL (1971) C-type virus in tumor tissue of a woolly monkey (*Lagothrix* spp) with fibrosarcoma. J Natl Cancer Inst 47:731–743

Ting CN, Rosenberg MP, Snow MP, Samuelson LC, Meisler MH (1992) Endogenous retroviral sequences are required for tissue-specific expression of a human salivary amylase gene. Genes Dev 6:1457–1465

Toyoshima K, Vogt PK (1969) Temperature sensitive mutants of an avian sarcoma virus. Virology 39:930–931

Twort FW (1915) An investigation on the nature of ultramicroscopic viruses. Lancet 2:1241–1243

Van Valen L (1973) A new evolutionary law. Evol Theor 1:1–30

Vargas A, Toufaily C, LeBellego F, Rassart É, Lafond J, Barbeau B (2011) Reduced expression of both syncytin 1 and syncytin 2 correlates with severity of preeclampsia. Reprod Sci 18: 1085–1091

Varmus HE (2017) How tumor virology evolved into cancer biology and transformed oncology. Annu Rev Cancer Biol 1:11.1–11.18 doi:10.1146/annurev-cancerbio-050216-034315

Venables P, Brookes SM, Griffiths D, Weiss RA, Boyd MT (1995) Abundance of an endogenous retroviral envelope protein in placental trophoblasts suggests a biological function. Virology 211:589–592
Vendramin R, Marine JC, Leucci E (2017) Non-coding RNAs: the dark side of nuclear-mitochondrial communication. EMBO J pii: e201695546. doi:10.15252/embj.201695546
Vogt M, Dulbecco R (1960) Virus-cell interaction with a tumor-producing virus. Proc Natl Acad Sci U S A 46:365–370
Vogt PK (1967) A virus released by "nonproducing" Rous sarcoma cells. Proc Natl Acad Sci U S A 58:801–808
Vogt PK (2002) Fortuitous convergences: the beginnings of JUN. Nat Rev Cancer 2:465–469
Vogt PK (2012) Retroviral oncogenes: a historical primer. Nat Rev Cancer 12:639–648
Vogt PK, Hu SS (1977) The genetic structure of RNA tumor viruses. Annu Rev Genet 11:203–238
Vogt PK, Ishizaki R (1965) Reciprocal patterns of genetic resistance to avian tumor viruses in two lines of chickens. Virology 26:664–672
Volz A, Sutter G (2017) Modified Vaccinia Virus Ankara: history, value in basic research, and current perspectives for vaccine development. Adv Virus Res 97:187–243. doi:10.1016/bs.aivir.2016.07.001
Warren JM, Simmons MP, Wu Z, Sloan DB (2016) Linear plasmids and the rate of sequence evolution in plant mitochondrial genomes. Genome Biol Evol 8(2):364–374. doi:10.1093/gbe/evw003
Weiss RA (1967) Spontaneous virus production from "non-virus producing" Rous sarcoma cells. Virology 32:19–23
Weiss RA (2006) The discovery of endogenous retroviruses. Retrovirology 2006(3):67
Weiss RA, Esparza J (2015) The prevention and eradication of smallpox: a commentary on Sloane (1755) 'An account of inoculation'. Philos Trans R Soc Lond B 370: pii: 20140378. doi:10.1098/rstb.2014.0378
Weiss RA, Kellam P (1997) Illicit viral DNA. Nature 390:235
Weiss RA, Vogt PK (2011) 100 years of Rous sarcoma virus. J Exp Med 208:2351–2355
Woese CR, Fox GE (1977) Phylogenetic structure of the prokaryotic domain: the primary kingdoms. Proc Natl Acad Sci U S A. 74:5088–5090
Xu W, Stadler CK, Gorman K, Jensen N, Kim D, Zheng H, Tang S, Switzer WM, Pye GW, Eiden MV (2013) An exogenous retrovirus isolated from koalas with malignant neoplasias in a US zoo. Proc Natl Acad Sci U S A 110:11547–11552. doi:10.1073/pnas.1304704110
Yenerall P, Zhou L (2012) Identifying the mechanisms of intron gain: progress and trends. Biol Direct 7:29. doi:10.1186/1745-6150-7-29
Zhdanov VM (1975) Integration of viral genomes. Nature 256:471–473

Virus-Host Gene Interactions Define HIV-1 Disease Progression

Daniela C. Monaco, Zachary Ende and Eric Hunter

Abstract In this chapter, we will review recent research on the virology of HIV-1 transmission and the impact of the transmitted virus genotype on subsequent disease progression. In most instances of HIV-1 sexual transmission, a single genetic variant, or a very limited number of variants from the diverse viral quasi-species present in the transmitting partner establishes systemic infection. Transmission involves both stochastic and selective processes, such that in general a minority variant in the donor is transmitted. While there is clear evidence for selection, the biological properties that mediate transmission remain incompletely defined. Nevertheless, the genotype of the transmitted founder virus, which reflects prior exposure to and escape from host immune responses, clearly influences disease progression. Some escape mutations impact replicative capacity, while others effectively cloak the virus from the newly infected host's immune response by preventing recognition. It is the balance between the impact of escape mutations on viral fitness and susceptibility to the host immunogenetics that defines HIV-1 disease progression.

Dedication: This chapter is dedicated to Peter K. Vogt in honor of his 50 years as an editor of Current Topics of Microbiology and Immunology, his continued contributions to science and 45 years of mentorship and friendship.

D.C. Monaco · Z. Ende · E. Hunter
Emory Vaccine Center, Yerkes National Primate Research Center,
954 Gatewood Road, Atlanta, GA 30329, USA

E. Hunter (✉)
Department of Pathology and Laboratory Medicine, Emory University,
954 Gatewood Road, Atlanta, GA 30329, USA
e-mail: ehunte4@emory.edu

Current Topics in Microbiology and Immunology (2017) 407:31–64
DOI 10.1007/82_2017_33

Published Online: 27 June 2017

Contents

1 Introduction

The current HIV/AIDS pandemic reflects a combination of sub-epidemics that vary in routes of transmission, prevalence and incidence rates, different at-risk populations, and viral subtypes. Overall in 2014, an estimated 36.7 million people were living with HIV-1, and 2.1 million new HIV-1 infections occurred globally (UNAIDS 2016). HIV-1, the causative agent of the AIDS pandemic and a member of the *lentivirus* genus of the family *Retroviridae*, continues to contribute to over 1 million deaths globally every year and has been coevolving in its only permissive host, humans, for many decades.

As with all retroviruses, HIV-1 integrates a copy of its genome into the chromosome of the host cell and establishes a chronic infection that is subject to immune control by the host; although, HIV-1 infects a subset of cells, $CD4^+$ T cells, that play a key role in orchestrating the immune response. Nevertheless, HIV-1 induces robust humoral and cellular immune responses, the latter of which can limit virus replication to varying degrees (Walker and Yu 2013). Genome-wide association studies have indicated that while genes in the human leukocyte antigen (HLA) locus account for the bulk of host control over virus replication, HLA can only account for approximately one-quarter of the observed variation in setpoint viremia (Fellay et al. 2007; McLaren et al. 2015). In this review, we will discuss recent data that suggest both control of viral replication and disease progression are the result of complex and evolving interplay between the virus that initiates infection and the host immune response.

2 HIV-1 Transmission

Heterosexual transmission of HIV-1 remains the predominant mode of transmission, particularly in sub-Saharan Africa, accounting for nearly 75% of new transmissions worldwide. In contrast, in developed countries of the Western

Hemisphere, men who have sex with men (MSM) continue to represent a majority of those newly acquiring HIV-1. Transmission via direct blood-to-blood contact through intravenous drug use (IDU) or contaminated blood transfusions constitutes a small percentage (<10%) of infections in the U.S., but transmission among IDUs is a significant contributor in parts of Southeast Asia and Eastern Europe. The risk of sexually transmitted infection by HIV-1 is highly dependent on the route of transmission; exposure in MSM via the rectum has an estimated infection probability ranging from 1 in 20–300, whereas for infection via the male genital tract the risk ranges from 1 in 300–7000. Infection in the female genital tract carries a risk of approximately 1 in 200–2000 (Hladik and McElrath 2008). Frequencies of transmission vary based on a number of factors that include the stage of infection and the level of viremia in the transmitting partner, with the risk of infection from patients with acute and early infection being significantly higher than that from those with established infection (Brenner et al. 2007; Miller et al. 2010; Powers et al. 2008; Wawer et al. 2005). This likely reflects the high viral loads (VL) observed in acute infection (Pilcher et al. 2007), lack of neutralizing antibody which may otherwise inactivate a majority of circulating virions in established infection, and essentially clonal amplification of a virus capable of initiating productive infection (see below). Other factors that have been shown to modulate the efficiency of sexual transmission include sexually transmitted diseases, particularly those that result in genital inflammation and ulcers, which can elevate HIV-1 shedding into the genital tract and increase the risk of infection 21-fold (Galvin and Cohen 2004); pregnancy during which a greater than two-fold increase in HIV-1 transmission has been observed (Gray et al. 2005); and circumcision, which in a series of clinical trials was shown to decrease transmission to the male by 60% (Auvert et al. 2005; Bailey et al. 2007; Gray et al. 2007). Transmission may also be biased by the viral subtype present in the population (Kiwanuka et al. 2009; Kamali et al. 2015).

3 Current Concepts of Genital Tract Infection and Systemic Spread

The nonhuman primate model of HIV-1 infection has proven to be a powerful tool for experimentally investigating the transmission event since it is not possible to study the earliest steps of sexual transmission in human subjects. It is well established that HIV-1 has evolved from a simian immunodeficiency virus (SIV_{cpz}) that naturally infects a species of chimpanzee, *Pan troglodytes troglodytes*, while HIV-2 is derived from another that naturally infects sooty mangabeys (*Cercocebus atys*; SIV_{sm}) (Hahn et al. 2000). While relatively benign in its natural host, SIV_{sm} can be adapted to replicate in and induce a rapid immunodeficiency disease in rhesus macaque monkeys similar to that induced by HIV-1 in humans (Paiardini et al. 2009; Klatt et al. 2012). Pathogenic variants isolated from rhesus macaques, referred to as $SIV_{mac,}$ have been particularly informative in defining steps involved in transmission of and systemic infection by primate lentiviruses when used in intra-vaginal and intra-rectal challenge models (Hatziioannou and Evans 2012; Evans and Silvestri 2013). Studies using high doses of SIV_{mac251} led to a model whereby SIV interaction with the cervicovaginal mucosa induces an innate response, amplified by recruitment of macrophages and plasmacytoid dendritic cells, that signals activated T cells to migrate to the site of infection. This increased availability of $CD4^+$ target cells allows amplification of infection from the initially infected cell, a partially activated mucosal $CD4^+$ T cell, to a level where virus or virus-infected cells can exit the mucosal tissue and travel to distal sites. These include regional lymph nodes and the gut-associated lymphoid tissue (GALT), where the bulk of early viral replication and T-cell depletion occurs. This model of a localized inflammatory response playing a role in facilitating transmission is supported by experiments demonstrating that vaginal infection could be inhibited by local application of glycerol monolaurate, an inhibitor of inflammation (Li et al. 2009). However, the need for local amplification of infection was recently questioned following a large study, where animals were infected intravaginally with a high dose of SIV_{mac251} and serially necropsied on days 0, 1, 3, 7, and 10. In this study, virus was occasionally detectable in tissues distal from the genital mucosa, such as the gastrointestinal tract, by day 1, and 89% (8 of 9) of animals had detectable levels of viral RNA in at least one distal tissue by day 3, suggesting that viral dissemination can occur rapidly (Barouch et al. 2016). However, the role of these early distal infections in the establishment of systemic infection remains unclear since plasma viremia was not detected until day 10 in a majority of the animals. It is also important to recognize that these high dose infections may not be directly comparable to human sexual transmission, where the inoculum of infectious virus is likely to be much lower.

Although a variety of potential initial targets of infection have been postulated, including resting $CD4^+$ T cells, macrophages and dendritic cells, Hope and colleagues recently demonstrated that Th17-lineage $CCR6^+$ $CD4^+$ T cells are the predominant targets of SIV during vaginal transmission by using a high titer, single-round non-replicating SIV construct that expresses luciferase and m-Cherry, (Stieh et al. 2014, 2016). This cell type plays a key role in maintaining the integrity of the gut mucosa and is rapidly depleted following both SIV and HIV-1 infection (Blaschitz and Raffatellu 2010; Dandekar et al. 2010; Cecchinato and Franchini 2010).

4 HIV-1 Transmission is Linked to a Genetic Bottleneck

The concept that transmission of HIV-1 involves a genetic bottleneck, in which one or a limited number of variants from the diverse population present in the transmitting partner establish productive infection in the uninfected partner, was first established in studies over two decades ago. By analyzing viral sequences from early time points in primary HIV-1 infection, as well as, in some cases, viral sequences from the donors of a small number of linked heterosexual, homosexual, and mother-to-child transmissions, these studies demonstrated that the virus population in the newly infected individual was much less diverse than that in the transmitting partner or mother (Wolinsky et al. 1992; Wolfs et al. 1992; Zhu et al. 1993; Zhang et al. 1993). Wolfs et al. also observed that in the two heterosexual transmissions described, the transmitted virus appeared to be a minor variant in the blood of the transmitters.

The difficulty of obtaining samples at early time points after infection, and from donors of linked HIV-1 transmission pairs, hindered the investigation of transmitted viral characteristics for over a decade. At this time a more in-depth analysis was performed using samples from heterosexual transmissions in a previously established cohort of serodiscordant couples in Lusaka, Zambia (Derdeyn et al. 2004). Derdeyn et al. sequenced almost 300 cloned HIV-1 Subtype C Envelope gene amplicons derived from eight heterosexual transmission pairs shortly after transmission, and performed genetic and phenotypic studies (Derdeyn et al. 2004). The strength of this study was that the donor and recipient pairs were confirmed to be epidemiologically linked (Trask et al. 2002). A strong genetic bottleneck was observed in each transmission pair, in that the *env* sequences derived from each linked recipient emanated from a single branch on their respective donor *env* sequence phylogenetic tree, arguing that a single transmitted founder (T/F) virus established infection in each case.

More definitive analyses have relied on the use of end-point dilution PCR [termed single genome amplification or SGA (Salazar-Gonzalez et al. 2008)] to amplify sequences from multiple single genomes present in plasma very early after primary infection of individuals followed by direct sequencing of the DNA amplicon. This approach, in contrast to bulk PCR followed by cloning, avoids sequencing errors introduced by the Taq polymerase, in vitro recombination induced by template switching during the PCR reaction, and non-proportional representation of sequences as a result of template resampling. Furthermore, it has been possible to define the number of T/F variants, and the nucleotide sequence of each variant, by applying a mathematical model of early virus evolution to the SGA sequences (Lee et al. 2009). This model assumes that, in the absence of immune selection, replicating genomes accumulate random mutations at a constant rate defined in large part by the error rate of the reverse transcriptase. Using this approach, Keele and colleagues showed that 78 out of 102 subjects with acute subtype B HIV-1 infection had evidence of systemic infection by a single virus, while the remaining 24 had been infected by approximately two to five viruses (Keele et al. 2008). Applying this same method to 20 subtype A and C heterosexual transmission pairs for whom multiple sequences from both partners were derived, Haaland et al. (2009) determined that a single T/F virus established infection in 90% of cases, while an analysis of 69 newly infected subtype C individuals from South Africa by Abrahams et al. (2009) showed that 78% involved single variant transmission. It is clear from such studies, that, in situations where multiple viruses initiate infection, the number of infecting variants does not follow a Poisson distribution, with a majority involving only two or three variants but occasionally more than five. This is inconsistent with each variant being transmitted independently with low probability (Abrahams et al. 2009). Most likely, in these cases, factors such as sexually transmitted infections and potentially, in young women, the use of hormonal contraceptives, lowered the barrier to transmission (Haaland et al. 2009; Sagar et al. 2004).

Modulation of the genetic bottleneck: Inflammatory responses to both the existing microbiome as well as to sexually transmitted infections clearly increase the frequency of HIV-1 transmission. Recent studies of young women in a South African cohort have shown that in those where the microbiome is deficient in *Lactobacillus* species, and yet diverse with *Prevotella* and *Gardnerella* species, there are increased genital pro-inflammatory cytokine concentrations (Anahtar et al. 2015). This ecologically diverse, *Lactobacillus* deficient microbiome was also associated with a higher risk of HIV-1 acquisition (Gosmann et al. 2017). Moreover, sexually transmitted infections (STIs), including herpes simplex type 2, which induce inflammation and ulcers in an uninfected partner, as well as similar infections in the transmitting partner, are known to increase the risk of transmission (Galvin and Cohen 2004).

Although not necessarily impacting the multiplicity of infection, viral load in the transmitting partner has been shown to modulate the likelihood of transmission. Studies in HIV-1 discordant couples have shown that partners with VL less than 10,000 copies/ml only rarely transmitted to their partners, while those with VL greater than 100,000 copies/ml transmitted much more frequently, with on average a 2.5-fold increase in risk with each $\log_{10}$ increase in VL (Quinn et al. 2000; Fideli et al. 2001). In Zambian discordant couples this increased risk was most evident in female to male (FTM) transmissions (Fideli et al. 2001), which may reflect higher VL in the genital tract of the transmitting partner (Pilcher et al. 2007) and a greater chance of virus reaching the genital mucosa. This is also consistent with NHP studies where both the frequency of infection and the number of transmitted variants increased with the dose of the inoculum (Liu et al. 2010). Moreover, in the macaque model, SIV in the plasma from animals in the acute stage of infection, where potentially neutralizing antibodies are absent, has a specific infectivity almost 100 times greater than that of virus in the plasma from chronically infected animals (Ma et al. 2009).

As highlighted by Joseph et al. (2015) and shown in Fig. 1, virus transmission can fail at multiple steps following inoculation onto a mucosal surface (Joseph et al. 2015). Thus, STIs could abrogate the barrier imposed by an intact mucosa by inducing breaks in the epithelial lining thereby allowing more viral variants to initiate infection in the mucosal tissue; alternatively, inflammation induced by genital infections could increase the availability of activated $CD4^{+}$ cells required to establish a spreading infection, in this way allowing infections that would have failed due to lack of target cells to expand. It is still not defined, under conditions of low multiplicity of infection, where the probability of infection is less than 1% per coital act, how many viruses initiate an abortive infection in the mucosa.

It has been possible to reproduce the transmission-linked genetic bottleneck in the nonhuman primate model of HIV-1 infection, where rhesus macaque monkeys

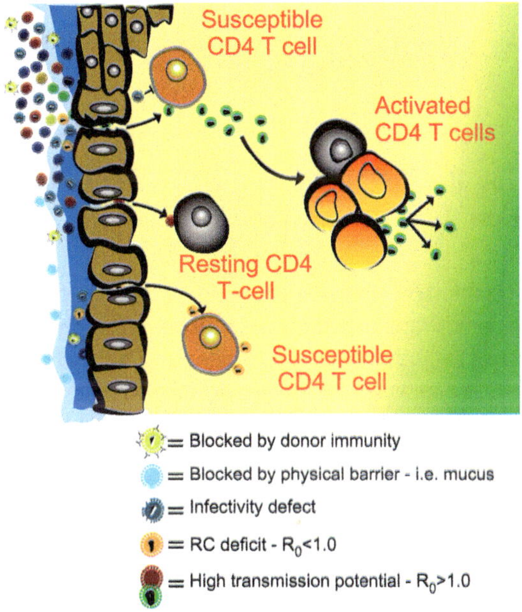

Fig. 1 Potential barriers to HIV transmission across the genital mucosa. During transmission, the virus encounters a number of barriers to infection including mucous and epithelial layers that can block access to target cells. However, viruses can penetrate host defenses through temporary breaks in the epithelium or dendritic cell sampling the mucosal environment. The entering viruses must interact with susceptible CD4+ CCR5+ T cells to propagate since entry into nonpermissive resting CD4+ T cells will result in nonproductive infection. Similarly, if the RC of the virus is unable to sustain a spreading infection (R0 < 1.0), even infection of susceptible cells will not result in dissemination. It is likely that viruses with a replicative advantage will outcompete those that replicate less efficiently. Initial target cells are most likely susceptible CD4+ CCR5+ T cells, a majority of which may be of a Th17 lineage, but infection of macrophages and dendritic cells has also been reported. These cells can replicate virus locally or traffic to local secondary lymphoid structures, though virus could also diffuse there directly. Once virus reaches local lymph nodes and disseminates throughout the body, specifically to the gut mucosa, viral load increases exponentially in the blood. Approximately 70– 80% of mucosal infections are established by single variants. Adapted from Ende and Hunter (Ende et al. 2017)

are challenged repeatedly with low doses of SIV intra-vaginally or intra-rectally. Macaques challenged multiple times via the rectal route with a quasi-species of SIV_{mac251} or SIV_{smE660} were found, using the same SGA approach as in infected people, to be infected with a limited number of genetic variants—a majority with a single variant (Keele et al. 2009). The kinetics of virus replication in these animals resembled that observed in acutely infected people (Fiebig et al. 2003). A similar genetic bottleneck, with predominantly single variants establishing infection, was observed when macaques were challenged intra-vaginally or through the penile route, although both were significantly less efficient. Increasing the challenge dose via the intra-rectal route above 10^7 viral RNA copies resulted in infection being established by multiple (>10) T/F variants, suggesting that in this model system the genetic bottleneck could be overcome by increasing the size of the input inoculum (Liu et al. 2010).

5 Evidence for Both Chance and Selection Influencing Transmission

The question of whether HIV-1 transmission is predominantly a stochastic process, where a single genetic variant establishes infection simply because of the low probability of transmission, has been the focus of much debate over the last several years. Alternatively, certain viral phenotypes, which confer enhanced transmissibility on a viral variant, could be selected for during transmission and systemic spread.

It is very clear that some aspects of the transmission process do involve chance: the genetic variant must be present in the genital fluid of the transmitting partner at the time of intercourse; it must interact with the genital or rectal mucosa; it must cross the epithelial barrier and infect a susceptible $CD4^+$ T cell; and it must have a sufficient number of secondary target cells for infection to spread and establish a localized and then a systemic infection [Fig. 1; (Joseph et al. 2015)]. Despite these stochastic aspects of transmission, there is strong evidence that selection pressure is applied and that viruses with specific traits are selected for during the transmission process.

HIV-1 encodes two envelope (Env) glycoproteins gp41 and gp120 that form hetero-trimers on the surface of the virus, where three molecules of the membrane-spanning gp41 anchor an equivalent number of gp120 molecules, through non-covalent associations, in the viral membrane (Hunter 1997). Initially synthesized as a single precursor gp160, which is proteolytically cleaved during transport to the cell surface, these two glycoproteins are critical for viral entry into a target cell. The surface protein, gp120, contains the receptor binding domains, interacting first with CD4 molecules that are expressed primarily on a subset of T cells, $CD4^+$ T cells, and macrophages. CD4 binding induces a conformational change in the trimer, leading to interaction with a co-receptor and further conformational changes that allow gp41 to mediate fusion of the viral and target cell membranes (Wilen et al. 2012). Most viruses utilize the chemokine receptor, CCR5 (Deng et al. 1996; Choe et al. 1996; Dragic et al. 1996; Berger et al. 1999; Feng et al. 1996; Alkhatib et al. 1996), as their co-receptor but during chronic infection viruses can evolve to utilize a second chemokine receptor, CXCR4 (Coetzer et al. 2008; Regoes and Bonhoeffer 2005).

Initial evidence for selection came from the finding that the bulk of newly transmitted viruses utilized the CCR5 co-receptor, even if CXCR4 viruses were present in the transmitting partner, providing the first evidence that CXCR4 tropic viruses were selected against, while CCR5 tropic viruses were selected for, in transmission. The discovery that persons at high risk of HIV-1 infection and homozygous for a deletion of 32 amino acids in their CCR5 gene were protected from acquiring HIV-1 by mucosal exposure, provided further evidence that CXCR4 viruses were selected against during HIV-1 transmission and that predominantly CCR5-tropic viruses could infect by this route (Liu et al. 1996; Dean et al. 1996; Zimmerman et al. 1997; Samson et al. 1996; Michael et al. 1997). The propensity for CCR5-tropism remains unexplained (Margolis and Shattock 2006), but could be due to target cell availability at portals of entry (Liu et al. 2014). This would be consistent with the observation that $CCR5^{-}CXCR4^{+}$ $CD4^{+}$ T cells are a minor population (<25%) in both the epidermis and dermis of inner and outer foreskin compared to $CCR5^{+}CXCR4^{-}$ and $CCR5^{+}CXCR4^{+}$ $CD4^{+}$ T cells (Liu et al. 2014), and that CCR5 expression is high on the surface of human vaginal epithelial $CD4^{+}$ T cells (Hladik et al. 2007). It is also consistent with observations in monkey challenge studies, where the number of $CCR5^{+}$ T target cells in the mucosa correlated with susceptibility to infection (Pandrea et al. 2008; Pandrea and Apetrei 2010).

The concept of selection during transmission has also been supported by phylogenetic analyses of HIV-1 sequences from heterosexual transmission pairs, which suggest that evolution during chronic infection may reduce transmissibility of the virus and favor transmission of earlier less-evolved variants from the transmitting partner. By calculating evolutionary distances of recipient and donor Env sequences to their most recent common ancestor for 10 subtype D and 10 subtype A linked transmission pairs, Sagar et al. provided the initial evidence that newly infecting (recipient) viruses were evolutionarily closer to the most recent common ancestor (MRCA) than transmitting partner (donor) viruses (Sagar et al. 2009). Consistent with the concept of transmission of less-evolved viruses, intra-host diversity was found to be greater than inter-host diversity in two separate cohorts infected with subtypes A or D and B (Redd et al. 2012; Alizon and Fraser 2013). Moreover, a longitudinal analysis of the *env* gene of viruses from donors prior to transmission, also found that the virus that established infection in the recipient more closely resembled earlier viruses in the donor than the viruses circulating at the time of transmission (Redd et al. 2012).

These observations have been supported by a more recent study that included genes outside of *env*. In an analysis of 137 epidemiologically linked clade C heterosexual transmission pairs, Carlson et al. demonstrated a selection bias in favor of cohort consensus amino acid residues and against non-consensus polymorphisms in Gag, Pol and Nef, suggesting a transmission advantage for variants with consensus amino acid residues in proteins outside of Env (Carlson et al. 2014). This was particularly obvious following deep sequencing of the donor quasi-species and the newly infecting viruses from five transmission pairs. As shown in Fig. 2, the transmission efficiency of non-consensus polymorphisms was reduced by approximately 20% compared to consensus amino acids for residues present at both high and low frequencies in the donor virus population. Non-consensus polymorphisms were predicted to reduce the structural stability of viral proteins, consistent with reduced in vivo replicative fitness in their presence. The selection pressure for consensus residues was influenced by gender, with female to male transmission imposing a greater selection bias on the virus than male to female transmission, suggesting that women are infected with less fit viruses than men. Interestingly, selection bias was reduced in men with genital ulcers and inflammation (GUI) and when the donor partner exhibited high viral load (Carlson et al. 2014). Both factors are known to increase the risk of infection and number of genetic variants transmitted (see above), demonstrating the interplay between factors that influence general susceptibility and stringency of the genetic bottleneck. Preferential transmission of viruses closer to consensus was confirmed over the full genome in the same cohort with six transmission pairs using SGA (Deymier et al. 2015). These findings suggest that diversification and adaptation to immune responses, which occur in the virus population during chronic infection, may hinder its ability to transmit. It is possible, therefore, that if a T/F virus establishes infection and remains in a reservoir that is produced sporadically or has a slow turnover rate, then these viruses over time will be present as minor variants in the host and will also retain the characteristics necessary to transmit again. Recent studies in the Amsterdam cohort, where the phylogenetic relationship of proviruses present in the viral reservoir following suppressive anti-retroviral treatment to viruses from several time points prior to treatment, are consistent with this hypothesis. In several individuals, viral reservoir sequences were PCR amplified that were highly related to viruses observed in acute/early infection, although in each case they were a minority of the population (Brodin et al. 2016).

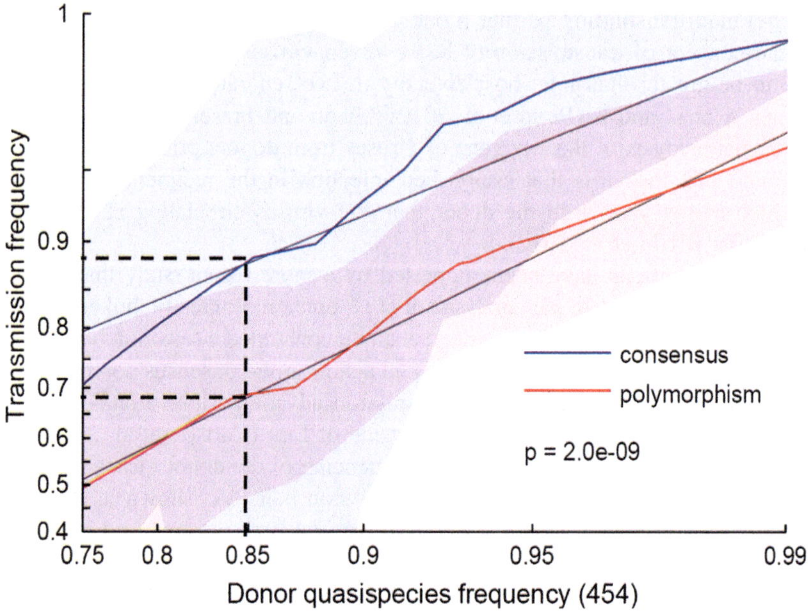

Fig. 2 Selection bias in heterosexual HIV-1 transmission. The odds that the transmitting partner's amino acid will be transmitted to the recipient is a function of the relative frequency of the amino acid in the donor virus quasi-species. The plot shows the empirical transmission probability (odds on a log10 scale) of a variant as a function of the relative in vivo frequency of the variant in the donor quasi-species, with a near 1-to-1 mapping for variants that match cohort consensus. In contrast, polymorphisms are uniformly less likely to be transmitted. Adapted from (Carlson et al. 2014)

Some of the first observations of a genetic bottleneck suggested that transmitted viruses were minor variants of the donor's plasma. In one of the initial studies that compared viral genetic diversity within and between partners in transmission pairs by analyzing virus from plasma, seminal fluid and seminal cells, the transmitted virus could be identified in both cell-free and cell-associated forms in the donor genital tract, and it was generally a minor variant of the genital tract (Zhu et al. 1996). A more recent analysis utilizing SGA of the V1–V4 region of Env from genital tract and plasma samples of eight subtype C infected heterosexual transmission pairs reported very similar results. It showed that, despite significant compartmentalization of viral genotypes with discrete populations within the genital compartment, the virus in the donor that most closely resembled the T/F was a minor variant, either of the genital tract or the plasma (Boeras et al. 2011). While genital tract enriched populations may be transient, and virus populations cannot be sampled precisely at the time of transmission (Anderson et al. 2010; Boeras et al. 2011), both studies do provide additional evidence in favor of HIV-1 selection during transmission.

6 Properties of the Transmitted/Founder Virus

Evidence of the genetic bottleneck and of selection during sexual transmission of HIV-1 has stimulated efforts to define biological characteristics of T/F viruses that could favor their transmission over a majority of the viruses that are circulating in the transmitting partner's quasi-species. To date, other than CCR5 tropism, no single trait has been consistently identified across the different studies and cohorts reported. In some respects, this could reflect the differences in the cohorts under study (e.g., MSM vs. heterosexual), as well as the stringency of the barriers to infection the virus must face (e.g., in the absence or presence of inflammation). Nevertheless, a number of properties have been linked to transmissibility and, while these have been addressed in recent reviews (Joseph et al. 2015; Ende et al. 2017, in press), will be summarized briefly here.

Co-receptor utilization: The observation that CCR5-tropic viruses are preferentially transmitted has been reproduced in most studies, including those where discrete T/F virus envelopes and full-length viruses were examined (Baalwa et al. 2013; Isaacman-Beck et al. 2009; Keele et al. 2008; Long et al. 2002; Parrish et al. 2013); however, this is not invariant, and infrequent CXCR4-tropic or dual-tropic transmitted strains have been observed. Although macrophages were once considered potential Trojan horses for carrying HIV-1 across the mucosa, it should be noted that T/F viral Envs mediate inefficient infection of macrophages and show a requirement for high levels of both CD4 and CCR5, with no evidence for preferential use of alternate coreceptors (Keele et al. 2008; Sagar et al. 2009; Salazar-Gonzalez et al. 2009; Isaacman-Beck et al. 2009; Alexander et al. 2010). These studies argue that neither infection of macrophages nor alternate coreceptor usage is advantageous for HIV-1 transmission.

Variable loop size and neutralization sensitivity: A major clue that transmission might select for traits other than co-receptor usage came from a comparison of the viral Env sequences from both partners of seven subtype C and one subtype G HIV-1 transmission pairs. It was found that within each pair, whether male-to-female or female-to-male, the newly transmitted viruses encoded statistically shorter, less glycosylated V1–V4 regions than their chronic counterparts (Derdeyn et al. 2004), raising the possibility that more compact envelope glycoproteins better interacted with critical target cells in the genital mucosa. The observation was confirmed using SGA in an additional 10 subtype C transmission pairs (Haaland et al. 2009). While similar results were obtained in studies of subtype A infected sex-workers in Kenya and subtype D and A transmission pairs from the Rakai district of Uganda (Chohan et al. 2005; Sagar et al. 2009), they have not been seen in most studies of recently transmitted subtype B HIV-1 (Chohan et al. 2005; Frost et al. 2005; Wilen et al. 2011). However, in a study comparing the SGA-derived Env sequences from 135 acutely infected and 140 chronically clade B HIV-infected individuals, statistically fewer N-linked glycosylation (PNLG) sites were found in the gp120s from early infection, with a trend toward fewer PNLG in the V1V2 loops and reduced V4 lengths (Gnanakaran et al. 2011).

An analysis of neutralization of subtype C heterosexually transmitted viruses demonstrated modestly increased sensitivity to antibodies in linked donor plasma taken near the time of transmission (Derdeyn et al. 2004; Deymier et al. 2015), though not to broadly neutralizing antibodies (Parrish et al. 2012) or pooled plasma (Derdeyn et al. 2004). It is possible that bound antibodies could enhance infection through capture by dendritic cell via Fc receptors in the mucosa of the new host, as has been reported for infected volunteers in the VAX004 vaccine trial, but this has not been demonstrated in non-vaccinated populations (Forthal et al. 2012). It is likely that donor antibody sensitivity reflects a surrogate marker of a different phenotype, such as mutational escape away from consensus, which in a recent study inversely correlated with donor antibody sensitivity over six subtype C transmission pairs (Deymier et al. 2015).

Interactions with the integrin $\alpha 4\beta 7$: $CD4^+$ T cells expressing $\alpha4\beta7$ home to mucosal sites, including the genital and gastrointestinal tract (Hawkins et al. 2000), and are highly susceptible to HIV-1 infection (Cicala et al. 2009). This susceptibility is likely facilitated by HIV-1 gp120 binding to $\alpha4\beta7$ via a motif in the second variable region, V2 (Arthos et al. 2008). A comparison of Envs from early in infection to later isolates from the same individual showed early high-affinity binding to $\alpha4\beta7$ that was lost over time. This appeared in part to be due to the absence of glycosylation at specific sites in V1 and V2 since mutation of these sites in the chronic envelopes increased $\alpha4\beta7$ binding (Nawaz et al. 2011). In addition, an analysis of viruses from the CAPRISA acute infection cohort from South Africa showed that dependence on $\alpha4\beta7$ for in vitro replication was high for T/F Env chimeras, particularly those encoding a P/SDI/V tri-peptide binding motif in the V2 region of gp120. This dependence on $\alpha4\beta7$ was lost during the first two months of infection, but regained at 39 months post infection for three individuals followed longitudinally (Richardson et al. 2015). An earlier comparison of subtype C T/F and chronic viruses had not observed differential inhibition of infection by a blocking antibody to α4β7 (Parrish et al. 2012); however, dissecting inhibition from antibody binding activation of cells has complicated the interpretation of this data.

Despite the mixed results of in vitro studies, administering a blocking antibody to α4β7 prior to SIV challenge in rhesus macaques decreased the number of animals infected and increased the number of challenges for infection to occur. Moreover, treated but infected animals showed a significant reduction in $CD4^+$ T cells loss in gut-associated lymphoid tissue (GALT) and evidence for limited trafficking of infection out of the genital mucosa (Byrareddy et al. 2014). Thus viruses with enhanced α4β7 affinity may possess increased transmissibility through the increased efficiency by which virus-infected cells are trafficked to the GALT.

Sensitivity to type I interferons: The innate immune response, in particular the production of type 1 interferons (IFN), is very important in a number of viral infections including lentiviruses (Doyle et al. 2015). Treatment of rhesus macaques with IFNα2 increased the number of challenges required to establish systemic SIV_{mac} (Sandler et al. 2014) and SHIV infection (Veazey et al. 2016). However, while IFNs are upregulated in the early stages of SIV infection in macaques (Abel et al. 2005), a recent large study of acute infection found that prior to day 10, when plasma viremia was apparent, SIV_{mac239} down-regulated the IFN response in cells it infected and instead stimulated an inflammasome response (Barouch et al. 2016).

Nevertheless, a number of recent papers have presented evidence that T/F variants are relatively resistant to IFN compared to viruses from chronic infection (Parrish et al. 2013; Foster et al. 2016; Iyer et al. 2017), although in a large study comparing T/F and chronic circulating viruses, subtype differences were observed. In this study, where subtype B T/F infectious molecular clones (IMCs) were less susceptible to IFNα, subtype C T/F, and chronic variants exhibited similar sensitivity to interferon (Parrish et al. 2013). In contrast, in a comparison of subtype B and C T/F variants and their matched 6-month post-infection and chronic infection counterparts, the T/F isolates were found to be more resistant to IFNα (Fenton-May et al. 2013). A specific restriction factor associated with co-receptor usage, IFN-induced transmembrane protein 1 (IFITM1), determined the resistance phenotype of the T/F and 6-month virus pairs (Foster et al. 2016), though VPU-tetherin interactions have also been implicated as major determinants of resistance for some of these variants (Kmiec et al. 2016). Both interferon-induced restriction factors IFITM1 and tetherin act at the plasma membrane and interact with the viral Env, which has been associated with IFN resistance in an investigation of SHIVs passaged in rhesus macaque cells in vitro (Boyd et al. 2016). In addition, a recent comparison of viral outgrowth isolates from eight transmission pairs, where the transmitting partner had viral loads exceeding 1×10^5 copies/ml, showed that all of the isolates from acute plasma were more resistant to both IFNα and IFNβ than those from the transmitting partner (Iyer et al. 2017).

However, not all studies have reported results consistent with these most recent observations. In an analysis of six subtype C transmission pairs, where IMCs representing the T/F virus from the newly infected partner and representative viruses from the transmitting partner were generated, replicative capacity (RC) was found to be positively associated with IFNα inhibition of replication, and a comparison of viruses matched for RC showed no consistent evidence of enhanced interferon resistance for T/F variants (Deymier et al. 2015). Furthermore, in an independent study, T/F isolates from 9 subtype B transmission pairs showed greater rather reduced sensitivity to IFN (Oberle et al. 2016), as did acute Env chimeras derived from 7 acute subtype B IDU infections when compared to chronic controls (Etemad et al. 2014). The conflicting results may stem from differences in the subjects under study, or the approaches taken to isolate infectious virus and assess IFN resistance. Additional studies where viruses from different HIV-1 subtypes and derived from both partners of transmission pairs very near the time of transmission are investigated will be critical to resolving the differences currently observed. Given the known impact of genital inflammation and ulcers in the uninfected partner, and VL in the chronically infected partner, on susceptibility to infection and the genetic bottleneck, it is likely these factors will need to be taken into account when comparing studies.

Infectivity and Replicative Capacity: One of the most compelling hypotheses regarding HIV transmission proposes that T/F variants replicate faster than other variants, granting a competitive advantage during the initial events of viral growth and dissemination (Shaw and Hunter 2012). This would also be compatible with the selection bias for consensus amino acid residues, which are predicted to increase structural stability and presumably function of the Gag and Pol proteins (Carlson et al. 2014). In general, studies have compared both infectivity and replication, in single and multi-round infection assays, respectively. Evidence that transmitted variants have enhanced infectivity is inconsistent and varies between cohorts and studies (Parrish et al. 2013; Selhorst et al. 2017b; Deymier et al. 2014; Oberle et al. 2016; Iyer et al. 2017), but is generally quite subtle (2-3fold) when it has been observed. Interestingly, an analysis of virus isolates and Env pseudoviruses from the CAPRISA 004 tenofovir gel trial showed that variants transmitted to women with genital inflammation were less infectious (Selhorst et al. 2017b). Genital ulceration and inflammation are known to increase the number of transmitting variants (Haaland et al. 2009) and decrease selection pressure (Carlson et al. 2014), consistent with this observation. For subtype B and C Env-pseudotyped (Isaacman-Beck et al. 2009; Oberle et al. 2016) and subtype B and C full-length (Parrish et al. 2013; Deymier et al. 2015; Yue et al. 2015; Ochsenbauer et al. 2012; Salazar-Gonzalez et al. 2009; Oberle et al. 2016) T/F variants, a 50–100-fold range of infectivity was found, similar to that seen in the non-transmitted variants. Chimeric viruses containing Gag from acute and early time points from subtype B (Brockman et al. 2010) and subtype C (Wright et al. 2010; Prince et al. 2012; Claiborne et al. 2015) strains also exhibited a broad range of RCs, which correlated well with those of IMCs derived from the same patients (Claiborne et al. 2015), suggesting that variants with relatively low and high RC in vitro have the potential to transmit. Similarly, most studies have not demonstrated a replicative advantage for transmitted variants measured in multiple round infections in vitro, instead showing transmission of viral variants with a range of RCs (Parrish et al. 2013; Deymier et al. 2015; Yue et al. 2015; Ochsenbauer et al. 2012; Salazar-Gonzalez et al. 2009; Oberle et al. 2016). Although, a recent study of eight subtype B and C transmission pairs did find that early isolates replicated somewhat (1.4-fold) higher than variants from matched donor partners (Iyer et al. 2017). Nevertheless, current data suggest that, in order to establish infection, the range of infectivity and RC, at least as measured in vitro, can be broad. However, as we will discuss below, the in vitro RC of the T/F virus can have a profound impact on both early pathogenesis and long-term trajectory of disease.

7 A Complex Interplay Between Host Immunity and Transmitted Virus Phenotype Defines Viral Control and Disease Progression

It is clear that traits of the transmitted virus influence the course of the disease. Despite human immunogenetic variability, high viral mutation rates leading to adaptation, and a restrictive genetic bottleneck during transmission, VL set-point in the donor correlates with that in the infected partner, although it is significantly modulated by the sex and immune response genes of the newly infected individual (Hecht et al. 2010; Yue et al. 2013). The heritability of VL suggests that disease progression itself can have a heritable component (Fraser et al. 2014) since VL is a strong predictor of disease progression (Mellors et al. 1995).

As we discuss in more detail below, the cellular immune response of a newly infected individual, programmed by their human leukocyte antigen (HLA) class I alleles, imposes selection pressures on the virus that result in the outgrowth of viruses with mutations in the controlling epitopes. The impact of these mutations on the T/F virus following transmission is complex. If these mutations are in epitopes normally recognized by the newly infected individuals class I alleles, they can reduce the ability of the immune system to control the replication of the virus and, therefore, result in enhanced disease progression (Carlson et al. 2016; Crawford et al. 2009; Monaco et al. 2016). On the other hand, although beneficial for the survival and propagation of the virus in the chronically infected partner, these mutations can negatively impact the replicative fitness of the virus when it is transmitted to an individual that does not share HLA-I alleles that recognize the epitope (Brockman et al. 2010; Chopera et al. 2008; Goepfert et al. 2008). The remainder of this review will discuss this complex interaction between the transmitted virus and its new host.

Host Control of Virus Replication: The cellular immune response plays a central role in controlling HIV-1 viral replication. In humans, the most compelling evidence in favor of this role comes from the consistent observation of a temporal association between the rapid decline in viremia during acute infection and the increase in numbers of HIV-specific cytotoxic T lymphocytes (CTLs) in the blood (Koup et al. 1994; Borrow et al. 1994). Studies in a nonhuman primate model of SIV or SHIV infection have provided the most direct evidence by depleting $CD8^+$ T cells at different stages of infection. During acute infection, transient depletion of $CD8^+$ T cells leads to increased plasma and cell-associated virus levels in both the peripheral blood and lymphoid tissues along with prolonged depletion of $CD4^+$ T cells and accelerated disease progression (Matano et al. 1998; Schmitz et al. 1999). Similarly, depletion of $CD8^+$ T cells during chronic infection results in a rapid increase in viremia that is again suppressed with the reappearance of SIV-specific CD8 + T cells (Schmitz et al. 1999; Jin et al. 1999). More recently, the role of CTLs in controlling viremia has also been demonstrated even in the presence of antiretroviral treatment (ART). Depletion of $CD8^+$ cells during ART significantly increased plasma VL and reconstitution of these cells was associated with re-establishment of viral control, providing a rationale for the administration of therapeutic vaccines in conjunction with ART (Cartwright et al. 2016).

Selection of CTL escape mutations: The changes in viremia associated with the presence or expansion of HIV-specific CTLs shows that these cells exert immunologic pressure on the virus, though they are still unable to clear or completely control the infection in the majority of infected individuals. A clear manifestation of this immunologic pressure is the appearance of escape mutations. These mutations have been identified both in acute and chronic infection and in all HIV-1 proteins, including accessory proteins, which indicates that the CTLs are constantly and widely targeting the virus during HIV-1 infection (Borrow et al. 1997; Goulder et al. 1997).

CTL escape mutations are thus able to release the pressure exerted by HIV-specific CTLs, and function at three different stages during the process of antigen presentation. First, these mutations can prevent the processing of the protein via the proteasome, abrogating the generation of the epitope before being loaded onto the HLA Class I molecule. Second, CTL escape mutations can compromise the loading of the HLA Class I molecule, which occurs in the endoplasmic reticulum before the loaded complex travels to the cell surface, by reducing its binding affinity for the epitope (Yokomaku et al. 2004). Finally, these mutations can reduce or prevent the interaction of the T Cell Receptor (TCR) with the HLA Class I-epitope complex (Iglesias et al. 2011).

While initial studies focused on mutations identified by subsequent increases in viremia in individuals harboring a particular HLA-I allele, CTL escape mutations have been identified more recently in population analyses using statistical methods. This approach is able to identify HIV-1 polymorphisms that are significantly more prevalent in individuals harboring a particular HLA-I allele, while at the same time correcting for covarying sites, linkage disequilibrium among HLA-I alleles, and mutations that could have arisen during the evolutionary history of the virus instead of in response to HLA-mediated immune pressure (Bhattacharya et al. 2007; Carlson et al. 2008).

Impact of CTL escape mutations on chronic and acute HIV-1 infection: CTL escape mutations are rapidly selected after the cytotoxic immune response is mounted (Fischer et al. 2010; Henn et al. 2012), and the time in which these mutations arise has been shown to be much faster in response to protective HLA alleles such as HLA B*57 or HLA B*27 (Roberts et al. 2015). Although these mutations confer a fitness advantage to the virus by preventing CTLs from targeting infected cells, they usually carry a significant replicative fitness cost (Martinez-Picado et al. 2006). However, because HIV-1 is a chronic disease, with continued virus replication, secondary or compensatory mutations frequently arise that reduce the impact of CTL escape mutations on replication, and allows for their maintenance (Brockman et al. 2007; Schneidewind et al. 2007).

Direct evidence for an impact on HIV-1 disease following transmission of CTL escape mutations has come from mother-to-child transmission (Goulder et al. 2001) as well as from serodiscordant couples (Goepfert et al. 2008; Crawford et al. 2009; Monaco et al. 2016; Carlson et al. 2016). In both cases, the viral quasi-species present in the donor at the time of transmission can be studied in order to distinguish transmitted mutations from those that arise early in the newly infected individual. Even though HLA B*27 and HLA B*57 are generally associated with protection against disease progression in HIV-infected individuals (Kaslow et al. 1996; Fellay et al. 2007; Kiepiela et al. 2007; Schneidewind et al. 2007), these earlier studies showed that transmitted CTL escape mutations, and the pre adapted epitopes associated with these alleles, are linked to rapid loss of control of viral replication and accelerated disease progression in HLA-matched individuals (Goulder et al. 2001; Crawford et al. 2009). These findings highlighted the fact that, when transmitted to HLA-matched individuals, CTL escape mutations abrogate protection against disease progression.

Population studies of the prevalence of CTL escape mutations indicate that these mutations are accumulating both over time (Cotton et al. 2014; Dilernia et al. 2008) and in relation to HLA allele prevalence (Kawashima et al. 2009). This phenomenon can lead to the circulation of viruses that are potentially already adapted to the HLA alleles in the HIV-1 negative population. While a study of the North American epidemic did not observe significant adaptation over time in circulating viruses over 20 years, this could have been related to the high HLA diversity and low genetic frequency of each allele in this population (Cotton et al. 2014). Since transmission of pre-adapted viruses compromises the ability of the CTL immune response to target these viruses in the newly infected individual, this raises the possibility that pre-adaptation of the T/F virus could have a significant impact on both viral control and disease progression.

This question was answered in recent studies of epidemiologically linked HIV-1 transmission pairs from a Zambian acute infection cohort, where the degree of preadaptation of the transmitted viruses was accurately estimated by focusing on polymorphisms present both in the donor and in the recipient. These polymorphisms were located both in positions statistically linked to the HLA alleles of the recipient and also in epitopes with a higher predicted HLA-I binding affinity. This analysis showed that approximately one-third of possible HLA-linked target sites were already adapted in T/F viruses and that this pre-adaptation compromised early immune recognition of the transmitted virus. Evaluation of CTL responses against adapted (meaning harboring HLA-linked mutations) and nonadapted epitopes across the entire Gag protein according to each individual's HLA alleles showed that adapted epitopes were significantly less recognized than nonadapted epitopes and, in the cases where an IFN-γ response was detected, it was of a lower magnitude. Moreover, in individuals that were infected with viruses where 50% of the HLA-linked target sites were pre-adapted, significantly fewer and less potent IFN-γ responses were elicited (Monaco et al. 2016). A parallel study, in which Carlson and colleagues used a mathematical model to estimate autologous adaptation of the transmitted virus, showed that CTLs targeting adapted epitopes had reduced antigen sensitivity and killing capacity when compared to those targeting nonadapted epitopes in both structural and accessory proteins (Carlson et al. 2016).

The degree of pre-adaptation in the Gag protein of a transmitted virus was also associated with higher viral loads and faster decline of $CD4^+$ T cells (Monaco et al. 2016), and this association was also true when the analysis was extended to Gag, Pol and Nef proteins (Carlson et al. 2016). However, this association with VL was stronger after accounting for polymorphisms that were not linked to the HLA molecules of the newly infected individual (nonassociated polymorphisms), which conversely associated with lower VL and slower decline of $CD4^+$ T cells. These results illustrate how HIV-1 disease progression is ultimately dictated by the balance between the two opposing forces of transmitted pre-adaptation, which determines the number and quality of epitopes effectively targeted—both initially and subsequently during the course of infection—and nonassociated polymorphisms, which likely lead to reduced viral replicative fitness (Monaco et al. 2016) (Fig. 3).

Importantly, the role of pre-adaptation and transmitted nonassociated polymorphisms on viral control and disease progression remained significant in the context of other factors known to influence clinical outcomes. In the case of disease

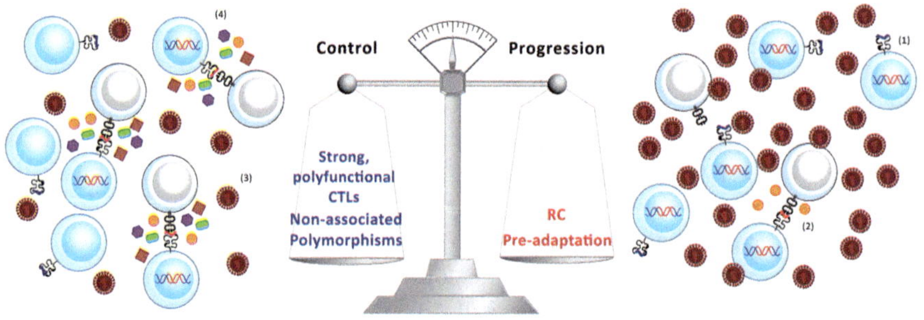

Fig. 3 The balance between transmitted viral characteristics and early CTL immune responses determines disease progression. High pre-adaptation and RC of the transmitted virus will tip the scale in favor of rapid disease progression. Pre-adaptation leads to impaired cytotoxic immune responses since the transmitted escape mutations prevent HLA molecules from presenting HIV epitopes (1) or lead to the presentation of epitopes that are poorly recognized by the TCR on the CTLs (2). High replication contributes with larger viral loads and proportions of infected cells, including the subsets associated with latency. On the other hand, large numbers of nonassociated polymorphisms and strong, poly-functional cytotoxic immune responses tip the scale in favor of slow disease progression or control. These polymorphisms impair virus replication (3) but may as well negatively impact innate immune responses while the activity of strong, poly-functional CTLs contributes to reducing viral load (4)

progression, a ratio between pre-adaptation and transmitted nonassociated polymorphisms was the strongest predictor of decline of $CD4^{+}$ T cells, followed by set-point VL and RC. The impact of pre-adaptation may reflect the reduced capacity of the immune system to target the virus both early in infection as well as late in infection, as the virus continues to adapt.

Taken together, these recent studies have significant implications for vaccine development. Immunogens harboring escape mutations may not be able to effectively prime an immune response. On the other hand, efficiently primed responses targeting epitopes where CTL escape mutations are frequently transmitted in a population may be less able to control those pre-adapted viruses. In this context, the use of immunogens that focus on inducing CTLs against conserved regions of the HIV-1 proteome where selection and transmission of CTL escape mutations are less frequent could help overcome these challenges.

Impact of immune selection on replicative capacity and subsequent disease progression: In contrast to the disease-enhancing effects of CTL escape mutations that represent preadaptation in the new host, CTL escape mutations transmitted to HLA mismatched individuals show the opposite effect by providing an advantage for the newly infected individual that is associated with a loss of replicative fitness. The fact that these detrimental mutations usually revert or are compensated for shortly after transmission is consistent with their impact on virus replication since natural selection favors viruses with higher replicative fitness (Crawford et al. 2007; Leslie et al. 2004; Schneidewind et al. 2009).

Since the most protective HLA alleles, HLA B*57 and B*27, target epitopes in the p24 capsid protein of HIV-1, initial studies focused on the impact of escape mutations in this region of the viral proteome on virus replication. Brockman et al. demonstrated that an escape mutation located in the B*57 TW10 epitope in Gag, $T_{242}N$, reduced replicative capacity of NL4-3 but this defect could be partially compensated by mutations at a second site in p24 (Brockman et al. 2007). Similarly, Crawford et al. demonstrated that the $A_{163}G$ escape mutation in the B*5703 KF11 epitope in Gag impaired replication but could be effectively compensated by an accompanying $S_{165}N$ mutation. In each case, individuals with mutations that compensated for the replication defect exhibited higher viral loads than those harboring just the escape mutation itself, correlating in vitro replication to in vivo levels of virus. Consistent with this, the in vitro RC of chimeric NL4-3 viruses encoding the *gag* gene (and a small region of protease) of viruses derived from individuals exhibiting elite control of VL was significantly lower than that of viruses derived from chronic progressors (Miura et al. 2009; Miura et al. 2010). In a large study of over 800 subtype B chronically infected individuals for whom Gag-Pro chimeras were constructed, a modest positive correlation was observed between RC and VL, while a negative correlation was observed between RC and $CD4^{+}$ T cell count, consistent with RC influencing $CD4^{+}$ T cell decline (Brockman et al. 2010). Similar results were observed in a South African subtype C cohort (Wright et al. 2010).

In two studies of subtype C acutely infected individuals, higher numbers of transmitted CTL escape mutations in Gag, but not in Nef, were associated with lower VLs and higher CD4$^+$ T cell counts in the newly infected HLA-mismatched recipient (Goepfert et al. 2008; Chopera et al. 2008). These data suggested that transmission of CTL escape mutations in the *gag* gene of the T/F virus might impact both viral control and disease progression through reduced RC. In an initial study of over 100 acutely infected Zambians, the RC conferred by transmitted subtype C gag sequences was moderately correlated with set-point VL. Moreover, individuals harboring low RC viruses exhibited significantly slower CD4$^+$ T cell decline during the first three years after transmission compared to those harboring high RC viruses (Prince et al. 2012). Studies with similarly constructed Gag-Pro chimeric viruses derived from recently infected individuals in South Africa also showed a trend toward more rapid CD4$^+$ T cell decline for high RC viruses (Wright et al. 2011).

A follow-up study also in acutely infected Zambians showed that RC of the transmitted variants predicted CD4 decline independently of other risk factors, such as VL and protective HLA alleles (Claiborne et al. 2015), a finding also confirmed in samples from the CAPRISA 004 trial (Selhorst et al. 2017a). The RCs of the Gag-Pro chimeric viruses studied correlated well with those of IMCs derived from the same patients (Claiborne et al. 2015). Infection with a high RC virus was associated with profound changes in immune function compared to infection with a low RC virus. For high RC viruses, high levels of inflammatory cytokines and markers of bacterial translocation were observed very early ($\sim$45 days after the estimated date of HIV-1 infection), as were activation/proliferation markers (CD38, HLA-DR, Ki-67) and markers of T-cell dysfunction (PD-1) in both CD4$^+$ and CD8$^+$ T cell populations. In addition, elevated levels of HIV-1 proviral DNA were observed in naïve and central memory CD4$^+$ T cells. All of these features appeared to contribute to more rapid disease progression (Claiborne et al. 2015), indicating that the replication phenotype of the T/F virus can have a profound impact on the earliest virus-host interactions that then define the trajectory of disease.

8 Conclusions

The past several years have seen significant progress in understanding the virology of both HIV-1 transmission and subsequent disease progression. It is established that in most instances of HIV-1 sexual transmission, a single T/F variant, or a limited number of variants, from the diverse quasi-species present in the transmitting partner, establishes systemic infection. What is less clear, in the context of the level of virus exposure common during sexual activity, is how many variants initiate localized replication, and how many of those may fail to propagate to more distal sites. Nor is it clear, given the different selection pressures in the presence and absence of mucosal infections and inflammation, for example, whether any one property, such as in vivo RC or reduced sensitivity to IFN, will set apart T/F viruses from their non-transmitted counterparts. Clearly, based on the spectrum of in vitro RC observed for T/F viruses, it seems likely that the bar for transmission can be quite low. It will be interesting to observe whether in vivo models such as the humanized mouse or in vitro mucosal tissue explant models will shed additional light on viral properties that facilitate transmission.

While selective bias does result in the partial loss of non-consensus polymorphisms (mutations) during transmission, a majority (many of which represent immune response escape mutations) are present in the T/F variant. Those that impact virus replication in a negative fashion, and result in viruses with low RC, can be highly protective to the newly infected host in terms of disease progression, by inducing lower levels of inflammation and immune dysfunction. They also result in lower levels of proviral DNA in naïve and memory $CD4^+$ T cells early in infection, which may mean smaller viral reservoirs post-ART and a greater opportunity for studies aimed at long-term remission (or cure). In contrast, mutations in epitopes that are normally targets for the immune response genes of the newly infected individual can have an opposite effect on disease progression by reducing immune recognition, target cell killing, and virus control. It is the balance of this complex interplay between the phenotype defined by viral genetics and host immunogenetics that ultimately defines the trajectory of HIV-1 disease in a newly infected individual, with the virus genotype playing a much greater role than previously envisaged.

References

Abel K, Rocke DM, Chohan B, Fritts L, Miller CJ (2005) Temporal and anatomic relationship between virus replication and cytokine gene expression after vaginal simian immunodeficiency virus infection. J Virol 79(19):12164–12172. doi:10.1128/JVI.79.19.12164-12172.2005 79/19/12164 [pii]

Abrahams MR, Anderson JA, Giorgi EE, Seoighe C, Mlisana K, Ping LH, Athreya GS, Treurnicht FK, Keele BF, Wood N, Salazar-Gonzalez JF, Bhattacharya T, Chu H, Hoffman I, Galvin S, Mapanje C, Kazembe P, Thebus R, Fiscus S, Hide W, Cohen MS, Karim SA, Haynes BF, Shaw GM, Hahn BH, Korber BT, Swanstrom R, Williamson C (2009) Quantitating the multiplicity of infection with human immunodeficiency virus type 1 subtype C reveals a non-poisson distribution of transmitted variants. J Virol 83(8):3556–3567. doi:10.1128/JVI.02132-08 JVI.02132-08 [pii]

Alexander M, Lynch R, Mulenga J, Allen S, Derdeyn CA, Hunter E (2010) Donor and recipient envs from heterosexual human immunodeficiency virus subtype C transmission pairs require high receptor levels for entry. J Virol 84(8):4100–4104. doi:10.1128/JVI.02068-09 JVI.02068-09 [pii]

Alizon S, Fraser C (2013) Within-host and between-host evolutionary rates across the HIV-1 genome. Retrovirology 10:49. doi:10.1186/1742-4690-10-49

Alkhatib G, Combadiere C, Broder CC, Feng Y, Kennedy PE, Murphy PM, Berger EA (1996) CC CKR5: a RANTES, MIP-1alpha, MIP-1beta receptor as a fusion cofactor for macrophage-tropic HIV-1. Science 272(5270):1955–1958

Anahtar MN, Byrne EH, Doherty KE, Bowman BA, Yamamoto HS, Soumillon M, Padavattan N, Ismail N, Moodley A, Sabatini ME, Ghebremichael MS, Nusbaum C, Huttenhower C, Virgin HW, Ndung'u T, Dong KL, Walker BD, Fichorova RN, Kwon DS (2015) Cervicovaginal bacteria are a major modulator of host inflammatory responses in the female genital tract. Immunity 42(5):965–976. doi:10.1016/j.immuni.2015.04.019

Anderson JA, Ping LH, Dibben O, Jabara CB, Arney L, Kincer L, Tang Y, Hobbs M, Hoffman I, Kazembe P, Jones CD, Borrow P, Fiscus S, Cohen MS, Swanstrom R (2010) HIV-1 Populations in Semen Arise through Multiple Mechanisms. PLoS Pathog 6(8):e1001053. doi:10.1371/journal.ppat.1001053

Arthos J, Cicala C, Martinelli E, Macleod K, Van Ryk D, Wei D, Xiao Z, Veenstra TD, Conrad TP, Lempicki RA, McLaughlin S, Pascuccio M, Gopaul R, McNally J, Cruz CC, Censoplano N, Chung E, Reitano KN, Kottilil S, Goode DJ, Fauci AS (2008) HIV-1 envelope protein binds to and signals through integrin alpha4beta7, the gut mucosal homing receptor for peripheral T cells. Nat Immunol 9(3):301–309. doi:10.1038/ni1566 ni1566 [pii]

Auvert B, Taljaard D, Lagarde E, Sobngwi-Tambekou J, Sitta R, Puren A (2005) Randomized, controlled intervention trial of male circumcision for reduction of HIV infection risk: the ANRS 1265 Trial. PLoS Med 2(11):e298. doi:10.1371/journal.pmed.0020298 05-PLME-RA-0310R1 [pii]

Baalwa J, Wang S, Parrish NF, Decker JM, Keele BF, Learn GH, Yue L, Ruzagira E, Ssemwanga D, Kamali A, Amornkul PN, Price MA, Kappes JC, Karita E, Kaleebu P, Sanders E, Gilmour J, Allen S, Hunter E, Montefiori DC, Haynes BF, Cormier E, Hahn BH, Shaw GM (2013) Molecular identification, cloning and characterization of transmitted/founder HIV-1 subtype A, D and A/D infectious molecular clones. Virology 436(1):33–48. doi:10.1016/j.virol.2012.10.009

Bailey RC, Moses S, Parker CB, Agot K, Maclean I, Krieger JN, Williams CF, Campbell RT, Ndinya-Achola JO (2007) Male circumcision for HIV prevention in young men in Kisumu, Kenya: a randomised controlled trial. Lancet 369(9562):643–656. doi:10.1016/S0140-6736(07)60312-2 S0140-6736(07)60312-2 [pii]

Barouch DH, Ghneim K, Bosche WJ, Li Y, Berkemeier B, Hull M, Bhattacharyya S, Cameron M, Liu J, Smith K, Borducchi E, Cabral C, Peter L, Brinkman A, Shetty M, Li H, Gittens C, Baker C, Wagner W, Lewis MG, Colantonio A, Kang HJ, Li W, Lifson JD, Piatak M Jr, Sekaly RP (2016) Rapid Inflammasome Activation following Mucosal SIV Infection of Rhesus Monkeys. Cell 165(3):656–667. doi:10.1016/j.cell.2016.03.021

Berger EA, Murphy PM, Farber JM (1999) Chemokine receptors as HIV-1 coreceptors: roles in viral entry, tropism, and disease. Annu Rev Immunol 17:657–700. doi:10.1146/annurev.immunol.17.1.657

Bhattacharya T, Daniels M, Heckerman D, Foley B, Frahm N, Kadie C, Carlson J, Yusim K, McMahon B, Gaschen B, Mallal S, Mullins JI, Nickle DC, Herbeck J, Rousseau C, Learn GH, Miura T, Brander C, Walker B, Korber B (2007) Founder effects in the assessment of HIV polymorphisms and HLA allele associations. Science 315(5818):1583–1586. doi:10.1126/science.1131528

Blaschitz C, Raffatellu M (2010) Th17 Cytokines and the Gut Mucosal Barrier. J Clin Immunol 30 (2):196–203. doi:10.1007/s10875-010-9368-7

Boeras DI, Hraber PT, Hurlston M, Evans-Strickfaden T, Bhattachary T, Giorgi EE, Mulenga J, Karita E, Korber BT, Allen S, Hart CE, Derdeyn CA, Hunter E (2011) Role of donor genital tract HIV-1 diversity in the transmission bottleneck. Proc Natl Acad Sci USA 108(46): E156–163. doi:10.1073/pnas.1103764108

Borrow P, Lewicki H, Hahn BH, Shaw GM, Oldstone MB (1994) Virus-specific CD8 + cytotoxic T-lymphocyte activity associated with control of viremia in primary human immunodeficiency virus type 1 infection. J Virol 68(9):6103–6110

Borrow P, Lewicki H, Wei X, Horwitz MS, Peffer N, Meyers H, Nelson JA, Gairin JE, Hahn BH, Oldstone MB, Shaw GM (1997) Antiviral pressure exerted by HIV-1-specific cytotoxic T lymphocytes (CTLs) during primary infection demonstrated by rapid selection of CTL escape virus. Nat Med 3(2):205–211

Boyd DF, Sharma A, Humes D, Cheng-Mayer C, Overbaugh J (2016) Adapting SHIVs In Vivo Selects for Envelope-Mediated Interferon-alpha Resistance. PLoS Pathog 12(7):e1005727. doi:10.1371/journal.ppat.1005727

Brenner BG, Roger M, Routy JP, Moisi D, Ntemgwa M, Matte C, Baril JG, Thomas R, Rouleau D, Bruneau J, Leblanc R, Legault M, Tremblay C, Charest H, Wainberg MA (2007) High rates of forward transmission events after acute/early HIV-1 infection. J Infect Dis 195 (7):951–959. doi:10.1086/512088 JID37441 [pii]

Brockman MA, Schneidewind A, Lahaie M, Schmidt A, Miura T, Desouza I, Ryvkin F, Derdeyn CA, Allen S, Hunter E, Mulenga J, Goepfert PA, Walker BD, Allen TM (2007) Escape and compensation from early HLA-B57-mediated cytotoxic T-lymphocyte pressure on human immunodeficiency virus type 1 Gag alter capsid interactions with cyclophilin A. J Virol 81(22):12608–12618. doi:10.1128/JVI.01369-07

Brockman MA, Brumme ZL, Brumme CJ, Miura T, Sela J, Rosato PC, Kadie CM, Carlson JM, Markle TJ, Streeck H, Kelleher AD, Markowitz M, Jessen H, Rosenberg E, Altfeld M, Harrigan PR, Heckerman D, Walker BD, Allen TM (2010) Early selection in Gag by protective HLA alleles contributes to reduced HIV-1 replication capacity that may be largely compensated for in chronic infection. J Virol 84(22):11937–11949. doi:10.1128/JVI.01086-10

Brodin J, Zanini F, Thebo L, Lanz C, Bratt G, Neher RA, Albert J (2016) Establishment and stability of the latent HIV-1 DNA reservoir. eLife 5. doi:10.7554/eLife.18889

Byrareddy SN, Kallam B, Arthos J, Cicala C, Nawaz F, Hiatt J, Kersh EN, McNicholl JM, Hanson D, Reimann KA, Brameier M, Walter L, Rogers K, Mayne AE, Dunbar P, Villinger T, Little D, Parslow TG, Santangelo PJ, Villinger F, Fauci AS, Ansari AA (2014) Targeting alpha4beta7 integrin reduces mucosal transmission of simian immunodeficiency virus and protects gut-associated lymphoid tissue from infection. Nat Med 20(12):1397–1400. doi:10.1038/nm.3715

Carlson JM, Brumme ZL, Rousseau CM, Brumme CJ, Matthews P, Kadie C, Mullins JI, Walker BD, Harrigan PR, Goulder PJ, Heckerman D (2008) Phylogenetic dependency networks: inferring patterns of CTL escape and codon covariation in HIV-1 Gag. PLoS Comput Biol 4(11):e1000225. doi:10.1371/journal.pcbi.1000225

Carlson JM, Schaefer M, Monaco DC, Batorsky R, Claiborne DT, Prince J, Deymier MJ, Ende ZS, Klatt NR, DeZiel CE, Lin TH, Peng J, Seese AM, Shapiro R, Frater J, Ndung'u T, Tang J, Goepfert P, Gilmour J, Price MA, Kilembe W, Heckerman D, Goulder PJ, Allen TM, Allen S, Hunter E (2014) HIV transmission. Selection bias at the heterosexual HIV-1 transmission bottleneck. Science 345 (6193):1254031. doi:10.1126/science.1254031

Carlson JM, Du VY, Pfeifer N, Bansal A, Tan VY, Power K, Brumme CJ, Kreimer A, DeZiel CE, Fusi N, Schaefer M, Brockman MA, Gilmour J, Price MA, Kilembe W, Haubrich R, John M, Mallal S, Shapiro R, Frater J, Harrigan PR, Ndung'u T, Allen S, Heckerman D, Sidney J, Allen TM, Goulder PJ, Brumme ZL, Hunter E, Goepfert PA (2016) Impact of pre-adapted HIV transmission. Nat Med 22(6):606–613. doi:10.1038/nm.4100

Cartwright EK, Spicer L, Smith SA, Lee D, Fast R, Paganini S, Lawson BO, Nega M, Easley K, Schmitz JE, Bosinger SE, Paiardini M, Chahroudi A, Vanderford TH, Estes JD, Lifson JD, Derdeyn CA, Silvestri G (2016) CD8(+) Lymphocytes Are Required for Maintaining Viral Suppression in SIV-Infected Macaques Treated with Short-Term Antiretroviral Therapy. Immunity 45(3):656–668. doi:10.1016/j.immuni.2016.08.018

Cecchinato V, Franchini G (2010) Th17 cells in pathogenic SIV infection of macaques. Current opinion in HIV and AIDS 5(2):141–145. doi:10.1097/COH.0b013e32833653ec

Choe H, Farzan M, Sun Y, Sullivan N, Rollins B, Ponath PD, Wu L, Mackay CR, LaRosa G, Newman W, Gerard N, Gerard C, Sodroski J (1996) The beta-chemokine receptors CCR3 and CCR5 facilitate infection by primary HIV-1 isolates. Cell 85(7):1135–1148

Chohan B, Lang D, Sagar M, Korber B, Lavreys L, Richardson B, Overbaugh J (2005) Selection for human immunodeficiency virus type 1 envelope glycosylation variants with shorter V1-V2 loop sequences occurs during transmission of certain genetic subtypes and may impact viral RNA levels. J Virol 79(10):6528–6531. doi:10.1128/JVI.79.10.6528-6531.2005 79/10/6528 [pii]

Chopera DR, Woodman Z, Mlisana K, Mlotshwa M, Martin DP, Seoighe C, Treurnicht F, de Rosa DA, Hide W, Karim SA, Gray CM, Williamson C, Team CS (2008) Transmission of

HIV-1 CTL escape variants provides HLA-mismatched recipients with a survival advantage. PLoS Pathog 4(3):e1000033. doi:10.1371/journal.ppat.1000033

Cicala C, Martinelli E, McNally JP, Goode DJ, Gopaul R, Hiatt J, Jelicic K, Kottilil S, Macleod K, O'Shea A, Patel N, Van Ryk D, Wei D, Pascuccio M, Yi L, McKinnon L, Izulla P, Kimani J, Kaul R, Fauci AS, Arthos J (2009) The integrin alpha4beta7 forms a complex with cell-surface CD4 and defines a T-cell subset that is highly susceptible to infection by HIV-1. Proc Natl Acad Sci U S A 106(49):20877–20882. doi:10.1073/pnas.0911796106 0911796106 [pii]

Claiborne DT, Prince JL, Scully E, Macharia G, Micci L, Lawson B, Kopycinski J, Deymier MJ, Vanderford TH, Nganou-Makamdop K, Ende Z, Brooks K, Tang J, Yu T, Lakhi S, Kilembe W, Silvestri G, Douek D, Goepfert PA, Price MA, Allen SA, Paiardini M, Altfeld M, Gilmour J, Hunter E (2015) Replicative fitness of transmitted HIV-1 drives acute immune activation, proviral load in memory CD4 + T cells, and disease progression. Proc Natl Acad Sci U S A 112(12):E1480–E1489. doi:10.1073/pnas.1421607112

Coetzer M, Nedellec R, Salkowitz J, McLaughlin S, Liu Y, Heath L, Mullins JI, Mosier DE (2008) Evolution of CCR5 use before and during coreceptor switching. J Virol 82(23):11758–11766. doi:10.1128/JVI.01141-08

Cotton LA, Kuang XT, Le AQ, Carlson JM, Chan B, Chopera DR, Brumme CJ, Markle TJ, Martin E, Shahid A, Anmole G, Mwimanzi P, Nassab P, Penney KA, Rahman MA, Milloy MJ, Schechter MT, Markowitz M, Carrington M, Walker BD, Wagner T, Buchbinder S, Fuchs J, Koblin B, Mayer KH, Harrigan PR, Brockman MA, Poon AF, Brumme ZL (2014) Genotypic and functional impact of HIV-1 adaptation to its host population during the North American epidemic. PLoS Genet 10(4):e1004295. doi:10.1371/journal.pgen.1004295

Crawford H, Prado JG, Leslie A, Hue S, Honeyborne I, Reddy S, van der Stok M, Mncube Z, Brander C, Rousseau C, Mullins JI, Kaslow R, Goepfert P, Allen S, Hunter E, Mulenga J, Kiepiela P, Walker BD, Goulder PJ (2007) Compensatory mutation partially restores fitness and delays reversion of escape mutation within the immunodominant HLA-B*5703-restricted Gag epitope in chronic human immunodeficiency virus type 1 infection. J Virol 81(15):8346–8351. doi:10.1128/JVI.00465-07

Crawford H, Lumm W, Leslie A, Schaefer M, Boeras D, Prado JG, Tang J, Farmer P, Ndung'u T, Lakhi S, Gilmour J, Goepfert P, Walker BD, Kaslow R, Mulenga J, Allen S, Goulder PJ, Hunter E (2009) Evolution of HLA-B*5703 HIV-1 escape mutations in HLA-B*5703-positive individuals and their transmission recipients. J Exp Med 206(4):909–921. doi:10.1084/jem.20081984

Dandekar S, George MD, Baumler AJ (2010) Th17 cells, HIV and the gut mucosal barrier. Curr Opin HIV AIDS 5(2):173–178. doi:10.1097/COH.0b013e328335eda3

Dean M, Carrington M, Winkler C, Huttley GA, Smith MW, Allikmets R, Goedert JJ, Buchbinder SP, Vittinghoff E, Gomperts E, Donfield S, Vlahov D, Kaslow R, Saah A, Rinaldo C, Detels R, O'Brien SJ (1996) Genetic restriction of HIV-1 infection and progression to AIDS by a deletion allele of the CKR5 structural gene. Hemophilia Growth and Development Study, Multicenter AIDS Cohort Study, Multicenter Hemophilia Cohort Study, San Francisco City Cohort, ALIVE Study. Science 273(5283):1856–1862

Deng H, Liu R, Ellmeier W, Choe S, Unutmaz D, Burkhart M, Di Marzio P, Marmon S, Sutton RE, Hill CM, Davis CB, Peiper SC, Schall TJ, Littman DR, Landau NR (1996) Identification of a major co-receptor for primary isolates of HIV-1. Nature 381(6584):661–666. doi:10.1038/381661a0

Derdeyn CA, Decker JM, Bibollet-Ruche F, Mokili JL, Muldoon M, Denham SA, Heil ML, Kasolo F, Musonda R, Hahn BH, Shaw GM, Korber BT, Allen S, Hunter E (2004) Envelope-constrained neutralization-sensitive HIV-1 after heterosexual transmission. Science 303(5666):2019–2022. doi:10.1126/science.1093137

Deymier MJ, Claiborne DT, Ende Z, Ratner HK, Kilembe W, Allen S, Hunter E (2014) Particle infectivity of HIV-1 full-length genome infectious molecular clones in a subtype C heterosexual transmission pair following high fidelity amplification and unbiased cloning. Virology 468–470:454–461. doi:10.1016/j.virol.2014.08.018

Deymier MJ, Ende Z, Fenton-May AE, Dilernia DA, Kilembe W, Allen SA, Borrow P, Hunter E (2015) Heterosexual Transmission of Subtype C HIV-1 Selects Consensus-Like Variants without Increased Replicative Capacity or Interferon-alpha Resistance. PLoS Pathog 11(9): e1005154. doi:10.1371/journal.ppat.1005154

Dilernia DA, Jones L, Rodriguez S, Turk G, Rubio AE, Pampuro S, Gomez-Carrillo M, Bautista CT, Deluchi G, Benetucci J, Lasala MB, Lourtau L, Losso MH, Perez H, Cahn P, Salomon H (2008) HLA-driven convergence of HIV-1 viral subtypes B and F toward the adaptation to immune responses in human populations. PLoS ONE 3(10):e3429. doi:10.1371/journal.pone.0003429

Doyle T, Goujon C, Malim MH (2015) HIV-1 and interferons: who's interfering with whom? Nat Rev Microbiol 13(7):403–413. doi:10.1038/nrmicro3449

Dragic T, Litwin V, Allaway GP, Martin SR, Huang Y, Nagashima KA, Cayanan C, Maddon PJ, Koup RA, Moore JP, Paxton WA (1996) HIV-1 entry into CD4 + cells is mediated by the chemokine receptor CC-CKR-5. Nature 381(6584):667–673. doi:10.1038/381667a0

Ende Z, Deymier M, Hunter E (2017) Virologic Aspects of Mucosal Transmission. In: Mestecky J, Veazey R (eds) Mucosal Iimmunology of HIV/SIV. Bentham Science, In Press

Etemad B, Gonzalez OA, White L, Laeyendecker O, Kirk GD, Mehta S, Sagar M (2014) Characterization of HIV-1 envelopes in acutely and chronically infected injection drug users. Retrovirology 11:106. doi:10.1186/s12977-014-0106-8

Evans DT, Silvestri G (2013) Nonhuman primate models in AIDS research. Curr Opin HIV AIDS 8(4):255–261. doi:10.1097/COH.0b013e328361cee8

Fellay J, Shianna KV, Ge D, Colombo S, Ledergerber B, Weale M, Zhang K, Gumbs C, Castagna A, Cossarizza A, Cozzi-Lepri A, De Luca A, Easterbrook P, Francioli P, Mallal S, Martinez-Picado J, Miro JM, Obel N, Smith JP, Wyniger J, Descombes P, Antonarakis SE, Letvin NL, McMichael AJ, Haynes BF, Telenti A, Goldstein DB (2007) A whole-genome association study of major determinants for host control of HIV-1. Science 317(5840):944–947. doi:10.1126/science.1143767 1143767 [pii]

Feng Y, Broder CC, Kennedy PE, Berger EA (1996) HIV-1 entry cofactor: functional cDNA cloning of a seven-transmembrane. G protein-coupled receptor. Science 272(5263):872–877

Fenton-May AE, Dibben O, Emmerich T, Ding H, Pfafferott K, Aasa-Chapman MM, Pellegrino P, Williams I, Cohen MS, Gao F, Shaw GM, Hahn BH, Ochsenbauer C, Kappes JC, Borrow P (2013) Relative resistance of HIV-1 founder viruses to control by interferon-alpha. Retrovirology 10:146. doi:10.1186/1742-4690-10-146

Fideli US, Allen SA, Musonda R, Trask S, Hahn BH, Weiss H, Mulenga J, Kasolo F, Vermund SH, Aldrovandi GM (2001) Virologic and immunologic determinants of heterosexual transmission of human immunodeficiency virus type 1 in Africa. AIDS Res Hum Retroviruses 17(10):901–910. doi:10.1089/088922201750290023

Fiebig EW, Wright DJ, Rawal BD, Garrett PE, Schumacher RT, Peddada L, Heldebrant C, Smith R, Conrad A, Kleinman SH, Busch MP (2003) Dynamics of HIV viremia and antibody seroconversion in plasma donors: implications for diagnosis and staging of primary HIV infection. AIDS 17(13):1871–1879. doi:10.1097/01.aids.0000076308.76477.b8

Fischer W, Ganusov VV, Giorgi EE, Hraber PT, Keele BF, Leitner T, Han CS, Gleasner CD, Green L, Lo CC, Nag A, Wallstrom TC, Wang S, McMichael AJ, Haynes BF, Hahn BH, Perelson AS, Borrow P, Shaw GM, Bhattacharya T, Korber BT (2010) Transmission of single HIV-1 genomes and dynamics of early immune escape revealed by ultra-deep sequencing. PLoS ONE 5(8):e12303. doi:10.1371/journal.pone.0012303

Forthal DN, Gabriel EE, Wang A, Landucci G, Phan TB (2012) Association of Fcgamma receptor IIIa genotype with the rate of HIV infection after gp120 vaccination. Blood 120(14):2836–2842. doi:10.1182/blood-2012-05-431361

Foster TL, Wilson H, Iyer SS, Coss K, Doores K, Smith S, Kellam P, Finzi A, Borrow P, Hahn BH, Neil SJ (2016) Resistance of Transmitted Founder HIV-1 to IFITM-Mediated Restriction. Cell Host Microbe 20(4):429–442. doi:10.1016/j.chom.2016.08.006

Fraser C, Lythgoe K, Leventhal GE, Shirreff G, Hollingsworth TD, Alizon S, Bonhoeffer S (2014) Virulence and pathogenesis of HIV-1 infection: an evolutionary perspective. Science 343 (6177):1243727. doi:10.1126/science.1243727

Frost SD, Liu Y, Pond SL, Chappey C, Wrin T, Petropoulos CJ, Little SJ, Richman DD (2005) Characterization of human immunodeficiency virus type 1 (HIV-1) envelope variation and neutralizing antibody responses during transmission of HIV-1 subtype B. J Virol 79(10):6523–6527. doi:10.1128/JVI.79.10.6523-6527.2005 79/10/6523 [pii]

Galvin SR, Cohen MS (2004) The role of sexually transmitted diseases in HIV transmission. Nat Rev Microbiol 2(1):33–42. doi:10.1038/nrmicro794

Gnanakaran S, Bhattacharya T, Daniels M, Keele BF, Hraber PT, Lapedes AS, Shen T, Gaschen B, Krishnamoorthy M, Li H, Decker JM, Salazar-Gonzalez JF, Wang S, Jiang C, Gao F, Swanstrom R, Anderson JA, Ping LH, Cohen MS, Markowitz M, Goepfert PA, Saag MS, Eron JJ, Hicks CB, Blattner WA, Tomaras GD, Asmal M, Letvin NL, Gilbert PB, Decamp AC, Magaret CA, Schief WR, Ban YE, Zhang M, Soderberg KA, Sodroski JG, Haynes BF, Shaw GM, Hahn BH, Korber B (2011) Recurrent Signature Patterns in HIV-1 B Clade Envelope Glycoproteins Associated with either Early or Chronic Infections. PLoS Pathog 7(9):e1002209. doi:10.1371/journal.ppat.1002209

Goepfert PA, Lumm W, Farmer P, Matthews P, Prendergast A, Carlson JM, Derdeyn CA, Tang J, Kaslow RA, Bansal A, Yusim K, Heckerman D, Mulenga J, Allen S, Goulder PJ, Hunter E (2008) Transmission of HIV-1 Gag immune escape mutations is associated with reduced viral load in linked recipients. J Exp Med 205(5):1009–1017. doi:10.1084/jem.20072457

Gosmann C, Anahtar MN, Handley SA, Farcasanu M, Abu-Ali G, Bowman BA, Padavattan N, Desai C, Droit L, Moodley A, Dong M, Chen Y, Ismail N, Ndung'u T, Ghebremichael MS, Wesemann DR, Mitchell C, Dong KL, Huttenhower C, Walker BD, Virgin HW, Kwon DS (2017) Lactobacillus-Deficient Cervicovaginal Bacterial Communities Are Associated with Increased HIV Acquisition in Young South African Women. Immunity 46(1):29–37. doi:10.1016/j.immuni.2016.12.013

Goulder PJ, Phillips RE, Colbert RA, McAdam S, Ogg G, Nowak MA, Giangrande P, Luzzi G, Morgan B, Edwards A, McMichael AJ, Rowland-Jones S (1997) Late escape from an immunodominant cytotoxic T-lymphocyte response associated with progression to AIDS. Nat Med 3(2):212–217

Goulder PJ, Brander C, Tang Y, Tremblay C, Colbert RA, Addo MM, Rosenberg ES, Nguyen T, Allen R, Trocha A, Altfeld M, He S, Bunce M, Funkhouser R, Pelton SI, Burchett SK, McIntosh K, Korber BT, Walker BD (2001) Evolution and transmission of stable CTL escape mutations in HIV infection. Nature 412(6844):334–338. doi:10.1038/35085576

Gray RH, Li X, Kigozi G, Serwadda D, Brahmbhatt H, Wabwire-Mangen F, Nalugoda F, Kiddugavu M, Sewankambo N, Quinn TC, Reynolds SJ, Wawer MJ (2005) Increased risk of incident HIV during pregnancy in Rakai, Uganda: a prospective study. Lancet 366 (9492):1182–1188. doi:10.1016/S0140-6736(05)67481-8

Gray RH, Kigozi G, Serwadda D, Makumbi F, Watya S, Nalugoda F, Kiwanuka N, Moulton LH, Chaudhary MA, Chen MZ, Sewankambo NK, Wabwire-Mangen F, Bacon MC, Williams CF, Opendi P, Reynolds SJ, Laeyendecker O, Quinn TC, Wawer MJ (2007) Male circumcision for HIV prevention in men in Rakai, Uganda: a randomised trial. Lancet 369(9562):657–666. doi:10.1016/S0140-6736(07)60313-4 S0140-6736(07)60313-4 [pii]

Haaland RE, Hawkins PA, Salazar-Gonzalez J, Johnson A, Tichacek A, Karita E, Manigart O, Mulenga J, Keele BF, Shaw GM, Hahn BH, Allen SA, Derdeyn CA, Hunter E (2009) Inflammatory genital infections mitigate a severe genetic bottleneck in heterosexual transmission of subtype A and C HIV-1. PLoS Pathog 5(1):e1000274. doi:10.1371/journal.ppat.1000274

Hahn BH, Shaw GM, De Cock KM, Sharp PM (2000) AIDS as a zoonosis: scientific and public health implications. Science 287(5453):607–614

Hatziioannou T, Evans DT (2012) Animal models for HIV/AIDS research. Nat Rev Microbiol 10 (12):852–867. doi:10.1038/nrmicro2911

Hawkins RA, Rank RG, Kelly KA (2000) Expression of mucosal homing receptor alpha4beta7 is associated with enhanced migration to the Chlamydia-infected murine genital mucosa in vivo. Infect Immun 68(10):5587–5594

Hecht FM, Hartogensis W, Bragg L, Bacchetti P, Atchison R, Grant R, Barbour J, Deeks SG (2010) HIV RNA level in early infection is predicted by viral load in the transmission source. AIDS 24(7):941–945. doi:10.1097/QAD.0b013e328337b12e

Henn MR, Boutwell CL, Charlebois P, Lennon NJ, Power KA, Macalalad AR, Berlin AM, Malboeuf CM, Ryan EM, Gnerre S, Zody MC, Erlich RL, Green LM, Berical A, Wang Y, Casali M, Streeck H, Bloom AK, Dudek T, Tully D, Newman R, Axten KL, Gladden AD, Battis L, Kemper M, Zeng Q, Shea TP, Gujja S, Zedlack C, Gasser O, Brander C, Hess C, Gunthard HF, Brumme ZL, Brumme CJ, Bazner S, Rychert J, Tinsley JP, Mayer KH, Rosenberg E, Pereyra F, Levin JZ, Young SK, Jessen H, Altfeld M, Birren BW, Walker BD, Allen TM (2012) Whole genome deep sequencing of HIV-1 reveals the impact of early minor variants upon immune recognition during acute infection. PLoS Pathog 8(3):e1002529. doi:10.1371/journal.ppat.1002529

Hladik F, McElrath MJ (2008) Setting the stage: host invasion by HIV. Nat Rev Immunol 8 (6):447–457. doi:10.1038/nri2302 nri2302 [pii]

Hladik F, Sakchalathorn P, Ballweber L, Lentz G, Fialkow M, Eschenbach D, McElrath MJ (2007) Initial events in establishing vaginal entry and infection by human immunodeficiency virus type-1. Immunity 26(2):257–270. doi:10.1016/j.immuni.2007.01.007 S1074-7613(07)00134-3 [pii]

Hunter E (1997) Viral Entry and Receptors. In: Coffin JM, Hughes SH, Varmus HE (eds) Retroviruses. Cold Spring Harbor (NY)

Iglesias MC, Almeida JR, Fastenackels S, van Bockel DJ, Hashimoto M, Venturi V, Gostick E, Urrutia A, Wooldridge L, Clement M, Gras S, Wilmann PG, Autran B, Moris A, Rossjohn J, Davenport MP, Takiguchi M, Brander C, Douek DC, Kelleher AD, Price DA, Appay V (2011) Escape from highly effective public CD8 + T-cell clonotypes by HIV. Blood 118(8):2138–2149. doi:10.1182/blood-2011-01-328781

Isaacman-Beck J, Hermann EA, Yi Y, Ratcliffe SJ, Mulenga J, Allen S, Hunter E, Derdeyn CA, Collman RG (2009) Heterosexual transmission of human immunodeficiency virus type 1 subtype C: Macrophage tropism, alternative coreceptor use, and the molecular anatomy of CCR5 utilization. J Virol 83(16):8208–8220. doi:10.1128/JVI.00296-09 JVI.00296-09 [pii]

Iyer SS, Bibollet-Ruche F, Sherrill-Mix S, Learn GH, Plenderleith L, Smith AG, Barbian HJ, Russell RM, Gondim MV, Bahari CY, Shaw CM, Li Y, Decker T, Haynes BF, Shaw GM, Sharp PM, Borrow P, Hahn BH (2017) Resistance to type 1 interferons is a major determinant of HIV-1 transmission fitness. Proc Natl Acad Sci U S A 114(4):E590–E599. doi:10.1073/pnas.1620144114

Jin X, Bauer DE, Tuttleton SE, Lewin S, Gettie A, Blanchard J, Irwin CE, Safrit JT, Mittler J, Weinberger L, Kostrikis LG, Zhang L, Perelson AS, Ho DD (1999) Dramatic rise in plasma viremia after CD8(+) T cell depletion in simian immunodeficiency virus-infected macaques. J Exp Med 189(6):991–998

Joseph SB, Swanstrom R, Kashuba AD, Cohen MS (2015) Bottlenecks in HIV-1 transmission: insights from the study of founder viruses. Nat Rev Microbiol 13(7):414–425. doi:10.1038/nrmicro3471

Kamali A, Price MA, Lakhi S, Karita E, Inambao M, Sanders EJ, Anzala O, Latka MH, Bekker LG, Kaleebu P, Asiki G, Ssetaala A, Ruzagira E, Allen S, Farmer P, Hunter E, Mutua G, Makkan H, Tichacek A, Brill IK, Fast P, Stevens G, Chetty P, Amornkul PN, Gilmour J, Partnership IAHP (2015) Creating an African HIV clinical research and prevention trials network: HIV prevalence, incidence and transmission. PLoS ONE 10(1):e0116100. doi:10.1371/journal.pone.0116100

Kaslow RA, Carrington M, Apple R, Park L, Munoz A, Saah AJ, Goedert JJ, Winkler C, O'Brien SJ, Rinaldo C, Detels R, Blattner W, Phair J, Erlich H, Mann DL (1996) Influence of combinations of human major histocompatibility complex genes on the course of HIV-1 infection. Nat Med 2(4):405–411

Kawashima Y, Pfafferott K, Frater J, Matthews P, Payne R, Addo M, Gatanaga H, Fujiwara M, Hachiya A, Koizumi H, Kuse N, Oka S, Duda A, Prendergast A, Crawford H, Leslie A, Brumme Z, Brumme C, Allen T, Brander C, Kaslow R, Tang J, Hunter E, Allen S, Mulenga J, Branch S, Roach T, John M, Mallal S, Ogwu A, Shapiro R, Prado JG, Fidler S, Weber J, Pybus OG, Klenerman P, Ndung'u T, Phillips R, Heckerman D, Harrigan PR, Walker BD, Takiguchi M, Goulder P (2009) Adaptation of HIV-1 to human leukocyte antigen class I. Nature 458(7238):641–645. doi:10.1038/nature07746

Keele BF, Giorgi EE, Salazar-Gonzalez JF, Decker JM, Pham KT, Salazar MG, Sun C, Grayson T, Wang S, Li H, Wei X, Jiang C, Kirchherr JL, Gao F, Anderson JA, Ping LH, Swanstrom R, Tomaras GD, Blattner WA, Goepfert PA, Kilby JM, Saag MS, Delwart EL, Busch MP, Cohen MS, Montefiori DC, Haynes BF, Gaschen B, Athreya GS, Lee HY, Wood N, Seoighe C, Perelson AS, Bhattacharya T, Korber BT, Hahn BH, Shaw GM (2008) Identification and characterization of transmitted and early founder virus envelopes in primary HIV-1 infection. Proc Natl Acad Sci U S A 105(21):7552–7557. doi:10.1073/pnas.0802203105 0802203105 [pii]

Keele BF, Li H, Learn GH, Hraber P, Giorgi EE, Grayson T, Sun C, Chen Y, Yeh WW, Letvin NL, Mascola JR, Nabel GJ, Haynes BF, Bhattacharya T, Perelson AS, Korber BT, Hahn BH, Shaw GM (2009) Low-dose rectal inoculation of rhesus macaques by SIVsmE660 or SIVmac251 recapitulates human mucosal infection by HIV-1. J Exp Med 206(5):1117–1134. doi:10.1084/jem.20082831 jem.20082831 [pii]

Kiepiela P, Ngumbela K, Thobakgale C, Ramduth D, Honeyborne I, Moodley E, Reddy S, de Pierres C, Mncube Z, Mkhwanazi N, Bishop K, van der Stok M, Nair K, Khan N, Crawford H, Payne R, Leslie A, Prado J, Prendergast A, Frater J, McCarthy N, Brander C, Learn GH, Nickle D, Rousseau C, Coovadia H, Mullins JI, Heckerman D, Walker BD, Goulder P (2007) CD8 + T-cell responses to different HIV proteins have discordant associations with viral load. Nat Med 13(1):46–53. doi:10.1038/nm1520 nm1520 [pii]

Kiwanuka N, Laeyendecker O, Quinn TC, Wawer MJ, Shepherd J, Robb M, Kigozi G, Kagaayi J, Serwadda D, Makumbi FE, Reynolds SJ, Gray RH (2009) HIV-1 subtypes and differences in heterosexual HIV transmission among HIV-discordant couples in Rakai. Uganda. AIDS 23 (18):2479–2484. doi:10.1097/QAD.0b013e328330cc08

Klatt NR, Silvestri G, Hirsch V (2012) Nonpathogenic simian immunodeficiency virus infections. Cold Spring Harb Perspect Med 2(1):a007153. doi:10.1101/cshperspect.a007153

Kmiec D, Iyer SS, Sturzel CM, Sauter D, Hahn BH, Kirchhoff F (2016) Vpu-Mediated Counteraction of Tetherin Is a Major Determinant of HIV-1 Interferon Resistance. mBio 7 (4). doi:10.1128/mBio.00934-16

Koup RA, Safrit JT, Cao Y, Andrews CA, McLeod G, Borkowsky W, Farthing C, Ho DD (1994) Temporal association of cellular immune responses with the initial control of viremia in primary human immunodeficiency virus type 1 syndrome. J Virol 68(7):4650–4655

Lee HY, Giorgi EE, Keele BF, Gaschen B, Athreya GS, Salazar-Gonzalez JF, Pham KT, Goepfert PA, Kilby JM, Saag MS, Delwart EL, Busch MP, Hahn BH, Shaw GM, Korber BT, Bhattacharya T, Perelson AS (2009) Modeling sequence evolution in acute HIV-1 infection. J Theor Biol 261(2):341–360. doi:10.1016/j.jtbi.2009.07.038 S0022-5193(09)00350-6 [pii]

Leslie AJ, Pfafferott KJ, Chetty P, Draenert R, Addo MM, Feeney M, Tang Y, Holmes EC, Allen T, Prado JG, Altfeld M, Brander C, Dixon C, Ramduth D, Jeena P, Thomas SA, St John A, Roach TA, Kupfer B, Luzzi G, Edwards A, Taylor G, Lyall H, Tudor-Williams G, Novelli V, Martinez-Picado J, Kiepiela P, Walker BD, Goulder PJ (2004) HIV evolution: CTL escape mutation and reversion after transmission. Nat Med 10(3):282–289. doi:10.1038/nm992nm992 [pii]

Li Q, Estes JD, Schlievert PM, Duan L, Brosnahan AJ, Southern PJ, Reilly CS, Peterson ML, Schultz-Darken N, Brunner KG, Nephew KR, Pambuccian S, Lifson JD, Carlis JV, Haase AT (2009) Glycerol monolaurate prevents mucosal SIV transmission. Nature 458(7241):1034–1038. doi:10.1038/nature07831 nature07831 [pii]

Liu R, Paxton WA, Choe S, Ceradini D, Martin SR, Horuk R, MacDonald ME, Stuhlmann H, Koup RA, Landau NR (1996) Homozygous defect in HIV-1 coreceptor accounts for resistance of some multiply-exposed individuals to HIV-1 infection. Cell 86(3):367–377

Liu J, Keele BF, Li H, Keating S, Norris PJ, Carville A, Mansfield KG, Tomaras GD, Haynes BF, Kolodkin-Gal D, Letvin NL, Hahn BH, Shaw GM, Barouch DH (2010) Low-dose mucosal simian immunodeficiency virus infection restricts early replication kinetics and transmitted virus variants in rhesus monkeys. J Virol 84(19):10406–10412. doi:10.1128/JVI.01155-10

Liu A, Yang Y, Liu L, Meng Z, Li L, Qiu C, Xu J, Zhang X (2014) Differential compartmentalization of HIV-targeting immune cells in inner and outer foreskin tissue. PLoS ONE 9(1):e85176. doi:10.1371/journal.pone.0085176

Long EM, Rainwater SM, Lavreys L, Mandaliya K, Overbaugh J (2002) HIV type 1 variants transmitted to women in Kenya require the CCR5 coreceptor for entry, regardless of the genetic complexity of the infecting virus. AIDS Res Hum Retroviruses 18(8):567–576. doi:10.1089/088922202753747914

Ma ZM, Stone M, Piatak M Jr, Schweighardt B, Haigwood NL, Montefiori D, Lifson JD, Busch MP, Miller CJ (2009) High specific infectivity of plasma virus from the pre-ramp-up and ramp-up stages of acute simian immunodeficiency virus infection. J Virol 83(7):3288–3297. doi:10.1128/JVI.02423-08 JVI.02423-08 [pii]

Margolis L, Shattock R (2006) Selective transmission of CCR5-utilizing HIV-1: the'gatekeeper' problem resolved? Nat Rev Microbiol 4(4):312–317. doi:10.1038/nrmicro1387

Martinez-Picado J, Prado JG, Fry EE, Pfafferott K, Leslie A, Chetty S, Thobakgale C, Honeyborne I, Crawford H, Matthews P, Pillay T, Rousseau C, Mullins JI, Brander C, Walker BD, Stuart DI, Kiepiela P, Goulder P (2006) Fitness cost of escape mutations in p24 Gag in association with control of human immunodeficiency virus type 1. J Virol 80(7):3617–3623. doi:10.1128/JVI.80.7.3617-3623.2006

Matano T, Shibata R, Siemon C, Connors M, Lane HC, Martin MA (1998) Administration of an anti-CD8 monoclonal antibody interferes with the clearance of chimeric simian/human immunodeficiency virus during primary infections of rhesus macaques. J Virol 72(1):164–169

McLaren PJ, Coulonges C, Bartha I, Lenz TL, Deutsch AJ, Bashirova A, Buchbinder S, Carrington MN, Cossarizza A, Dalmau J, De Luca A, Goedert JJ, Gurdasani D, Haas DW, Herbeck JT, Johnson EO, Kirk GD, Lambotte O, Luo M, Mallal S, van Manen D, Martinez-Picado J, Meyer L, Miro JM, Mullins JI, Obel N, Poli G, Sandhu MS, Schuitemaker H, Shea PR, Theodorou I, Walker BD, Weintrob AC, Winkler CA, Wolinsky SM, Raychaudhuri S, Goldstein DB, Telenti A, de Bakker PI, Zagury JF, Fellay J (2015) Polymorphisms of large effect explain the majority of the host genetic contribution to variation of HIV-1 virus load. Proc Natl Acad Sci U S A 112(47):14658–14663. doi:10.1073/pnas.1514867112

Mellors JW, Kingsley LA, Rinaldo CR Jr, Todd JA, Hoo BS, Kokka RP, Gupta P (1995) Quantitation of HIV-1 RNA in plasma predicts outcome after seroconversion. Ann Intern Med 122(8):573–579

Michael NL, Chang G, Louie LG, Mascola JR, Dondero D, Birx DL, Sheppard HW (1997) The role of viral phenotype and CCR-5 gene defects in HIV-1 transmission and disease progression. Nat Med 3(3):338–340

Miller WC, Rosenberg NE, Rutstein SE, Powers KA (2010) Role of acute and early HIV infection in the sexual transmission of HIV. Curr Opin HIV AIDS 5(4):277–282. doi:10.1097/COH.0b013e32833a0d3a 01222929-201007000-00004 [pii]

Miura T, Brockman MA, Brumme ZL, Brumme CJ, Pereyra F, Trocha A, Block BL, Schneidewind A, Allen TM, Heckerman D, Walker BD (2009) HLA-associated alterations in replication capacity of chimeric NL4-3 viruses carrying gag-protease from elite controllers of human immunodeficiency virus type 1. J Virol 83(1):140–149. doi:10.1128/JVI.01471-08

Miura T, Brumme ZL, Brockman MA, Rosato P, Sela J, Brumme CJ, Pereyra F, Kaufmann DE, Trocha A, Block BL, Daar ES, Connick E, Jessen H, Kelleher AD, Rosenberg E, Markowitz M, Schafer K, Vaida F, Iwamoto A, Little S, Walker BD (2010) Impaired replication capacity of acute/early viruses in persons who become HIV controllers. J Virol 84 (15):7581–7591. doi:10.1128/JVI.00286-10

Monaco DC, Dilernia DA, Fiore-Gartland A, Yu T, Prince JL, Dennis KK, Qin K, Schaefer M, Claiborne DT, Kilembe W, Tang J, Price MA, Farmer P, Gilmour J, Bansal A, Allen S, Goepfert P, Hunter E (2016) Balance between transmitted HLA preadapted and nonassociated polymorphisms is a major determinant of HIV-1 disease progression. J Exp Med 213 (10):2049–2063. doi:10.1084/jem.20151984

Nawaz F, Cicala C, Van Ryk D, Block KE, Jelicic K, McNally JP, Ogundare O, Pascuccio M, Patel N, Wei D, Fauci AS, Arthos J (2011) The genotype of early-transmitting HIV gp120 s promotes alphabeta-reactivity, revealing alphabetaCD4 + T cells as key targets in mucosal transmission. PLoS Pathog 7(2):e1001301. doi:10.1371/journal.ppat.1001301

Oberle CS, Joos B, Rusert P, Campbell NK, Beauparlant D, Kuster H, Weber J, Schenkel CD, Scherrer AU, Magnus C, Kouyos R, Rieder P, Niederost B, Braun DL, Pavlovic J, Boni J, Yerly S, Klimkait T, Aubert V, Trkola A, Metzner KJ, Gunthard HF, Swiss HIVCS (2016) Tracing HIV-1 transmission: envelope traits of HIV-1 transmitter and recipient pairs. Retrovirology 13(1):62. doi:10.1186/s12977-016-0299-0

Ochsenbauer C, Edmonds TG, Ding H, Keele BF, Decker J, Salazar MG, Salazar-Gonzalez JF, Shattock R, Haynes BF, Shaw GM, Hahn BH, Kappes JC (2012) Generation of transmitted/founder HIV-1 infectious molecular clones and characterization of their replication capacity in CD4 T lymphocytes and monocyte-derived macrophages. J Virol 86(5):2715–2728. doi:10.1128/JVI.06157-11

Paiardini M, Pandrea I, Apetrei C, Silvestri G (2009) Lessons learned from the natural hosts of HIV-related viruses. Annu Rev Med 60:485–495. doi:10.1146/annurev.med.60.041807.123753

Pandrea I, Apetrei C (2010) Where the wild things are: pathogenesis of SIV infection in African nonhuman primate hosts. Curr HIV/AIDS Rep 7(1):28–36. doi:10.1007/s11904-009-0034-8

Pandrea I, Onanga R, Souquiere S, Mouinga-Ondeme A, Bourry O, Makuwa M, Rouquet P, Silvestri G, Simon F, Roques P, Apetrei C (2008) Paucity of CD4 + CCR5 + T cells may prevent transmission of simian immunodeficiency virus in natural nonhuman primate hosts by breast-feeding. J Virol 82(11):5501–5509. doi:10.1128/JVI.02555-07

Parrish NF, Wilen CB, Banks LB, Iyer SS, Pfaff JM, Salazar-Gonzalez JF, Salazar MG, Decker JM, Parrish EH, Berg A, Hopper J, Hora B, Kumar A, Mahlokozera T, Yuan S, Coleman C, Vermeulen M, Ding H, Ochsenbauer C, Tilton JC, Permar SR, Kappes JC, Betts MR, Busch MP, Gao F, Montefiori D, Haynes BF, Shaw GM, Hahn BH, Doms RW (2012) Transmitted/founder and chronic subtype C HIV-1 use CD4 and CCR5 receptors with equal efficiency and are not inhibited by blocking the integrin alpha4beta7. PLoS Pathog 8(5): e1002686. doi:10.1371/journal.ppat.1002686

Parrish NF, Gao F, Li H, Giorgi EE, Barbian HJ, Parrish EH, Zajic L, Iyer SS, Decker JM, Kumar A, Hora B, Berg A, Cai F, Hopper J, Denny TN, Ding H, Ochsenbauer C, Kappes JC, Galimidi RP, West AP Jr, Bjorkman PJ, Wilen CB, Doms RW, O'Brien M, Bhardwaj N, Borrow P, Haynes BF, Muldoon M, Theiler JP, Korber B, Shaw GM, Hahn BH (2013) Phenotypic properties of transmitted founder HIV-1. Proc Natl Acad Sci U S A 110(17):6626–6633. doi:10.1073/pnas.1304288110

Pilcher CD, Joaki G, Hoffman IF, Martinson FE, Mapanje C, Stewart PW, Powers KA, Galvin S, Chilongozi D, Gama S, Price MA, Fiscus SA, Cohen MS (2007) Amplified transmission of HIV-1: comparison of HIV-1 concentrations in semen and blood during acute and chronic infection. AIDS 21(13):1723–1730. doi:10.1097/QAD.0b013e3281532c82

Powers KA, Poole C, Pettifor AE, Cohen MS (2008) Rethinking the heterosexual infectivity of HIV-1: a systematic review and meta-analysis. Lancet Infect Dis 8(9):553–563. doi:10.1016/S1473-3099(08)70156-7

Prince JL, Claiborne DT, Carlson JM, Schaefer M, Yu T, Lahki S, Prentice HA, Yue L, Vishwanathan SA, Kilembe W, Goepfert P, Price MA, Gilmour J, Mulenga J, Farmer P, Derdeyn CA, Tang J, Heckerman D, Kaslow RA, Allen SA, Hunter E (2012) Role of

transmitted Gag CTL polymorphisms in defining replicative capacity and early HIV-1 pathogenesis. PLoS Pathog 8(11):e1003041. doi:10.1371/journal.ppat.1003041
Quinn TC, Wawer MJ, Sewankambo N, Serwadda D, Li C, Wabwire-Mangen F, Meehan MO, Lutalo T, Gray RH (2000) Viral load and heterosexual transmission of human immunodeficiency virus type 1. Rakai Project Study Group. N Engl J Med 342(13):921–929. doi:10.1056/NEJM200003303421303
Redd AD, Collinson-Streng AN, Chatziandreou N, Mullis CE, Laeyendecker O, Martens C, Ricklefs S, Kiwanuka N, Nyein PH, Lutalo T, Grabowski MK, Kong X, Manucci J, Sewankambo N, Wawer MJ, Gray RH, Porcella SF, Fauci AS, Sagar M, Serwadda D, Quinn TC (2012) Previously transmitted HIV-1 strains are preferentially selected during subsequent sexual transmissions. J Infect Dis 206(9):1433–1442. doi:10.1093/infdis/jis503
Regoes RR, Bonhoeffer S (2005) The HIV coreceptor switch: a population dynamical perspective. Trends Microbiol 13(6):269–277. doi:10.1016/j.tim.2005.04.005
Richardson SI, Gray ES, Mkhize NN, Sheward DJ, Lambson BE, Wibmer CK, Masson L, Werner L, Garrett N, Passmore JA, Karim QA, Karim SS, Williamson C, Moore PL, Morris L (2015) South African HIV-1 subtype C transmitted variants with a specific V2 motif show higher dependence on alpha4beta7 for replication. Retrovirology 12:54. doi:10.1186/s12977-015-0183-3
Roberts HE, Hurst J, Robinson N, Brown H, Flanagan P, Vass L, Fidler S, Weber J, Babiker A, Phillips RE, McLean AR, Frater J, investigators St (2015) Structured observations reveal slow HIV-1 CTL escape. PLoS genetics 11 (2):e1004914. doi:10.1371/journal.pgen.1004914
Sagar M, Lavreys L, Baeten JM, Richardson BA, Mandaliya K, Ndinya-Achola JO, Kreiss JK, Overbaugh J (2004) Identification of modifiable factors that affect the genetic diversity of the transmitted HIV-1 population. AIDS 18(4):615–619. 00002030-200403050-00005 [pii]
Sagar M, Laeyendecker O, Lee S, Gamiel J, Wawer MJ, Gray RH, Serwadda D, Sewankambo NK, Shepherd JC, Toma J, Huang W, Quinn TC (2009) Selection of HIV variants with signature genotypic characteristics during heterosexual transmission. J Infect Dis 199(4):580–589. doi:10.1086/596557
Salazar-Gonzalez JF, Bailes E, Pham KT, Salazar MG, Guffey MB, Keele BF, Derdeyn CA, Farmer P, Hunter E, Allen S, Manigart O, Mulenga J, Anderson JA, Swanstrom R, Haynes BF, Athreya GS, Korber BT, Sharp PM, Shaw GM, Hahn BH (2008) Deciphering human immunodeficiency virus type 1 transmission and early envelope diversification by single-genome amplification and sequencing. J Virol 82(8):3952–3970. doi:10.1128/JVI.02660-07 JVI.02660-07 [pii]
Salazar-Gonzalez JF, Salazar MG, Keele BF, Learn GH, Giorgi EE, Li H, Decker JM, Wang S, Baalwa J, Kraus MH, Parrish NF, Shaw KS, Guffey MB, Bar KJ, Davis KL, Ochsenbauer-Jambor C, Kappes JC, Saag MS, Cohen MS, Mulenga J, Derdeyn CA, Allen S, Hunter E, Markowitz M, Hraber P, Perelson AS, Bhattacharya T, Haynes BF, Korber BT, Hahn BH, Shaw GM (2009) Genetic identity, biological phenotype, and evolutionary pathways of transmitted/founder viruses in acute and early HIV-1 infection. J Exp Med 206(6):1273–1289. doi:10.1084/jem.20090378 jem.20090378 [pii]
Samson M, Libert F, Doranz BJ, Rucker J, Liesnard C, Farber CM, Saragosti S, Lapoumeroulie C, Cognaux J, Forceille C, Muyldermans G, Verhofstede C, Burtonboy G, Georges M, Imai T, Rana S, Yi Y, Smyth RJ, Collman RG, Doms RW, Vassart G, Parmentier M (1996) Resistance to HIV-1 infection in caucasian individuals bearing mutant alleles of the CCR-5 chemokine receptor gene. Nature 382(6593):722–725. doi:10.1038/382722a0
Sandler NG, Bosinger SE, Estes JD, Zhu RT, Tharp GK, Boritz E, Levin D, Wijeyesinghe S, Makamdop KN, del Prete GQ, Hill BJ, Timmer JK, Reiss E, Yarden G, Darko S, Contijoch E, Todd JP, Silvestri G, Nason M, Norgren RB Jr, Keele BF, Rao S, Langer JA, Lifson JD, Schreiber G, Douek DC (2014) Type I interferon responses in rhesus macaques prevent SIV infection and slow disease progression. Nature 511(7511):601–605. doi:10.1038/nature13554
Schmitz JE, Kuroda MJ, Santra S, Sasseville VG, Simon MA, Lifton MA, Racz P, Tenner-Racz K, Dalesandro M, Scallon BJ, Ghrayeb J, Forman MA, Montefiori DC, Rieber EP, Letvin NL,

Reimann KA (1999) Control of viremia in simian immunodeficiency virus infection by CD8 + lymphocytes. Science 283(5403):857–860

Schneidewind A, Brockman MA, Yang R, Adam RI, Li B, Le Gall S, Rinaldo CR, Craggs SL, Allgaier RL, Power KA, Kuntzen T, Tung CS, LaBute MX, Mueller SM, Harrer T, McMichael AJ, Goulder PJ, Aiken C, Brander C, Kelleher AD, Allen TM (2007) Escape from the dominant HLA-B27-restricted cytotoxic T-lymphocyte response in Gag is associated with a dramatic reduction in human immunodeficiency virus type 1 replication. J Virol 81(22):12382–12393. doi:10.1128/JVI.01543-07

Schneidewind A, Brumme ZL, Brumme CJ, Power KA, Reyor LL, O'Sullivan K, Gladden A, Hempel U, Kuntzen T, Wang YE, Oniangue-Ndza C, Jessen H, Markowitz M, Rosenberg ES, Sekaly RP, Kelleher AD, Walker BD, Allen TM (2009) Transmission and long-term stability of compensated CD8 escape mutations. J Virol 83(8):3993–3997. doi:10.1128/JVI.01108-08

Selhorst P, Combrinck C, Ndabambi N, Ismail SD, Abrahams MR, Lacerda M, Samsunder N, Garrett N, Abdool Karim Q, Abdool Karim SS, Williamson C (2017a) Replication Capacity of Viruses from Acute Infection Drives HIV-1 Disease Progression. J Virol 91 (8). doi:10.1128/JVI.01806-16

Selhorst P, Masson L, Ismail SD, Samsunder N, Garrett N, Mansoor LE, Abdool Karim Q, Abdool Karim SS, Passmore JS, Williamson C (2017b) Cervicovaginal Inflammation Facilitates Acquisition of Less Infectious HIV Variants. Clin Infect Dis 64(1):79–82. doi:10.1093/cid/ciw663

Shaw GM, Hunter E (2012) HIV transmission. Cold Spring Harb Perspect Med 2 (11). doi:10.1101/cshperspect.a006965

Stieh DJ, Maric D, Kelley ZL, Anderson MR, Hattaway HZ, Beilfuss BA, Rothwangl KB, Veazey RS, Hope TJ (2014) Vaginal challenge with an SIV-based dual reporter system reveals that infection can occur throughout the upper and lower female reproductive tract. PLoS Pathog 10(10):e1004440. doi:10.1371/journal.ppat.1004440

Stieh DJ, Matias E, Xu H, Fought AJ, Blanchard JL, Marx PA, Veazey RS, Hope TJ (2016) Th17 Cells Are Preferentially Infected Very Early after Vaginal Transmission of SIV in Macaques. Cell Host Microbe 19(4):529–540. doi:10.1016/j.chom.2016.03.005

Trask SA, Derdeyn CA, Fideli U, Chen Y, Meleth S, Kasolo F, Musonda R, Hunter E, Gao F, Allen S, Hahn BH (2002) Molecular epidemiology of human immunodeficiency virus type 1 transmission in a heterosexual cohort of discordant couples in Zambia. J Virol 76(1):397–405

UNAIDS (2016) Global AIDS Update 2016

Veazey RS, Pilch-Cooper HA, Hope TJ, Alter G, Carias AM, Sips M, Wang X, Rodriguez B, Sieg SF, Reich A, Wilkinson P, Cameron MJ, Lederman MM (2016) Prevention of SHIV transmission by topical IFN-beta treatment. Mucosal Immunol 9(6):1528–1536. doi:10.1038/mi.2015.146

Walker BD, Yu XG (2013) Unravelling the mechanisms of durable control of HIV-1. Nat Rev Immunol 13(7):487–498. doi:10.1038/nri3478

Wawer MJ, Gray RH, Sewankambo NK, Serwadda D, Li X, Laeyendecker O, Kiwanuka N, Kigozi G, Kiddugavu M, Lutalo T, Nalugoda F, Wabwire-Mangen F, Meehan MP, Quinn TC (2005) Rates of HIV-1 transmission per coital act, by stage of HIV-1 infection, in Rakai. Uganda. J Infect Dis 191(9):1403–1409. doi:10.1086/429411 JID33445 [pii]

Wilen CB, Parrish NF, Pfaff JM, Decker JM, Henning EA, Haim H, Petersen JE, Wojcechowskyj JA, Sodroski J, Haynes BF, Montefiori DC, Tilton JC, Shaw GM, Hahn BH, Doms RW (2011) Phenotypic and immunologic comparison of clade B transmitted/founder and chronic HIV-1 envelope glycoproteins. J Virol 85(17):8514–8527. doi:10.1128/JVI.00736-11

Wilen CB, Tilton JC, Doms RW (2012) HIV: Cell Binding and Entry. Cold Spring Harbor Perspectives in Medicine 2(8):a006866. doi:10.1101/cshperspect.a006866

Wolfs TF, Zwart G, Bakker M, Goudsmit J (1992) HIV-1 genomic RNA diversification following sexual and parenteral virus transmission. Virology 189(1):103–110

Wolinsky SM, Wike CM, Korber BT, Hutto C, Parks WP, Rosenblum LL, Kunstman KJ, Furtado MR, Munoz JL (1992) Selective transmission of human immunodeficiency virus type-1 variants from mothers to infants. Science 255(5048):1134–1137

Wright JK, Brumme ZL, Carlson JM, Heckerman D, Kadie CM, Brumme CJ, Wang B, Losina E, Miura T, Chonco F, van der Stok M, Mncube Z, Bishop K, Goulder PJ, Walker BD, Brockman MA, Ndung'u T (2010) Gag-protease-mediated replication capacity in HIV-1 subtype C chronic infection: associations with HLA type and clinical parameters. J Virol 84 (20):10820–10831. doi:10.1128/JVI.01084-10

Wright JK, Novitsky V, Brockman MA, Brumme ZL, Brumme CJ, Carlson JM, Heckerman D, Wang B, Losina E, Leshwedi M, van der Stok M, Maphumulo L, Mkhwanazi N, Chonco F, Goulder PJ, Essex M, Walker BD, Ndung'u T (2011) Influence of Gag-protease-mediated replication capacity on disease progression in individuals recently infected with HIV-1 subtype C. J Virol 85(8):3996–4006. doi:10.1128/JVI.02520-10

Yokomaku Y, Miura H, Tomiyama H, Kawana-Tachikawa A, Takiguchi M, Kojima A, Nagai Y, Iwamoto A, Matsuda Z, Ariyoshi K (2004) Impaired processing and presentation of cytotoxic-T-lymphocyte (CTL) epitopes are major escape mechanisms from CTL immune pressure in human immunodeficiency virus type 1 infection. J Virol 78(3):1324–1332

Yue L, Prentice HA, Farmer P, Song W, He D, Lakhi S, Goepfert P, Gilmour J, Allen S, Tang J, Kaslow RA, Hunter E (2013) Cumulative impact of host and viral factors on HIV-1 viral-load control during early infection. J Virol 87(2):708–715. doi:10.1128/JVI.02118-12

Yue L, Pfafferott KJ, Baalwa J, Conrod K, Dong CC, Chui C, Rong R, Claiborne DT, Prince JL, Tang J, Ribeiro RM, Cormier E, Hahn BH, Perelson AS, Shaw GM, Karita E, Gilmour J, Goepfert P, Derdeyn CA, Allen SA, Borrow P, Hunter E (2015) Transmitted virus fitness and host T cell responses collectively define divergent infection outcomes in two HIV-1 recipients. PLoS Pathog 11(1):e1004565. doi:10.1371/journal.ppat.1004565

Zhang LQ, MacKenzie P, Cleland A, Holmes EC, Brown AJ, Simmonds P (1993) Selection for specific sequences in the external envelope protein of human immunodeficiency virus type 1 upon primary infection. J Virol 67(6):3345–3356

Zhu T, Mo H, Wang N, Nam DS, Cao Y, Koup RA, Ho DD (1993) Genotypic and phenotypic characterization of HIV-1 patients with primary infection. Science 261(5125):1179–1181

Zhu T, Wang N, Carr A, Nam DS, Moor-Jankowski R, Cooper DA, Ho DD (1996) Genetic characterization of human immunodeficiency virus type 1 in blood and genital secretions: evidence for viral compartmentalization and selection during sexual transmission. J Virol 70 (5):3098–3107

Zimmerman PA, Buckler-White A, Alkhatib G, Spalding T, Kubofcik J, Combadiere C, Weissman D, Cohen O, Rubbert A, Lam G, Vaccarezza M, Kennedy PE, Kumaraswami V, Giorgi JV, Detels R, Hunter J, Chopek M, Berger EA, Fauci AS, Nutman TB, Murphy PM (1997) Inherited resistance to HIV-1 conferred by an inactivating mutation in CC chemokine receptor 5: studies in populations with contrasting clinical phenotypes, defined racial background, and quantified risk. Mol Med 3(1):23–36

Sequencing the Biology of Entry: The Retroviral *env* Gene

Ronald Swanstrom, William D. Graham and Shuntai Zhou

Abstract The surface envelope protein of any virus is major determinant of the host cell that is infected and as a result a major determinant of viral pathogenesis. Retroviruses have a single surface protein named Env. It is a trimer of heterodimers and is responsible for binding to the host cell receptor and mediating fusion between the viral and host membranes. In this review we will discuss the history of the discovery of the avian leukosis virus (ALV) and human immunodeficiency virus type 1 (HIV-1) Env proteins and their receptor specificity, comparing the many differences but having some similarities. Much of the progress in these fields has relied on viral genetics and genetic polymorphisms in the host population. A special feature of HIV-1 is that its persistent infection in its human host, to the point of depleting its favorite target cells, allows the virus to evolve new entry phenotypes to expand its host range into several new cell types. This variety of entry phenotypes has led to confusion in the field leading to the major form of entry phenotype of HIV-1 being overlooked until recently. Thus an important part of this story is the description and naming of the most abundant entry form of the virus: R5 T cell-tropic HIV-1.

Contents

Dedication: Sometimes the cart does come before the horse. A few heartfelt words about knowing Peter Vogt for forty years are included at the end.

R. Swanstrom (✉) · W.D. Graham · S. Zhou
Department of Biochemistry and Biophysics, Lineberger Cancer Center,
University of North Carolina, Chapel Hill, NC 27599, USA
e-mail: ron_swanstrom@med.unc.edu

Current Topics in Microbiology and Immunology (2017) 407:65–82
DOI 10.1007/82_2017_35

Published Online: 09 July 2017

1 Introduction

Retroviruses are sloppy. One can marvel at the regularity of the surface envelope protein of flaviviruses and alphaviruses where the proteins outside of the membrane are organized with the icosahedral shell inside of the membrane. One can envy the high density of envelope proteins that decorate the surface of coronaviruses and influenza virus. While some retroviruses show-off a high density of their surface envelope protein (Env), for others inclusion of the Env protein almost seems like an afterthought, given the apparent low density of Env on the virus surface (Zhu et al. 2006; Martin et al. 2016). In the case of HIV-1, the Env protein appears on the surface of the cell and if not quickly incorporated into a budding virion it is cycled back off the surface (Rowell et al. 1995; Sauter et al. 1996), presumably to avoid marking the cell as infected, a biologic example of "use it or lose it."

All retroviruses encode an Env protein, representing a primordial strategy for how at least some viral particles are able to fuse their membranes with the target cellular membrane. There are cellular proteins that can fuse membranes, for example the SNAREs, but the details of how they accomplish this are sufficiently different that it is hard to argue they are the origin of the viral envelope proteins like retroviruses use. Ironically it is easier to make the opposite claim. The cellular protein syncytin 1 is responsible for fusing cells to make a multicellular placenta in primates. Syncytin 1 is derived from an endogenous retrovirus whose *env* gene is now developmentally regulated. When expressed it fuses adjacent cells in the same way a viral membrane is fused to the cellular membrane; similar capture events appear to have happened in other eutherian mammals (Cornelis et al. 2013). The current lack of a true cellular protein that acts like the retroviral Env protein, and the fact that the retroviral Env protein functions in the same way as the influenza viral protein (and the surface protein of other distant viruses too), points to an early evolution of this type of membrane fusion capacity in enveloped viruses with this gene evolving with viruses as they generated different lineages. It is also likely that this gene has been passed among different viral lineages followed by distinctive evolution within the lineages to retain the basic fusion mechanism but with greatly varying sequence.

There are several universal points about this class of viral entry proteins (Fig. 1). First, they are type 1 transmembrane (TM) proteins with an N-terminal signal sequence that threads them into the endoplasmic reticulum and a stop-transfer sequence that stays in the membrane near the C terminus of the protein. This gives a luminal/outside N terminus of the protein and leaves the C terminus "inside" of the cell, which later becomes the inside of the virus particle when it buds from the cell taking some of the cellular membrane as its envelope. Second, there is an obligatory trimerization of the protein to stabilize its structure. Third, each subunit of the trimer is cleaved by a host protein (furin or furin-like) in the late Golgi compartment to create an extracellular component and a TM component, both of which are retained in the trimer complex and with each other. Fourth, the extracellular component (called SU for surface component in retrovirus nomenclature) of the cleaved protein is responsible for interacting with the host receptor(s), while the TM component exposes a hydrophobic stretch of amino acids (fusion peptide) at its new N terminus that is able to insert into the host membrane to facilitate fusion of host and viral membranes. There is a signal transduction event between the SU and TM components of the protein that occurs when the SU component binds its receptor thus defining the moment when fusion to the host cell must occur. There are diverse host cell receptors for viruses with this type of surface envelope protein, and the protein domains that bind the receptor and transduce the signal have lost any semblance of a common ancestor. And fifth, in contrast to the diversity of SU protein function the fusion mechanism is highly

Fig. 1 The retroviral Env protein. The viral Env protein exists as a trimer of heterodimers (SU/TM) embedded in the viral membrane (*parallel wavy lines*). The TM/transmembrane protein is *orange* with the lighter shade of orange indicating the cytoplasmic domain, the *dark blue portion* indicating the membrane-spanning domain, the *gray regions* representing the heptad repeats (hr1 and hr2) involved in the formation of the six helix bundle, and the *red portion* indicating the N-terminal fusion peptide. The *green region* represents the receptor binding/surface/SU protein, with the cleft indicating the receptor-binding region. The *purple box* indicates the region for HIV-1 that rearranges after binding CD4 to form the coreceptor binding site

conserved, in function if not in sequence. Insertion of the TM N-terminal fusion peptide into the host cell membrane is linked to the formation of a proximal trimer of a heptad repeat. This is followed by a more distal heptad repeat that folds up on the N-terminal heptad repeat to form a hairpin. Both of these heptad repeats (hr) are in the extracellular domain of TM such that the N-terminal repeat (hr1) is adjacent to the fusion peptide inserted into the host membrane and the C-terminal heptad repeat (hr2) is adjacent to the viral membrane-spanning portion of TM; the juxtaposition of the heptad repeats in the hairpin brings the two membranes together to promote fusion, bringing the inside of the virus particle into the inside of the target host cell.

There are several other curious points to mention about viruses that use this type of viral entry protein. First, the somewhat loose relationship between the TM viral envelope protein and the viral capsid inside of the viral envelope allows other proteins to be incorporated as TM proteins into the virus particle. Cellular proteins can be incorporated and there is an ongoing interest as to whether such proteins can alter the biology of the virus particle. Second, other viral proteins can be incorporated into the virus envelope which now gives the virus a host range specificity not encoded in its own genome, a phenomenon known as pseudotyping. Third, the virus-producing cell still expresses the cellular receptor. Since the virus does not what its envelope proteins interacting with the host cell receptor on a cell that is already infected, viruses use a variety of strategies of lowering the amount of receptor on the surface of the infected cell and regulating the fusogenicity of the envelope protein until the virus particle has budded. The overall effect of either down-regulating the receptor or tying up the receptor with newly synthesized Env protein prevents the infected cell from getting super-infected (usually), a phenomenon known as interference.

In this review we will focus on the SU domain of two very different retroviruses: avian leukosis virus (ALV) and human immunodeficiency virus type 1 (HIV-1). We will examine what sequence evolution can tell us about the proteins, we will look at their receptors and the conformational changes induced by binding to the receptor, and we will explore how these proteins can change host range to target different cell types. Some commonalities will emerge, but it will also be striking how different these two groups of viruses have solved the challenge of host cell targeting.

2 Avian Leukosis Virus (ALV)

It was around 1900 that enough evidence was accumulating to make the concept of a virus tenable. At that time most of the evidence was based on identifying a transmissible disease-causing agent that could be passed through a filter that would retain bacteria. There was an ongoing discussion as to whether these filterable agents could be very very small bacteria (but still independently living), toxins, or viruses that could only replicate within a host cell. As viruses became synonymous with filterable agents the list of viral diseases grew. There were no tools to measure a virus directly so disease was the only readout initially. Thus Ellermann and Bang described the first ALV-associated disease with the description of avian leukemia in 1908. Within a few years Peyton Rous would get credit for describing the first tumor-causing virus, Rous sarcoma virus, in 1910, for which he would get the Nobel Prize in 1966.

There was of course an underlying biology of these viruses that could not have been guessed in 1910, with genetic heterogeneity in the chickens defining susceptibility and in the virus changing receptors, and even endogenous viruses in the chicken genome. These phenomena were all revealed slowly, initially as unexpected observations that had to be reproduced enough so that a description could be formulated that could then be used to support suggestions of underlying mechanisms.

2.1 History of Viruses and Susceptibility

When disease is the only readout progress is slow. In a retrospective published by Harry Rubin in 2011, he reviewed the role new assays played in developing an understanding of the viruses, not just the disease (Rubin 2011). In 1938 Keogh showed that (tumor-like) lesions could be produced on the chorioallantoic membrane (CAM) of a chicken egg providing a semi-quantitative alternative to titering virus in chickens. It took another 20 years until Prince discovered that the high number of false negatives (i.e., eggs that gave no lesions) was a genetic trait of the chicken strains used, defining a sensitive and resistant allele for the virus used, and the first measure of the viral receptor. It was also around this time that Howard Temin, then a graduate student collaborating with Harry Rubin, perfected the focus-forming assay with an agar overlay to allow titration of the transforming activity of RSV. Each egg could be turned into multiple plates of chicken embryo fibroblasts (CEFs) and read much easier than counting lesions on CAM.

With a new, more powerful, assay the genetics of the virus started to come into focus. A subset of embryos that should have given cells that were susceptible to transformation by RSV were in fact resistant, defining a resistance-inducing factor (RIF). For a number of years avian leukosis viruses had been studied for their natural infection of chickens, and it turned out that the RIF was the result of the occasional embryo that was already infected with an ALV. This revealed the phenomenon of interference, the mechanism being that when a cell is already expressing a viral envelope protein it removes or engages its normal receptor precluding infection with a new virus with the same receptor specificity.

It had been 50 years of research and experimental passaging of Rous sarcoma virus by this time so it should not be surprising that different virus stocks had acquired different properties and even different passenger viruses. RSV is unique in that it can be propagated as a replication competent virus, in contrast to most other acutely transforming retroviruses that carry deletion of part of the viral genome with the acquisition of the cellular oncogene. But RSV will delete the *v-src* gene to give rise to a non-acutely transforming, replication competent Rous-associated virus (RAV), again essentially a standard ALV. However, the isolation of the first RAV (RAV-1) with its associated interference properties was followed by the isolation of a second RAV (RAV-2) with distinct interference properties. The clear implication was that these two viruses used different host receptors. Furthermore, the isolation of defective forms of the transforming virus component (defective for replication) allowed rescue of the transforming genome with any ALV helper virus, with the transforming component now having the cell infectivity properties of the helper virus (i.e., a pseudotype). The combination of cell susceptibility to infection and transformation, and the ability to test viruses for the use of the same or different receptors through interference led to the identification of virus subgroups, specifically subgroups A, B, C, D, and E (with J coming later). Conversely it was possible to identify the genetic loci for susceptibility in chickens, the presumed receptors which were named *tva*, *tvb*, and *tvc* (with *tvb* serving a receptor for subgroups D and E ALV). Thus great strides in the genetics of the virus and the host were accomplished using the focus-forming assay. These insights into the biology of the virus and the host set the stage for the coming tools of cloning and sequencing.

2.2 The Viral Env Gene/Protein

The development of electron microscopy led to the ability to "see" viruses and it became apparent that viruses had complex structures. Advances in growing and purifying the virus, and in protein analysis allowed for the identification of proteins associated with the virus particle, and presumably encoded by the viral genome. In the early 1970s a small group of investigators, including Peter Vogt, had characterized virion proteins for RSV, including virion-associated glycoproteins (Duesberg et al. 1970; Robinson et al. 1970). By 1971 the combination of EM, virus purification, radioactive labeling of proteins, and protease treatment was used to show that it was the virion glycoprotein that was on the exterior of the viral envelope (Rifkin and Compans 1971). With viral genetics providing recombinants with altered host range it became possible to map the location of the glycoprotein within the viral genome using the pre-sequencing tool of assessing patterns of RNase T1-resistant oligonucleotides displayed using 2D electrophoresis, placing the *env* gene at the 3′ end of the ALV genome (Joho et al. 1975; Coffin and Billiter 1976; Wang et al. 1976).

2.3 Sequencing, Cloning, and Sequencing

In February of 1975 The Asilomar Conference on Recombinant DNA was held to consider the risks and rewards of using recombinant DNA tools and cloning. With guidelines in place a new era of biology began. Virologists had to become at least mediocre bacteriologists capable of growing phage and plasmids and isolating biological clones. Those up to the task were rewarded with a new view of genes and genomes. With the availability of unlimited amounts of cloned DNA, sequencing tools quickly followed and gene and inferred protein sequences were revealed. Since DNA is relatively homogeneous in its chemical properties, in contrast to its information properties, virtually anything could be cloned with the order determined by the investigator's interest. For those interested in viruses with RNA genomes the challenge was greater, while for retrovirologists most went after the DNA form of the viral genome. However, in an early tour de force, virtually the entire RSV genome was sequenced as cDNA products that were made using random primers, purified reverse transcriptase, and viral genomic RNA (Schwartz et al. 1983). The authors noted in their 1983 paper that "at the time this project was initiated, molecular cloning of RSV was prohibited." The placement of the *env* gene, upstream of the *v-src* gene, was proven by comparing the amino-terminal protein sequences of the SU (gp85) and TM (gp37) Env protein subunits and placing them on the nucleotide sequence (Hunter et al. 1983). The strain of RSV that was sequenced carried a subgroup C *env* gene providing the first view of the sequence of this protein.

One of the funny things about looking at sequences is that you learn some things from looking at the first sequence of a gene, but you learn a whole lot of information that was not available by looking at the first sequence when you get to look at the second sequence, hopefully a similar but not identical sequence. Specifically you learn about which regions are identical (or nearly identical) and which are different. For evolutionary differences there are two considerations: first is a relatively uninteresting evolutionary drift that occurs with evolutionary distance; more interesting are the sequences that rapidly evolve due to strong selective pressure where the differences can be linked to changes in biological function. Thus the second and third sequences reported, for subgroup B and E *env* genes, revealed conserved and variable regions that could only be interpreted as the protein framework and two regions of variability as determinants of receptor specificity, named host range 1 and 2 or hr 1 and 2 (Dorner et al. 1985). The analysis of subgroups A and D *env* genes (Bova et al. 1986, 1988) reinforced this idea.

2.4 Receptors

The molecular cloning of receptors was a difficult endeavor. In cloning the viral genome hybridization probes made as radioactive cDNA from viral genomic RNA could be used to find the desired clones. However, even though the receptors had been characterized genetically and named (*tva, tvb*, etc.), the only assay for the receptors was a biological assay. After an effort of several years the first ALV receptor was cloned for subgroup A viruses and identified as being related to the low density lipoprotein receptor (Bates et al. 1993). Next the receptor for subgroup B and D viruses was cloned and recognized as a member of the TNFR protein family (Brojatsch et al. 1996), then the receptor for subgroup E was identified as another member of the TNFR family (Adkins et al. 2001), and finally the subgroup C receptor was identified as a member of the butyrophilin protein family (Elleder et al. 2005). The lesson learned is that viruses pick out receptors using a logic that as yet escapes us. The polymorphisms in the receptor/susceptibility loci, in part expanded in the population by breeding but still from naturally existing alleles, suggests selection against having functional receptors by the host. This is analogous to the presence of xenotropic MLV endogenous viruses, i.e., an endogenous virus that can not infect its host (or that the host lost the functional receptor after the endogenous provirus was fixed). Since evolution is not functionally blind, it must be the case that there is a primary receptor and at least a low level of interaction with other surface proteins, as different from each other as they are, to allow the evolution of different subgroups with different receptor specificities. The ultimate "receptor switch" then provides the strong selective pressure for rapid sequence change in the hr1 and hr2 regions.

2.5 Subgroup J, Biology in Real Time

Humans are overly egocentric such that we do not readily conceive of the next thing until it happens as a surprise. There will always be new viral variants that appear with a slightly different constellation of properties that allow the next epidemic. An unknown primate lentivirus was percolating along in chimpanzees and we became aware of it only when it turned into a worldwide epidemic in humans. Thus we should not be surprised that just as we were getting comfortable with the biology of ALV, and ways to avoid it in chicken flocks, a new subgroup appeared, subgroup J. As reviewed in Payne and Nair (2012) a new pathogenic subgroup of ALV was identified in the late 1980s, subsequently shown to be a recombinant between ALV and the *env* gene of an endogenous retroviral allele. The receptor for this virus was cloned and shown to be a Na+/H+ exchanger type 1 protein (Chai and Bates 2006), thus adding to the seemingly non sequitur list of cellular proteins capable of serving as receptors.

3 HIV-1 Env Proteins: Still Trying to Get It Right

HIV-1 was discovered around the time other retroviral genomes were just coming into the hands of cloners. Given its novelty as the second human retrovirus, and certainly the scarier of the two, the understanding of the framework of its biology occurred quickly. An important early observation was that CD4 on the surface of T cells was a receptor important for cell entry, demonstrated by blocking viral infection with an anti-CD4 antibody (Dalgleish et al. 1984; Klatzmann et al. 1984). This is a fundamental property of HIV-1 as no CD4-independent virus has ever been isolated, although it has been possible to select for a CD4-independent SIV in cell culture (Swanstrom et al. 2016). However, after identifying CD4 as a receptor for HIV-1, the entry field took a confusing twist that is still confounding us today. In a way that was analogous to some chickens/eggs/CEFs being resistant to infection (for lack of a functional receptor), certain strains of HIV-1 could not grow on cell lines that clearly expressed CD4, even though all viral isolates could grow in PBMCs. Thirty years later we are still trying to overcome the initial interpretation of these results that now need to be updated.

The identification of CD4 as a receptor and the use of transformed T cell lines that express CD4 (usually derived from a leukemia) make perfect sense. The ability of some strains of virus to grow in the T cell lines earned them the name T cell-tropic. Those that failed to grow had to be something else. Macrophages express a low level of CD4 and can support a low level of entry for most isolates and overt replication for some isolates. This was enough for the isolates that could not grow in the T cell lines to earn the name macrophage-tropic. Thus HIV-1 isolates fell into these two nice categories, T cell-tropic and macrophage-tropic. Unfortunately both names are inappropriate for what they were trying to describe.

This problem of misidentification became exacerbated with the important discovery that HIV-1, unlike most of the other viruses we know about, has a second receptor now called the co-receptor. A coreceptor was identified for one of the "T cell-tropic" viruses which was the chemokine receptor CXCR4 (Feng et al. 1996). By analogy the "macrophage-tropic" viruses should use a similar molecule, which was quickly identified as the chemokine receptor CCR5 (Alkhatib et al. 1996; Choe et al. 1996; Deng et al. 1996; Doranz et al. 1996; Dragic et al. 1996). Viruses using these coreceptors were named X4 and R5 viruses and the dual entry phenotype of HIV-1 was further engrained as X4 T cell-tropic and R5 macrophage-tropic. If the names were inappropriate before this only made it worse. In fact there are three types of HIV-1 whose entry phenotypes are most accurately described as a jumble of these two inappropriate names.

The missing piece in this story was apparent to a few investigators who came to understand that the "R5 macrophage-tropic" viruses were not homogeneous. Some were effective at using a low density of CD4 while most were inefficient at entering cells that displayed a low density of CD4 (Kabat et al. 1994; Platt et al. 1998). Furthermore, the viruses capable of using a low density of CD4 were most reliably found in the brain late in infection (Gorry et al. 2002; Peters et al. 2004; Dunfee et al. 2006; Martin-Garcia et al. 2006; Peters et al. 2006; Duenas-Decamp et al. 2009; Schnell et al. 2011). Since macrophages have a low density of CD4 (CD4 has no function in macrophages) (Lee et al. 1999; Joseph et al. 2014), it is a small leap to suggest that the viruses that can use a low density of CD4 to enter cells are in fact the ones that have evolved to become macrophage-tropic. We used a cell line (Affinofile cells) designed to allow regulated control of CD4 levels (Johnston et al. 2009) to explore these issues at length. After analyzing between 100 and 200 *env* genes cloned from a variety of sources, we came to the conclusion that R5 macrophage-tropic viruses are rare (see for example Ping et al. 2013). This left the majority of R5 viruses as distinct from R5 macrophage-tropic viruses and without a name.

As discussed later, X4 viruses evolve from R5 viruses so the default version of HIV-1 has to be an R5 virus. But if most of these isolates do not enter macrophages efficiently then where are these viruses growing? Of course the answer is T cells and they are rightly called T cell-tropic, or more descriptively R5 T cell-tropic. However, we started this discussion with the idea that only X4 viruses were growing in T cells, or more accurately T cell lines. For those who are good at puzzles the answer is probably clear. The CD4+ T cells lines used to grow HIV-1 in the early days expressed CXCR4 but not CCR5, a point that at the time was meaningless since the concept of the coreceptor did not exist. However, just as chickens, eggs, and CEFs can lack functional receptors for certain subgroups of ALV, these cell lines were heterogeneous for expression of the coreceptors, although in general most of these transformed cell lines express CXCR4 and not CCR5. The better analogy comes with the observation that about 10% of northern Europeans carry a CCR5 allele with an inactivating mutation, meaning that about 1% of this population is resistant to infection by the most common form of HIV-1, i.e., R5 T cell-tropic virus (Dean et al. 1996; Samson et al. 1996; Liu et al. 1996). Such individuals can be infected with an X4 T cell-tropic virus (Theodorou et al. 1997; Michael et al. 1998), although transmission of this virus is rare. It should be noted that when we have examined X4 viruses where the *env* genes were isolated without tissue culture passage we find that they require a high density of CD4 so that they are still appropriately called X4 T cell-tropic (M. Bednar and R.S., in preparation).

To summarize HIV-1 entry phenotypes thus far: there are three entry phenotypes, R5 T cell-tropic, R5 macrophage-tropic, and X4 T cell-tropic. The default form of HIV-1 is R5 T cell-tropic, using the CCR5 coreceptor and requiring a high density of CD4 for entry. X4 T cell-tropic viruses evolve late in disease with a switch in coreceptor use. In an analogous way macrophage-tropic viruses also evolve late in disease (in a T cell-poor environment) and gain the ability to enter cells with a low density of CD4 more efficiently. Thus the vexing legacy of the early entry work is that in naming two types of entry phenotypes the one that was excluded is the one that is actually the predominant form of HIV-1: R5 T cell-tropic.

3.1 The Two Evolutionary Variants of R5 T Cell-Tropic Viruses

The transmitted form of HIV-1 and the form that is found in the blood throughout most of the infection is the R5 T cell-tropic form of HIV-1 (using CCR5 as the coreceptor and requiring a high density of CD4 for efficient entry) (Ochsenbauer et al. 2012; Ping et al. 2013). That means that the other forms must evolve from the R5 T cell-tropic form (Fig. 2). With the discovery of coreceptors it was possible to make the link with the CXCR4-using version of HIV-1 as appearing late in infection

(Connor et al. 1997; Brumme et al. 2005; Moyle et al. 2005). As noted above for the evolution of different subgroups of ALV (probably on a much longer time scale), the R5 T cell-tropic version of HIV-1 must interact at least a little bit with CXCR4 to allow that evolutionary pathway to occur. At the extreme, the adaptation is so complete that the virus losses the ability to interact with CCR5, however most viruses never get that far in their evolution so the typical virus seen is termed dual-tropic, meaning it can use both CXCR4 and CCR5. We have observed that these dual-tropic viruses have a reduced affinity for CCR5, as measured by increased sensitivity to a CCR5 antagonist, suggesting that the dual-tropic viruses are more X4 than R5 (M. Bednar and R.S., in preparation). We have also observed that X4 viruses require a high density of CD4 for efficient entry, retaining their X4 T cell-tropic moniker. Since X4 viruses evolve late in disease we can speculate that either they require an immunodeficient host to evolve or that they evolve when the preferred target cells, CD4+ CCR5+ T cells, become limiting. One theory is that CD4+ CXCR4+ naive T cells support the replication of X4 viruses (Ribeiro et al. 2006).

The second evolutionary variant of HIV-1 is of course the macrophage tropic variant. Where the change in coreceptor use is more dramatic and easy to measure, the change in CD4 use is less dramatic and harder to measure (Joseph et al. 2014; Arrildt et al. 2015), accounting for much of the confusion about this phenotype. These viruses evolve in a T cell-poor environment, especially in the CNS (Sturdevant et al. 2015). In addition to their ability to efficiently enter cells with a low density of CD4 there is a tightly linked phenotype of increased sensitivity to neutralization by soluble CD4 (Arrildt et al. 2015). It appears these viruses are primed to undergo the fusion conformation cascade with fewer interactions with CD4.

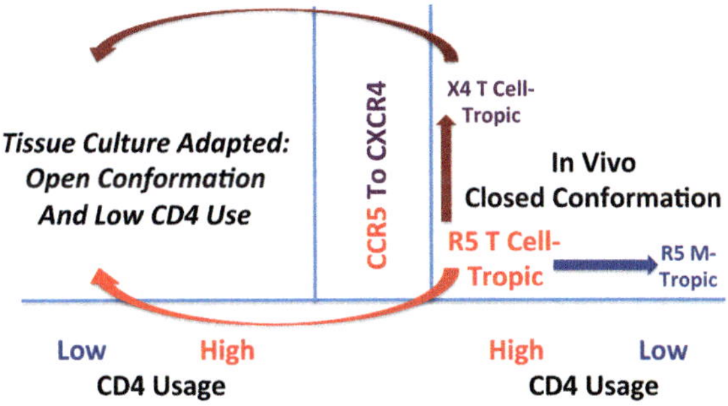

Fig. 2 Pathways for the evolution of the HIV-1 Env protein entry phenotype. The major entry phenotype form for HIV-1 is the R5 T cell-tropic form. It uses CCR5 as the coreceptor, but requires a high density of CD4, as is found on T cells, for efficient entry. In vivo it evolves to switch coreceptor to use CXCR4. Alternatively, it can evolve to use a low density of CD4 to enter cells such as macrophages, which have a density of CD4 about 25-fold lower than that found on T cells. Also in vivo, these viruses are found in a closed conformation, i.e., resistant to neutralization, especially to antibodies targeting the epitopes exposed after binding CD4. In cell culture the virus follows another evolutionary pathway in which an open conformation is generated allowing the use of a low density of CD4 for entry. Presumably this enables more rapid entry under culture conditions. This can happen for both the X4 form of the virus and the R5 T cell-tropic form of the virus and should be considered an artifact of tissue culture adaptation

3.2 A Fourth Entry Phenotype Is an Artifact of Tissue Culture Adaptation

It is common practice in studying a virus to grow the virus in tissue culture. For most of the viral functions this is probably not a bad idea, at least in the short term. However, given that many attenuated viral vaccines were developed simply by passaging an isolate in culture we have to acknowledge that passage in culture does put the virus under a very different selective pressure compared to what it experiences in vivo. The strongest selective pressure in cell culture is likely to occur on the viral entry phenotype. There is no wrong cell to infect given that the culture is largely homogeneous, so viral replication in cell culture becomes a race to see who can enter cells most quickly. HIV-1 seems to be especially susceptible to a serious cell culture artifact that is as yet not fully appreciated.

As noted above the HIV-1 Env protein goes through a conformational change when binding the CD4 receptor (McDougal et al. 1986; White et al. 2010; Munro et al. 2014). For most viruses this would lead to insertion of the fusion peptide followed by membrane fusion. However, for HIV-1 there is an extra step. The first conformational change is to create and expose the CCR5 binding site. Binding to CCR5 then triggers insertion of the fusion peptide, formation of the six helix bundle, and membrane fusion between the host and viral membranes. Thus the HIV-1 Env protein trimer must transition through a lot of conformational space to carry out its job. When examining the antibodies from people infected with HIV-1 there seem to be abundant antibodies to epitopes that are created after binding to CD4. This is likely the selective pressure that keeps the trimer in a "closed" conformation where these epitopes are either covered or not yet even formed. The transient exposure of these epitopes after CD4 engagement leaves too little space at the surface of the cell for these antibodies to be effective (Labrijn et al. 2003). However, in cell culture no such antibody selective pressure exists.

An important observation was the realization that passage of HIV-1 in tissue culture led to a highly neutralization sensitive form of the Env protein where it was now sensitive to these CD4 binding-induced epitopes (Moore et al. 1995). The conformation associated with this state has become known as the "open" conformation. The important thing to realize is that a virus with its Env protein in an open conformation is a tissue culture artifact, such viruses are not found in vivo (Mascola et al. 1996; Harris et al. 2011). In the race to grow the fastest in cell culture the incorporation of mutations that dispense with the closed conformation and allow the virus to skip most of the conformational change induced by binding CD4 are selected, although these viruses are still CD4-dependent (Fig. 2).

The next idea we will discuss is partly data-driven and partly prediction. Tissue culture adaptation has another phenotype that causes confusion with macrophage tropism; like macrophage-tropic viruses, tissue culture-adapted viruses are also able to use a low density of CD4 to enter cells (Kabat et al. 1994). The important distinction to make is that tissue culture adaptation selects for an open conformation while macrophage tropism does not. There do seem to be some structural changes in the Env protein associated with macrophage tropism, but these do not go as far as representing the tissue culture-adapted open conformation (Arrildt et al. 2015). This is an important distinction to make because these two forms appear to share the feature of being able to use CD4 at a low density. We would predict that the pathway to low CD4 use in the context of tissue culture adaptation is distinct from the more relevant pathway that the virus uses in vivo to become macrophage-tropic. In developing a relevant understanding of macrophage tropism it will be important to rigorously avoid confounding information that comes from tissue culture-adapted viruses.

Acknowledgements Our own work is supported by the National Institutes of Health. We thank many lab members who contributed to the ideas in this review. We also thank Adrienne Swanstrom for editorial assistance.

References

Adkins HB, Blacklow SC, Young JA (2001) Two functionally distinct forms of a retroviral receptor explain the nonreciprocal receptor interference among subgroups B, D, and E avian leukosis viruses. J Virol 75(8):3520–3526. doi:10.1128/JVI.75.8.3520-3526.2001

Alkhatib G, Combadiere C, Broder CC, Feng Y, Kennedy PE, Murphy PM, Berger EA (1996) CC CKR5: a RANTES, MIP-1 alpha, MIP-1 beta receptor as a fusion cofactor for macrophage-tropic HIV-1. Science 272(5270):1955–1958. doi:10.1126/science.272.5270.1955

Arrildt KT, LaBranche CC, Joseph SB, Dukhovlinova EN, Graham WD, Ping L-H, Schnell G, Sturdevant CB, Kincer LP, Mallewa M, Heyderman RS, Van Rie A, Cohen MS, Spudich S, Price RW, Montefiori DC, Swanstrom R (2015) Phenotypic correlates of HIV-1 macrophage tropism. J Virol 89(22):11294–11311. doi:10.1128/jvi.00946-15

Bates P, Young JA, Varmus HE (1993) A receptor for subgroup A Rous sarcoma virus is related to the low density lipoprotein receptor. Cell 74(6):1043–1051

Bova CA, Manfredi JP, Swanstrom R (1986) Env genes of avian retroviruses: nucleotide sequence and molecular recombinants define host range determinants. Virology 152(2):343–354

Bova CA, Olsen JC, Swanstrom R (1988) The avian retrovirus env gene family: molecular analysis of host range and antigenic variants. J Virol 62(1):75–83

Brojatsch J, Naughton J, Rolls MM, Zingler K, Young JA (1996) CAR1, a TNFR-related protein, is a cellular receptor for cytopathic avian leukosis-sarcoma viruses and mediates apoptosis. Cell 87(5):845–855

Brumme ZL, Goodrich J, Mayer HB, Brumme CJ, Henrick BM, Wynhoven B, Asselin JJ, Cheung PK, Hogg RS, Montaner JS, Harrigan PR (2005) Molecular and clinical epidemiology of CXCR4-using HIV-1 in a large population of antiretroviral-naive individuals. J Infect Dis 192(3):466–474. doi:10.1086/431519

Chai N, Bates P (2006) Na+/H+ exchanger type 1 is a receptor for pathogenic subgroup J avian leukosis virus. Proc Natl Acad Sci U S A 103 (14):5531–5536. doi:10.1073/pnas.0509785103

Choe H, Farzan M, Sun Y, Sullivan N, Rollins B, Ponath PD, Wu L, Mackay CR, LaRosa G, Newman W, Gerard N, Gerard C, Sodroski J (1996) The beta-chemokine receptors CCR3 and CCR5 facilitate infection by primary HIV-1 isolates. Cell 85(7):1135–1148. doi:10.1016/s0092-8674(00)81313-6

Coffin JM, Billeter MA (1976) A physical map of the Rous sarcoma virus genome. J Mol Biol 100 (3):293–318

Connor RI, Sheridan KE, Ceradini D, Choe S, Landau NR (1997) Change in coreceptor use correlates with disease progression in HIV-1–infected individuals. J Exp Med 185(4):621–628

Cornelis G, Heidmann O, Degrelle SA, Vernochet C, Lavialle C, Letzelter C, Bernard-Stoecklin S, Hassanin A, Mulot B, Guillomot M, Hue I, Heidmann T, Dupressoir A (2013) Captured retroviral envelope syncytin gene associated with the unique placental structure of higher ruminants. Proc Natl Acad Sci U S A 110(9):E828–E837. doi:10.1073/pnas.1215787110

Dalgleish AG, Beverley PCL, Clapham PR, Crawford DH, Greaves MF, Weiss RA (1984) The CD4 (T4) antigen is an essential component of the receptor for the AIDS retrovirus. Nature 312 (5996):763−767. doi:10.1038/312763a0

Dean M, Carrington M, Winkler C, Huttley GA, Smith MW, Allikmets R, Goedert JJ, Buchbinder SP, Vittinghoff E, Gomperts E, Donfield S, Vlahov D, Kaslow R, Saah A, Rinaldo C, Detels R, Obrien SJ (1996) Genetic restriction of HIV-1 infection and progression to AIDS by a deletion allele of the CKR5 structural gene. Science 273(5283):1856–1862. doi:10.1126/science.273.5283.1856

Deng HK, Liu R, Ellmeier W, Choe S, Unutmaz D, Burkhart M, DiMarzio P, Marmon S, Sutton RE, Hill CM, Davis CB, Peiper SC, Schall TJ, Littman DR, Landau NR (1996) Identification of a major co-receptor for primary isolates of HIV-1. Nature 381(6584):661–666. doi:10.1038/381661a0

Doranz BJ, Rucker J, Yi Y, Smyth RJ, Samson M, Peiper SC, Parmentier M, Collman RG, Doms RW (1996) A dual-tropic primary HIV-1 isolate that uses fusin and the beta-chemokine receptors CKR-5, CKR-3, and CKR-2b as fusion cofactors. Cell 85(7):1149–1158. doi:10.1016/s0092-8674(00)81314-8

Dorner AJ, Stoye JP, Coffin JM (1985) Molecular basis of host range variation in avian retroviruses. J Virol 53(1):32–39

Dragic T, Litwin V, Allaway GP, Martin SR, Huang Y, Nagashima KA, Cayanan C, Maddon PJ, Koup RA, Moore JP, Paxton WA (1996) HIV-1 entry into CD4+ cells is mediated by the chemokine receptor CC-CKR-5. Nature 381(6584):667–673. doi:10.1038/381667a0

Duenas-Decamp MJ, Peters PJ, Burton D, Clapham PR (2009) Determinants flanking the CD4 binding loop modulate macrophage tropism of human immunodeficiency virus type 1 R5 envelopes. J Virol 83(6):2575–2583. doi:10.1128/jvi.02133-08

Duesberg PH, Martin GS, Vogt PK (1970) Glycoprotein components of avian and murine RNA tumor viruses. Virology 41(4):631–646

Dunfee RL, Thomas ER, Gorry PR, Wang JB, Taylor J, Kunstman K, Wolinsky SM, Gabuzda D (2006) The HIV Env variant N283 enhances macrophage tropism and is associated with brain infection and dementia. Proc Natl Acad Sci U S A 103(41):15160–15165. doi:10.1073/pnas.0605513103

Elleder D, Stepanets V, Melder DC, Senigl F, Geryk J, Pajer P, Plachy J, Hejnar J, Svoboda J, Federspiel MJ (2005) The receptor for the subgroup C avian sarcoma and leukosis viruses, Tvc, is related to mammalian butyrophilins, members of the immunoglobulin superfamily. J Virol 79(16):10408–10419. doi:10.1128/JVI.79.16.10408-10419.2005

Feng Y, Broder CC, Kennedy PE, Berger EA (1996) HIV-1 entry cofactor: functional cDNA cloning of a seven-transmembrane, G protein-coupled receptor. Science 272(5263):872–877

Gorry PR, Taylor J, Holm GH, Mehle A, Morgan T, Cayabyab M, Farzan M, Wang H, Bell JE, Kunstman K, Moore JP, Wolinsky SM, Gabuzda D (2002) Increased CCR5 affinity and reduced CCR5/CD4 dependence of a neurovirulent primary human immunodeficiency virus type 1 isolate. J Virol 76(12):6277–6292. doi:10.1128/jvi.76.12.6277-6292.2002

Harris A, Borgnia MJ, Shi D, Bartesaghi A, He H, Pejchal R, Kang Y, Depetris R, Marozsan AJ, Sanders RW, Klasse PJ, Milne JLS, Wilson IA, Olson WC, Moore JP, Subramaniam S (2011) Trimeric HIV-1 glycoprotein gp140 immunogens and native HIV-1 envelope glycoproteins display the same closed and open quaternary molecular architectures. Proc Natl Acad Sci U S A 108(28):11440–11445. doi:10.1073/pnas.1101414108

Hunter E, Hill E, Hardwick M, Bhown A, Schwartz DE, Tizard R (1983) Complete sequence of the Rous sarcoma virus env gene: identification of structural and functional regions of its product. J Virol 46(3):920–936

Johnston SH, Lobritz MA, Nguyen S, Lassen K, Delair S, Posta F, Bryson YJ, Arts EJ, Chou T, Lee B (2009) A quantitative affinity-profiling system that reveals distinct CD4/CCR5 usage patterns among human immunodeficiency virus type 1 and simian immunodeficiency virus strains. J Virol 83(21):11016–11026. doi:10.1128/JVI.01242-09

Joho RH, Billeter MA, Weissmann C (1975) Mapping of biological functions on RNA of avian tumor viruses: location of regions required for transformation and determination of host range. Proc Natl Acad Sci U S A 72(12):4772–4776

Joseph SB, Arrildt KT, Swanstrom AE, Schnell G, Lee B, Hoxie JA, Swanstrom R (2014) Quantification of entry phenotypes of macrophage-tropic HIV-1 across a wide range of CD4 densities. J Virol 88(4):1858–1869. doi:10.1128/JVI.02477-13

Kabat D, Kozak SL, Wehrly K, Chesebro B (1994) Differences in CD4 dependence for infectivity of laboratory-adapted and primary patient isolates of human immunodeficiency virus type 1. J Virol 68(4):2570–2577

Klatzmann D, Champagne E, Chamaret S, Gruest J, Guetard D, Hercend T, Gluckman JC, Montagnier L (1984) T-lymphocyte T4 molecule behaves as the receptor for human retrovirus LAV. Nature 312(5996):767–768

Labrijn AF, Poignard P, Raja A, Zwick MB, Delgado K, Franti M, Binley J, Vivona V, Grundner C, Huang CC, Venturi M, Petropoulos CJ, Wrin T, Dimitrov DS, Robinson J, Kwong PD, Wyatt RT, Sodroski J, Burton DR (2003) Access of antibody molecules to the conserved coreceptor binding site on glycoprotein gp120 is sterically restricted on primary human immunodeficiency virus type 1. J Virol 77(19):10557–10565

Lee B, Sharron M, Montaner LJ, Weissman D, Doms RW (1999) Quantification of CD4, CCR5, and CXCR4 levels on lymphocyte subsets, dendritic cells, and differentially conditioned monocyte-derived macrophages. Proc Natl Acad Sci U S A 96(9):5215–5220

Liu R, Paxton WA, Choe S, Ceradini D, Martin SR, Horuk R, MacDonald ME, Stuhlmann H, Koup RA, Landau NR (1996) Homozygous defect in HIV-1 coreceptor accounts for resistance of some multiply-exposed individuals to HIV-1 infection. Cell 86(3):367–377. doi:10.1016/s0092-8674(00)80110-5

Martin JL, Cao S, Maldonado JO, Zhang W, Mansky LM (2016) Distinct particle morphologies revealed through comparative parallel analyses of retrovirus-like particles. J Virol 90 (18):8074–8084. doi:10.1128/JVI.00666-16

Martin-Garcia J, Cao W, Varela-Rohena A, Plassmeyer ML, Gonzalez-Scarano F (2006) HIV-1 tropism for the central nervous system: brain-derived envelope glycoproteins with lower CD4 dependence and reduced sensitivity to a fusion inhibitor. Virology 346(1):169–179

Mascola JR, Snyder SW, Weislow OS, Belay SM, Belshe RB, Schwartz DH, Clements ML, Dolin R, Graham BS, Gorse GJ, Keefer MC, McElrath MJ, Walker MC, Wagner KF, McNeil JG, McCutchan FE, Burke DS (1996) Immunization with Envelope Subunit Vaccine Products Elicits Neutralizing Antibodies against Laboratory-Adapted but Not Primary Isolates of Human Immunodeficiency Virus Type 1. J Infect Dis 173(2):340–348

McDougal JS, Nicholson JK, Cross GD, Cort SP, Kennedy MS, Mawle AC (1986) Binding of the human retrovirus HTLV-III/LAV/ARV/HIV to the CD4 (T4) molecule: conformation dependence, epitope mapping, antibody inhibition, and potential for idiotypic mimicry. J Immunol 137(9):2937–2944

Michael NL, Nelson JA, KewalRamani VN, Chang G, O'Brien SJ, Mascola JR, Volsky B, Louder M, White GC, 2nd, Littman DR, Swanstrom R, O'Brien TR (1998) Exclusive and persistent use of the entry coreceptor CXCR4 by human immunodeficiency virus type 1 from a subject homozygous for CCR5 delta32. J Virol 72(7):6040–6047

Moore JP, Cao YZ, Qing L, Sattentau QJ, Pyati J, Koduri R, Robinson J, Barbas CF, Burton DR, Ho DD (1995) Primary isolates of human immunodeficiency virus type 1 are relatively resistant to neutralization by monoclonal antibodies to gp120, and their neutralization is not predicted by studies with monomeric gp120. J Virol 69(1):101–109

Moyle GJ, Wildfire A, Mandalia S, Mayer H, Goodrich J, Whitcomb J, Gazzard BG (2005) Epidemiology and predictive factors for chemokine receptor use in HIV-1 infection. J Infect Dis 191(6):866–872. doi:10.1086/428096

Munro JB, Gorman J, Ma X, Zhou Z, Arthos J, Burton DR, Koff WC, Courter JR, Smith AB, Kwong PD, Blanchard SC, Mothes W (2014) Conformational dynamics of single HIV-1 envelope trimers on the surface of native virions. Science 346(6210):759–763. doi:10.1126/science.1254426

Ochsenbauer C, Edmonds TG, Ding HT, Keele BF, Decker J, Salazar MG, Salazar-Gonzalez JF, Shattock R, Haynes BF, Shaw GM, Hahn BH, Kappes JC (2012) Generation of transmitted/founder HIV-1 infectious molecular clones and characterization of their replication capacity in CD4 T lymphocytes and monocyte-derived macrophages. J Virol 86(5):2715–2728

Payne LN, Nair V (2012) The long view: 40 years of avian leukosis research. Avian Pathol 41 (1):11–19. doi:10.1080/03079457.2011.646237

Peters PJ, Bhattacharya J, Hibbitts S, Dittmar MT, Simmons G, Bell J, Simmonds P, Clapham PR (2004) Biological analysis of human immunodeficiency virus type 1 R5 envelopes amplified from brain and lymph node tissues of AIDS patients with neuropathology reveals two distinct tropism phenotypes and identifies envelopes in the brain that confer an enhanced tropism and fusigenicity for macrophages. J Virol 78(13):6915–6926. doi:10.1128/jvi.78.13.6915-6926.2004

Peters PJ, Sullivan WM, Duenas-Decamp MJ, Bhattacharya J, Ankghuambom C, Brown R, Luzuriaga K, Bell J, Simmonds P, Ball J, Clapham PR (2006) Non-macrophage-tropic human immunodeficiency virus type 1 R5 envelopes predominate in blood, lymph nodes, and semen: Implications for transmission and pathogenesis. J Virol 80(13):6324–6332. doi:10.1128/jvi.02328-05

Ping LH, Joseph SB, Anderson JA, Abrahams MR, Salazar-Gonzalez JF, Kincer LP, Treurnicht FK, Arney L, Ojeda S, Zhang M, Keys J, Potter EL, Chu H, Moore P, Salazar MG, Iyer S, Jabara C, Kirchherr J, Mapanje C, Ngandu N, Seoighe C, Hoffman I, Gao F, Tang Y, Labranche C, Lee B, Saville A, Vermeulen M, Fiscus S, Morris L, Karim SA, Haynes BF, Shaw GM, Korber BT, Hahn BH, Cohen MS, Montefiori D, Williamson C, Swanstrom R (2013) Comparison of viral Env proteins from acute and chronic infections with subtype C human immunodeficiency virus type 1 identifies differences in glycosylation and CCR5 utilization and suggests a new strategy for immunogen design. J Virol 87(13):7218–7233. doi:10.1128/JVI.03577-12

Platt EJ, Wehrly K, Kuhmann SE, Chesebro B, Kabat D (1998) Effects of CCR5 and CD4 cell surface concentrations on infections by macrophagetropic isolates of human immunodeficiency virus type 1. J Virol 72(4):2855–2864

Ribeiro RM, Hazenberg MD, Perelson AS, Davenport MP (2006) Naive and memory cell turnover as drivers of CCR5-to-CXCR4 tropism switch in human immunodeficiency virus type 1: implications for therapy. J Virol 80(2):802–809. doi:10.1128/JVI.80.2.802-809.2006

Rifkin D, Compans RW (1971) Identification of the spike proteins of Rous sarcoma virus. Virology 46(2):485–489

Robinson WS, Hung P, Robinson HL, Ralph DD (1970) Proteins of avian tumor viruses with different coat antigens. J Virol 6(5):695–698

Rowell JF, Stanhope PE, Siliciano RF (1995) Endocytosis of endogenously synthesized HIV-1 envelope protein. Mechanism and role in processing for association with class II MHC. J Immunol 155(1):473–488

Rubin H (2011) The early history of tumor virology: Rous, RIF, and RAV. Proc Natl Acad Sci U S A 108(35):14389–14396. doi:10.1073/pnas.1108655108

Samson M, Libert F, Doranz BJ, Rucker J, Liesnard C, Farber CM, Saragosti S, Lapoumeroulie C, Cognaux J, Forceille C, Muyldermans G, Verhofstede C, Burtonboy G, Georges M, Imai T, Rana S, Yi YJ, Smyth RJ, Collman RG, Doms RW, Vassart G, Parmentier M (1996) Resistance to HIV-1 infection in Caucasian individuals bearing mutant alleles of the CCR-5 chemokine receptor gene. Nature 382(6593):722–725. doi:10.1038/382722a0

Sauter MM, Pelchen-Matthews A, Bron R, Marsh M, LaBranche CC, Vance PJ, Romano J, Haggarty BS, Hart TK, Lee WM, Hoxie JA (1996) An internalization signal in the simian immunodeficiency virus transmembrane protein cytoplasmic domain modulates expression of envelope glycoproteins on the cell surface. J Cell Biol 132(5):795–811

Schnell G, Joseph S, Spudich S, Price RW, Swanstrom R (2011) HIV-1 replication in the central nervous system occurs in two distinct cell types. PLoS Pathog 7(10):e1002286. doi:10.1371/journal.ppat.1002286

Schwartz DE, Tizard R, Gilbert W (1983) Nucleotide sequence of Rous sarcoma virus. Cell 32 (3):853–869

Sturdevant CB, Joseph SB, Schnell G, Price RW, Swanstrom R, Spudich S (2015) Compartmentalized replication of R5 T cell-tropic HIV-1 in the central nervous system early in the course of infection. Plos Pathog 11(3):e1004720. doi:10.1371/journal.ppat.1004720

Swanstrom AE, Haggarty B, Jordan APO, Romano J, Leslie GJ, Aye PP, Marx PA, Lackner AA, Del Prete GQ, Robinson JE, Betts MR, Montefiori DC, LaBranche CC, Hoxie JA (2016) Derivation and characterization of a CD4-independent, non-CD4-tropic simian immunodeficiency virus. J Virol 90(10):4966–4980. doi:10.1128/jvi.02851-15

Theodorou I, Meyer L, Magierowska M, Katlama C, Rouzioux C (1997) HIV-1 infection in an individual homozygous for CCR5 delta 32. Seroco Study Group. Lancet 349(9060):1219–1220

Wang LH, Duesberg P, Mellon P, Vogt PK (1976) Distribution of envelope-specific and sarcoma-specific nucleotide sequences from different parents in the RNAs of avian tumor virus recombinants. Proc Natl Acad Sci U S A 73(4):1073–1077

White TA, Bartesaghi A, Borgnia MJ, Meyerson JR, de la Cruz MJ, Bess JW, Nandwani R, Hoxie JA, Lifson JD, Milne JLS (2010) Molecular architectures of trimeric SIV and HIV-1 envelope glycoproteins on intact viruses: strain-dependent variation in quaternary structure. PLoS pathog 6(12):e1001249

Zhu P, Liu J, Bess J, Jr., Chertova E, Lifson JD, Grise H, Ofek GA, Taylor KA, Roux KH (2006) Distribution and three-dimensional structure of AIDS virus envelope spikes. Nature 441 (7095):847–852. doi:10.1038/nature04817

Personal Perspective from RS I first met Peter Vogt in 1975. I was a new postdoctoral fellow in the Bishop/Varmus lab at UCSF where we participated in the "West Coast RNA Tumor Virus Meeting." It was a group that met several times each year with somewhat changing characters but with a core of Peter's lab, Mike's and Harold's lab, Peter Duesberg's lab, and Steve Martin's lab. I joined this group just after the discovery of *v-src* sequences in the normal vertebrate genome. The race was on to genetically define new viral oncogenes and to test their host cell origin, and to try to understand what the functions of the protein products were and how they could change to induce cellular transformation and cancer. At least from my perspective Peter Vogt was the one who brought the biology of these viruses to this larger group, providing the starting point for the important molecular understanding of the transforming genes that followed. Peter has been a full partner in this field, fittingly spending these last years searching for small molecule inhibitors of the proteins encoded in the same transforming genes he helped discover.

Peter's personal story of how he became an academic researcher and his long successful career are inspiring but not mine to tell. In my time with Peter I always found him to be incredibly thoughtful and generous. He is truly a gentle man. He has a deep-seated love of art that led him to be a painter in his own right. I have two Peter Vogt paintings, one of a rose demonstrating an eye for capturing simple beauty. The other is a street scene from Pasadena where he lived while on the faculty at USC Medical School. Pasadena played an important role for me in my extended family when I was young so to be able to capture part of that memory as seen through Peter's eyes is a special treat. True to Peter's nature, both of these paintings (and many others) were donated to USC to allow them to be sold to raise funds for graduate education.

I think Peter enjoys people in general and he is good spirits. I stand with a large group of people who have seen Peter as a role model for scientific excellence and personal humility. I have always enjoyed his company given the endlessly interesting things he pursues. One of the things I have noticed over the years is that PIs frequently have their personality reflected through the people in their lab. Thus it is easy to point to many of Peter's trainees as scientifically rigorous and genuinely nice people. I am sure Klaus and Eric take pride in being able to share some of their feelings toward Peter by creating this volume in his honor.

I met Eric Hunter when he was a postdoctoral fellow with Peter, Eric forever being slightly older than I am. It is a friendship that has lasted over 40 years. When Eric asked if I would contribute a chapter there was really only one answer. In 1990 Peter approached me to coedit CTMI volume 157 "Retroviruses: Strategies of Replication." In trying to round out my own list of authors I asked Eric to write a chapter on "Retrovirus Envelope Glycoproteins" which he agreed to do with me as a coauthor. Now the shoe is on the other foot, but as I look back on that old chapter it is clear that we did not bring the breadth to the topic that (mostly) Eric and I managed as younger writers.

When I met Peter it was a time when young people were joining the field with the hope of getting to work on their own new oncogene. Alas, some of us were just virologists. It is fun to use the miracle of PubMed to be reminded of Peter's work before I met him. The foundational work that set the stage for the discovery of cellular oncogenes also provided the underlying biology for some of the molecular studies I did. The introduction of molecular cloning and sequencing allowed us to ask how the sequences of the ALV strains differed given their host range differences. A few years later another generation of postdocs, Paul Bates and John Young, would build on the same biology to clone the cellular receptors for some of these strains to provide a complete view of how the virus chooses which cells to infect.

Along the way the oncogene field and the retrovirus field largely split. The next retrovirus challenge made itself known with the discovery of HIV. I have taken as our charge in this new review to revisit the biology of viral entry for the avian retroviruses that brought us *v-src*, *v-myc*, *v-myb*, etc., and today threaten new ALV pandemics. We have paired those lessons with the ones learned from HIV-1 infecting human cells. It is perhaps an odd pairing but simply reflects the twists and turns of a typical scientist following related biological phenomena. Finally, this effort is offered with affection and admiration to one of my heroes.

Infectious Agents in Bovine Red Meat and Milk and Their Potential Role in Cancer and Other Chronic Diseases

Harald zur Hausen, Timo Bund and Ethel-Michele de Villiers

Abstract Red meat and dairy products have frequently been suggested to represent risk factors for certain cancers, chronic neurodegenerative diseases, and autoimmune and cardiovascular disorders. This review summarizes the evidence and investigates the possible involvement of infectious factors in these diseases. The isolation of small circular single-stranded DNA molecules from serum and dairy products of Eurasian Aurochs (*Bos taurus*)-derived cattle, obviously persisting as episomes in infected cells, provides the basis for further investigations. Gene expression of these agents in human cells has been demonstrated, and frequent infection of humans is implicated by the detection of antibodies in a high percentage of healthy individuals. Epidemiological observations suggest their relationship to the development multiple sclerosis, to heterophile antibodies, and to N-glycolylneuraminic acid (Neu5Gc) containing cell surface receptors.

Contents

Dedication This manuscript is dedicated to Dr. Peter Vogt, San Diego, whose pioneering studies stimulated, encouraged and influenced a large number of young scientists and greatly contributed to the progress of Virology and Cell Biology.

H. zur Hausen (✉) · T. Bund · E.-M. de Villiers (✉)
Deutsches Krebsforschungszentrum, Im Neuenheimer Feld 280, 69120 Heidelberg, Germany
e-mail: zurhausen@dkfz.de

E.-M. de Villiers
e-mail: e.devilliers@dkfz.de

T. Bund
e-mail: t.bund@dkfz.de

Current Topics in Microbiology and Immunology (2017) 407:83–116
DOI 10.1007/82_2017_3

Published Online: 28 March 2017

1 Increased Cancer Risk After Red Meat or Dairy Product Consumption

An increased risk has been reported for several cancers after prolonged consumption of red or processed meat (Table 1). A large number of cohort studies consistently concluded an increase of approximately 15–30% for a specific group of cancers, colorectal carcinomas (reviewed in zur Hausen 2012; zur Hausen and de Villiers 2015). A small number of additional studies failed to confirm this increase. For other cancers, the data are less consistent, although in recent years the number of positive reports seems to have increased (Table 1).

The list of cancer types (shown in Table 1) linked to red meat and milk consumption is not complete. Individual reports have also been published for liver, gall bladder, bladder, and thyroid cancers and, in addition, for multiple myelomas. Yet, by far the most consistent reports concern colorectal cancers (zur Hausen 2012; zur Hausen and de Villiers 2015). A number of responsible factors specifically in red meat have been incriminated (reviewed in Alisson-Silva et al. 2016). They include most prominently chemical carcinogens (aromatic hydrocarbons and others) arising in the preparatory steps for meat consumption, heme iron, and the composition of the colon microbiome (reviewed in Sugimura et al. 2004; Fonseca-Nunes et al. 2014). Many of these studies, however, disregarded that countries with a high level of red meat consumption (e.g., Mongolia, China, and Bolivia) reveal remarkably low incidence rates of this cancer. Long-time consumption of fish or poultry in the diet does not lead to an increased incidence of colon cancer, although the same chemical mutagens arise in preparatory steps for fish and chicken consumption as in red meat (Knize et al. 1999; Kao et al. 2014; Hernández et al. 2015). Several studies attempted to analyze the potential carcinogenicity of heme iron but failed to support its carcinogenic function (reviewed in Andersen and Vogel 2015). More recently, specific bacteria in the colonic microbial flora have been suggested to play a role in colon cancer development, specifically *Fusobacterium nucleatum* and

Table 1 Cancer risk factors linked to nutrition, aerosols, and infections

Type of cancer	Biologic risk factor	Consistency of observation	Reference
Colon	Red meat	High	Norat et al. (2005), Huxley et al. (2009), Bouvard et al. (2015), Lippi et al. (2016)
Breast	Red meat, milk	Moderate	Taylor et al. (2007), Linos et al. (2008), Farvid et al. (2015), [a]Kabat et al. 2009)
Lung	Red meat, milk, slaughterhouse aerosols	Moderate	Durusoy et al. (2006), Xue et al. 2014), Ji et al. (2015), Lippi et al. (2016)
Prostate	Milk, red meat	Moderate	Kolonel (2001), Bosetti et al. (2004), Amin et al. (2008)
Pancreas	Red meat	Low	Larsson and Wolk (2012), [a]Rohrmann et al. (2013)
Esophagus	Red meat	Low	Choi et al. (2013), Zhang (2013), Lippi et al. (2016)
Stomach	Red meat (*Helicobacter pylori* Epstein–Barr virus)	Low	Matos and Brandani (2002), González and López-Carrillo (2010), Song et al. (2014), Epplein et al. (2014), Lippi et al. (2016)
Non-Hodgkin's lymphoma	Red meat	Low	Cross and Lim (2006), Caini et al. 2016)
Hodgkin's disease	Animal exposure, milk, red meat	Low	Khuder et al. (1999), Martin et al. (2005a), Epstein et al. (2015)

[a]Did not confirm data

Streptococcus bovis (reviewed in Boleij et al. 2011; Han 2015). These agents are also found in dental plaques and seem to contribute to other inflammatory reactions, not necessarily linked to cancer development (Flynn et al. 2016).

An intriguing hypothesis has been presented by Ajit Varki and his colleagues (Samraj et al. 2015). They demonstrated that the sialic acid, N-glycolylneuraminic acid (Neu5Gc), not synthesized in humans, is bioavailable in various diets and in particular in red meat and dairy products. Its incorporation into cell membranes of humans can evoke an immune response. This "xeno-autoimmunization" represents an immune reaction directed against Neu5Gc and may cause chronic inflammations at expressed sites. The hydrolysis of the sialic acid N-acetylneuraminic acid (Neu5Ac) into Neu5Gc is directed by the gene cytidine monophospho-N-acetylneuraminic acid hydrolase (CMAH). A specific deletion of exons in the CMAH gene apparently selectively occurred in humans and in members of the *Mustelidae* and *Pinnipedia* genera (Ng et al. 2014), inactivating the function of CMAH. A deletion at a different site of the gene coding for this enzyme has been identified in New World monkeys (Springer et al. 2014). Human anti-Neu5Gc antibodies appear during infancy and correlate with weaning and exposure to dietary Neu5Gc. However, dietary Neu5Gc alone cannot elicit anti-Neu5Gc antibodies in mice with a human-like Neu5Gc deficiency (Taylor et al. 2010). It should be mentioned here, but will be discussed later (Sect. 8.3) that Hanganutziu–Deicher and Paul–Bunnell heterophilic antibodies reveal reactivations directed against Neu5Gc. These observations are interesting, but as such difficult to interpret in view of the low colon cancer rate in Mongolia, where red meat consumption is particularly high (Maytsetseg and Riichiro 2006).

The global epidemiology of colorectal cancer, and in particular its very low incidence in populations not consuming beef (Hindu population in India), suggested to us that a beef-specific factor might contribute to the colon cancer risk (zur Hausen 2012) (see Sect. 6).

In the following, we summarize data in support of the previously postulated zoonotic infectious components in the origin of specific cancers and several chronic diseases (zur Hausen 2001, 2009; zur Hausen and de Villiers 2015).

2 Risk Reduction for Specific Cancers and Two Chronic Diseases by Long-Time Breast-Feeding

It became apparent that infections with several infectious agents linked to oncogenesis occur early in life and result in long-time latency prior to cancer development (in average between 5 and 60 years—zur Hausen and de Villiers 2014a). Several protective events acting during the first two years of life and reducing the risk for certain cancers and chronic diseases have been described. Besides the effect of frequent, mainly respiratory infections during this period (zur Hausen and de

Table 2 Protective effect of prolonged breast-feeding (>6 months) for specific cancers and chronic neurological diseases

Disease	Consistency of observation	Reference
Acute lymphoblastic leukemia	Moderate	Bener et al. (2001), Guise et al. (2005), Martin et al. (2005a), McNally and Parker (2006), Greenop et al. (2015)
Hodgkin's disease	Moderate	Martin et al. (2005a), McNally and Parker (2006), Bener et al. (2008)
Neuroblastoma	Low	Daniels et al. (2002), Martin et al. (2005a), Ferris i Tortajada et al. (2005)
Premenopausal breast cancer	Low	[a]Michels et al. (2001), Abou Dakn et al. (2003), Martin et al. (2005b), [a]Wise et al. (2009)
Multiple sclerosis	Moderate	Pisacane et al. (1994), Conradi et al. (2013), Graves et al. (2014), Ragnedda et al. (2015)
Diabetes types 1 and 2	Low	Ip et al. (2007), Gouveri et al. (2011), Stuebe (2009)

[a]Did not confirm data

Villiers 2014b), long-time breast-feeding has often been cited as a protective effect for the risk reduction (Table 2).

The protective effect is particularly noteworthy in view of the demonstration of specific glycoconjugates in human milk which block the binding of several infectious agents to their receptors (e.g., rota-, noro-, human immunodeficiency viruses and *Candida albicans*) (Peterson et al. 1998; Etzold and Bode 2014; Gonia et al. 2015). Although the presence of similar oligosaccharides has been noted in milk of several other mammalian species (Urashima et al. 2009; Takimori et al. 2011), their concentration in human milk seems to be particularly high. *Disialyl-lacto-N-tetraose* is most abundant in human milk, but missing in milk from most other species (Monti et al. 2015). The same accounts for *2'-fucosyllactose* and *3'-fucosyllactose* which specifically block binding of norovirus-like particles to human histo-blood group antigens (Weichert et al. 2016). Bovine milk contains less diverse free oligosaccharides than human milk and significant less fucosylated glycans (Peterson et al. 2013). Peterson et al. put the suggestion forward that bovine milk glycoconjugates have evolved specifically to protect young calves from infection by bovine pathogens.

Interferon gamma is consistently produced during the prolonged breast-feeding period (Lohman-Payne et al. 2012) and may, among other cytokines, also contribute to the observed risk reduction. The role of specific immunoglobulins is probably limited to the early period after delivery and decreases with the gradual maturation of the gastrointestinal tract and the immune system of the newborn (Zhang et al. 2016a).

This protective effect is schematically outlined in Fig. 1.

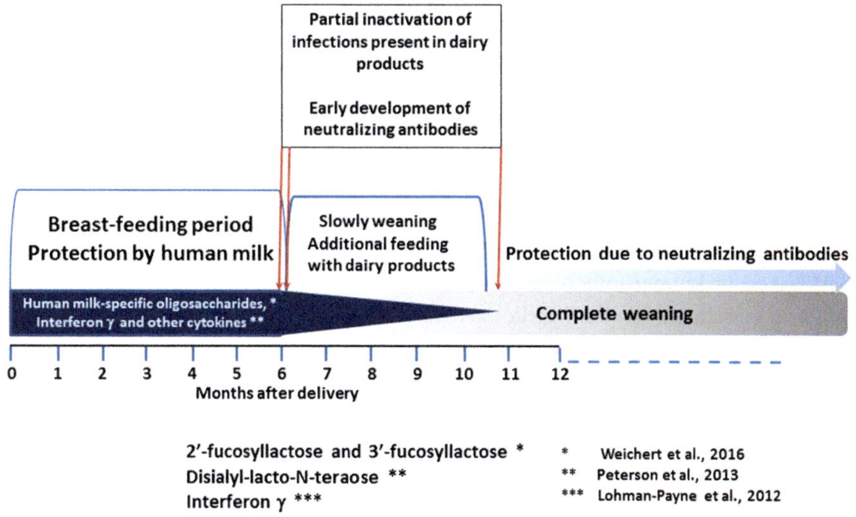

Fig. 1 Protective factors in human milk present during the first 6 months of breast-feeding. Three sugars, 2′-fucosyllactose, 3′-fucosyllactose, and disialyl-lacto-N-tetraose, seem to be selectively enriched in human milk. Their binding to glycoprotein cell surface receptors emerges as a protective mechanism against various potentially dangerous infections

This model assumes that a more matured immune system, as existing in the second half of the first year of life, will be better equipped to cope with and neutralize infections. Suspected infectious agents should be present in low concentrations in dairy products which are gradually added during the weaning period. The data for premenopausal breast cancer are somewhat more controversial: At least two follow-up studies failed to confirm this effect (Michels et al. 2001; Wise et al. 2009). This may relate, however, to the difficulties in ascribing the late outcome of a disease to infectious events occurring in the first year of life. An explanation was previously postulated for the repeatedly reported protective effect of non-specific infections during the first 1–2 years of life, reducing the risk for tumors mentioned in Table 2. Prenatal infections of potentially oncogenic agents will presumably result in immune tolerance of the adaptive immune system. Activation of innate immune functions, however, by non-specific postnatal infections could explain this protective effect (zur Hausen and de Villiers 2014b).

3 Are Diet-Attributed Factors Increasing Cancer Risks Human-Specific?

The analysis of cancer risks in other mammalian species poses a few problems: Life expectancy is commonly low in wild animals, and those suffering from chronic diseases will be an easy prey of predators or succumb more readily to parasitic infections. Domestic animals have in part prolonged life spans which accounts in particular for carnivores (dogs and cats). Herbivores (e.g., cattle, pigs, sheep, and goat) are frequently kept for limited time periods before being slaughtered and prepared for consumption. Exceptions are horses or animals kept in zoological gardens. Analyses are further complicated by genetic factors and in part by substantial race differences in cancer incidence (Graf et al. 2016). Thus, reliable comparisons of cancer rates between carnivores and herbivores are difficult to obtain, except for tumors already occurring at young age or by comparing aged horses with old carnivores.

Several predominantly hematopoietic, but also skin cancers in both groups have been linked to known infectious agents (e.g., feline and bovine leukemias, bovine Theileria infections, bovine papillomavirus, and others). This hampers any analysis of dietary factors as cancer risks for carnivores and herbivores.

Yet, according to Munson and Moresco (2007), "*differences in cancer prevalence between carnivores and herbivores and between captive and wild animals are striking and support the hypotheses that diet and reproductive history are major risk factors.*" At least one tumor type, cancer of the mammary gland, occurs far more frequently in carnivores than in herbivores (Munson and Moresco 2007; Owston et al. 2008; Merlo et al. 2008). Only single reports describe this cancer in herbivores (e.g., McElroy and Bassett 2010). Gastrointestinal cancer appears to be relatively rare in both groups. In comparison with humans, this may in part relate to differences in the microbiome. *Postmortem* analyses of cancers in 241 old horses (older than 15 years) identified neoplastic disease in 18.7%, most commonly diagnosed as squamous cell carcinoma, lymphoma, or melanoma (Hendrix and Newkirk 2014; Miller et al. 2016; Luethy et al. 2016).

It is likely that cats and dogs, as well as carnivores in zoological gardens, receive a diet enriched in red meat. Although a larger number of individual reports document cancers in wild carnivores, there is a notable absence of similar reports for wild herbivores, as well as for whales, dolphins, seals, and sea lions.

In general, the scarcity of solid epidemiological data on cancer incidence in animals does presently not permit the conclusion that risks linked to red meat consumption are solely confined to the human race, certainly not as long as we do not have conclusive evidence for the responsible factor(s).

4 Increased Risk for Neurodegenerative Diseases After Red Meat or Dairy Product Consumption

Besides cancers, several chronic neurodegenerative diseases have also repeatedly been linked to dairy product consumption or to a diet rich in red meat (Table 3).

Bovine transmissible spongiform encephalopathy (TSE) has clearly been identified as a zoonosis, transmitted to humans via consumption of bovine meat and organs (Diack et al. 2014). It can also be transmitted to other ruminants, cats, and

Table 3 Neurodegenerative diseases suspected to be linked to meat and milk consumption

Neurodegenerative disease	Risk factor	Consistency of observation	Reference
Bovine transmissible spongiform encephalopathy (TSE)	Meat or consumption of animal organs	High	Baron et al. (2007), Simmons et al. (2008), Davenport et al. (2015)
Multiple sclerosis	Milk, farming, vitamin D deficiency, herpesvirus reactivation	High	Butcher (1976), Malosse et al. (1992), Malosse and Perron (1993), Sepcic et al. (1993), Winer et al. (2001), Otaegui et al. (2007), Munger et al. (2011)
Parkinson's disease	Milk consumption	Moderate	Park et al. (2005), Chen et al. (2007), Grant (2013), Jiang et al. (2014), Abbot et al. (2016)
Amyotrophic lateral sclerosis	Milk, meat, farming	Moderate	Kang et al. (2014), Felmus et al. (1976), Huisman et al. (2015)
Alzheimer's disease	Red meat, obesity	Low	Gu et al. (2010), Chen et al. (2016), Pedditizi et al. (2016)

rodents (Leunda et al. 2013). Transmission of scrapie (another TSE in sheep) to humans has not been reported, although a recent publication claimed a successful transmission to a cynomolgus macaque (Comoy et al. 2015).

TSEs have been postulated to result from misfolding of prion proteins, initiating their aggregation and catalyzing the same specific misfolding events in newly synthesized prion molecules, with detrimental effects for the respective neurons (Prusiner 1982; Harrison et al. 2001; Małolepsza 2008). A controversial claim has been made by *Manuelidis* and co-workers arguing in favor of a role of a more conventional infectious event in TSE induction after isolating two types of small presumably single-stranded circular DNA molecules (*Slow Progressing Hidden Infections of variable(X) latency* "*Sphinx*") from murine neuronal tissue culture cells infected with materials from different experimentally induced TSEs (Manuelidis 2011, 2013; Botsios and Manuelidis 2016).

Consumption of milk and to a lesser degree of red meat has been repeatedly considered as an important risk factor for multiple sclerosis (MS), Parkinson's disease (PD), and amyotrophic lateral sclerosis (ALS) (Table 3). Milk consumption, farming, vitamin D deficiency, and a reactivated herpesvirus have been claimed as important risks for MS. A model postulating an interaction between a latent bovine milk factor, reactivated latent herpesvirus DNA, and vitamin D deficiency has recently been published (zur Hausen 2015). For MS, events early in life seem to be responsible for the development of the disease later in life (references in Table 2).

As previously mentioned, milk consumption has also been repeatedly reported as a risk factor for Parkinson's disease and amyotrophic lateral sclerosis, although slightly less consistent than for MS. Even less data exist for Alzheimer's disease, here pointing to red meat uptake and obesity as risk factors.

5 Increased Risk for Cardiovascular or Autoimmune Diseases After Red Meat or Dairy Product Consumption

Several reports point to dietary factors, red meat, and dairy products as potential risks for cardiovascular disorders, arterial sclerosis, Crohn's disease, and ulcerative colitis. Still, the data are relatively scarce. They are even less convincing for diabetes type 2 and type 1 and still require further epidemiological investigation (Table 4).

The "*patchy*" occurrence of lesions in colitis ulcerosa and Crohn's disease (Fig. 2) (Price and Morson 1975) and to a certain degree also in cardiovascular disease (Meyer et al. 1994) may raise the suspicion that dual infections in initiating cells, similar to multiple sclerosis, could play a role in the etiology of these conditions (see Sect. 8.1).

Table 4 Cardiovascular disorders, Crohn's disease, and diabetes type 2 in relation to diet

Cardiovascular or autoimmune diseases	Risk factor	Consistency of observation	Reference
Cardiovascular disorders, atherosclerosis	Red meat consumption	Moderate	Pan et al. (2012), Micha et al. (2012), Larsson and Orsini (2014), Battaglia Richi et al. (2015)
Crohn's disease and colitis ulcerosa	Vitamin D deficiency, animal proteins, and farm dairy products	Moderate	Battaglia Richi et al. (2015), Bovalino et al. (2016), Cantarna and Mahon (2004), D'Souza et al. (2008), Jantchou et al. (2010), Juste (2010), Hou et al. (2011)
Diabetes type 2	Red meat	Low	Pan et al. (2011), Micha et al. (2012), Battaglia Richi et al. (2015)

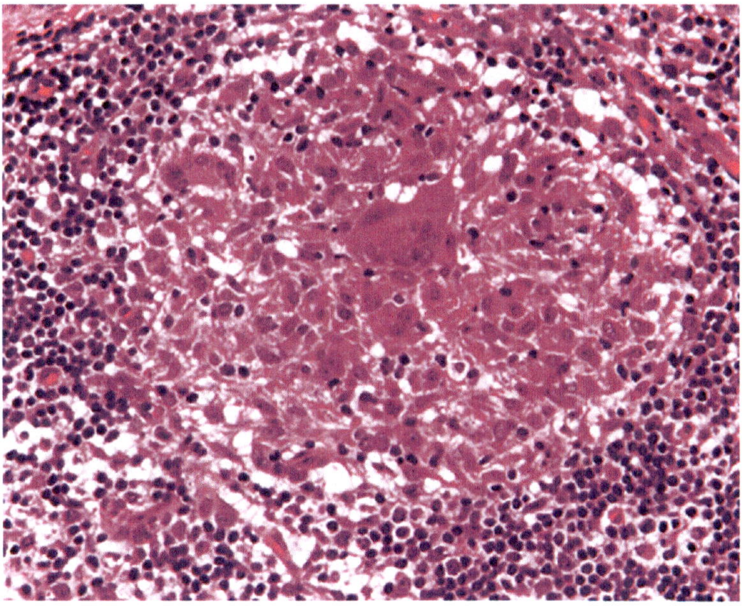

Fig. 2 Patchy appearance of Crohn's disease lesion. Intense inflammatory reaction (*Source* Machado et al. 2010)

6 Are the Observed Effects for Meat and Dairy Product Consumption Species-Specific for Eurasian Bos Taurus-Derived Dairy Cattle?

The majority of reviews dealing with the risk of meat consumption in relation to specific cancers or chronic diseases globally generalized "*red meat*" as beef, lamb, goat, horse, and pork meat consumed raw, fried, broiled, or processed (reviewed in zur Hausen 2012). Only recently did a few publications assess the risk of subtypes of meat specifically for colorectal cancer (reviewed in zur Hausen 2012; zur Hausen and de Villiers 2015; Aykan 2015; Lippi et al. 2016; Carr et al. 2016). Beef emerged as a prevailing risk factor, although lamb/mutton was also occasionally quoted as a risk. A few studies even attributed risk to pork consumption, even though parts of the world with a very high consumption of pork (e.g., China and Pacific Islands) reveal a relatively low incidence of colorectal cancer (zur Hausen and de Villiers 2015).

As indicated in Sect. 1, further analyses (zur Hausen and de Villiers 2015) revealed that the consumption of beef per se would not suffice to explain the regional differences in colon cancer incidence: Mongolia with the endogenous Yak (*Bos mutus* and *Bos grunniens*) and Zebu (*Bos indicus*)-derived Chinese Yellow cattle populations, Equatorial Africa, where in particular Zebu-crossbreeds with endogenous cattle formed the Watusi cattle, and in Bolivia with an almost pure-bred Zebu cattle population, all reveal relatively low colon cancer rates. This seems to reflect a better adaptation of these cattle to harsh environmental conditions. Interestingly, different susceptibilities for specific infectious events have been noted among ungulates and even races of cattle (Dechamma et al. 1998; Sharma et al. 2011).

Based on these considerations, we published a model, postulating a role of infectious events in colon carcinogenesis (zur Hausen and de Villiers 2015). Our model took into account that none of the presently known infectious factors linked to human cancers is by itself sufficient for cancer induction, but requires genetic or epigenetic modifications in specific cellular genes or cellular signaling cascades involved in protection against uncontrolled proliferation (zur Hausen and de Villiers 2014a).

Interestingly, in most parts of the world, the incidence of colon and breast cancers seems to be correlated (reviewed in zur Hausen and de Villiers 2015), although the breast cancer rate usually exceeds that of colon cancer. A few exceptions, however, exist: In Japan and Korea, the incidence of colon cancer increased rapidly 15–20 years after the Second World War (Japan) or after the Korean War (Korea). This increase surmounted that of breast cancer and has been linked to changes in dietary habits accompanying large imports of red meat (beef and pork) mainly from the USA. Conversely, breast cancer incidence increased more rapidly in India, where during the past few decades large dairy cattle farms have been established. In this country, children are commonly initiated on bovine milk after 3 months of age (Mayuri et al. 2012).

Breast cancer (mainly premenopausal), prostate cancer, and lung cancer have repeatedly been linked to dairy product consumption, although not as consistently as red meat consumption to colon cancer (Table 1). Some follow-up studies even failed to confirm this link (e.g., Kabat et al. 2009). One of the problems may stem from the fact that a putative infectious agent linked to colon cancer will reach its target cells (colon cells) free of intervening immune reactions. In contrast, reaching target cells in breast, in prostate, and probably also in the lung requires systemic transfer, most likely via the blood or the lymphatic system (in lung cancer possibly also by aerosols). This would raise the probability of immune reactions after repeated exposures to the same agent. Thus, the initial infection, most likely followed by systemic spreading via viremia or, even more likely, by the transfer via permissive or semipermissive cells of the hematopoietic system, will result in probably decade-long latent infections of cells in specific target organs. As a consequence, in particular for factors in milk and dairy products, the initial events will date back to the early period of life, most likely to the period of weaning. Subsequent immunologic interactions will probably neutralize further identical or closely related infectious events. In this case, epidemiological follow-up studies of dietary habits in adults will not be very informative and may be the reason for the reported inconsistencies for the discussed tumor types (zur Hausen and de Villiers 2015).

Based on these considerations and on the existence of a viral factor causing mammary cancer in mice, we postulated a "*milk factor*" for human breast cancer, differing from a "*meat factor*" for colon cancer (zur Hausen and de Villiers 2015). Yet, both factors should originate from the same host, *Bos taurus*-derived Eurasian dairy cattle.

7 Evidence for Episomally Persisting Infectious Factors Transmissible from Dairy Cattle to Humans

The mentioned epidemiological evidence prompted experiments to analyze 120 individual serum samples from healthy dairy cows, provided by the Veterinary Institute of the University of Leipzig. In addition, 4 samples of commercially available dairy milk, 30 samples from healthy human blood donors, 30 sera from multiple sclerosis patients, as well as 13 autopsy (brain) samples from patients with MS, were analyzed. Rolling-circle amplification resulted in the isolation of 18 circular single-stranded DNA molecules, ranging between 1084 and 2958 nucleotides (Lamberto et al. 2014; Funk et al. 2014; Gunst et al. 2014; Whitley et al. 2014).

Fourteen isolates originated from cattle/cow sera or milk and 4 isolates from brain and serum samples from MS patients. The isolates were grouped based on nucleotide sequence relatedness to known organisms. Twelve isolates are assigned to 2 groups based on their nucleotide similarity to Sphinx 1.76 (Group 1) and

Sphinx 2.36 (Group 2), respectively (Manuelidis et al. 2011). These latter isolates originated from infectious cultures and from brain samples of transmissible spongiform encephalopathy (TSE). In view of the bovine origin of both of our groups, we labeled them as bovine meat and milk factors (BMMF) groups 1 and 2. None of these DNA isolates was found in blood samples of 30 healthy human blood donors.

Group 1 (BMMF1) consists of 4 isolates from dairy cows (*cow milk isolate (CMI)*) (CMI1.252, CMI2.214, CMI3.168, and CMI4.158), 5 isolates from bovine sera (*healthy cow blood isolate (HCBI)*) (HCBI3.108, HCBI4.296, HCBI5.173, HCBI6.159, and HCBI6.252), and 2 isolates (*MS biopsy isolate (MSBI)*) (MSBI1.176 and MSBI2.176) from a single MS brain sample (Whitley et al. 2014) (Fig. 3). Interestingly, this group harbors 2 subgroups with sizes of the genomes of CMI3.168, CMI4.158, HCBI3.108, HCBI5.173, and HCBI6.159 ranging between 1086 and 1766 nucleotides, whereas the genome sizes of CMI1.252, CMI2.214, HCBI4.296, and HCBI6.252 vary between 2148 and 2958 nucleotides.

The main open reading frame (ORF) of all isolates encodes for an initiator replication (repB) protein. Whereas the repB proteins of CMI1.252 and HCBI6.252 are identical, they and others share 80–98% amino acid identity between isolates. Exceptions are HCBI4.296, HCBI5.173, and MSBI2.176 sharing 50–68% amino acid identity to the rep protein of MSBI1.176 and Sphinx 1.76 (Fig. 4).

In the subgroup with larger genomes, CMI1.252 and HCBI6.252 are almost identical in nucleotide sequence and have a second large ORF encoding for a putative transcription-regulating protein (219aa). Similarly, a second ORF in the larger HCBI4.296 may encode a protein (225aa) with similarity to mobilization/plasmid recombination proteins (Whitley et al. 2014). The exact functions of these putative proteins are presently under investigation. The BMMF1 isolates all share about 70% nucleotide identity to plasmids of *Acinetobacter baumanii* as determined by BLAST analyses. HCBI5.173 shares this degree of identity not only to plasmids of *A. baumanii*, but similarly to plasmids of *Psychrobacter* spp.

Another common feature between all isolates in BMMF1 is the iteron-like tandem repeat region which may constitute binding sites for the rep protein (Chattoraj 2000). Each isolate harbors at least 3 direct repeats of 22 nt each, plus a partial repeat of 17/18 nt. CMI4.158 has 4 tandem repeats. CLUSTAL alignment of these regions displays their distinct nucleotide identities, although single-nucleotide differences between the first repeat and the respective subsequent repeats are seen in MSBI1.176, CMI3.168, and HCBI3.108 (Fig. 5). The repeat regions of HCBI4.296 and HCBI5.173 differ considerably (CMI2.214 and MSBI2.176 to a lesser extent) from those of the other isolates—this mirrors the large differences in nucleotide sequences of the complete genomes. Interestingly, despite these diversities, all the BMMF1 isolates have a conserved core palindromic sequence located upstream of the repeats. A similar conserved palindromic structure has been suggested to play a role in replication initiation (Dziewit et al. 2013). The palindromic sequence of the majority of BMMF1 isolates is 5′-TAAATGCTTTTA-3′ (12 nt) located 41–52 nt upstream of the repeat region. The exceptions are CMI4.158 with 4 additional

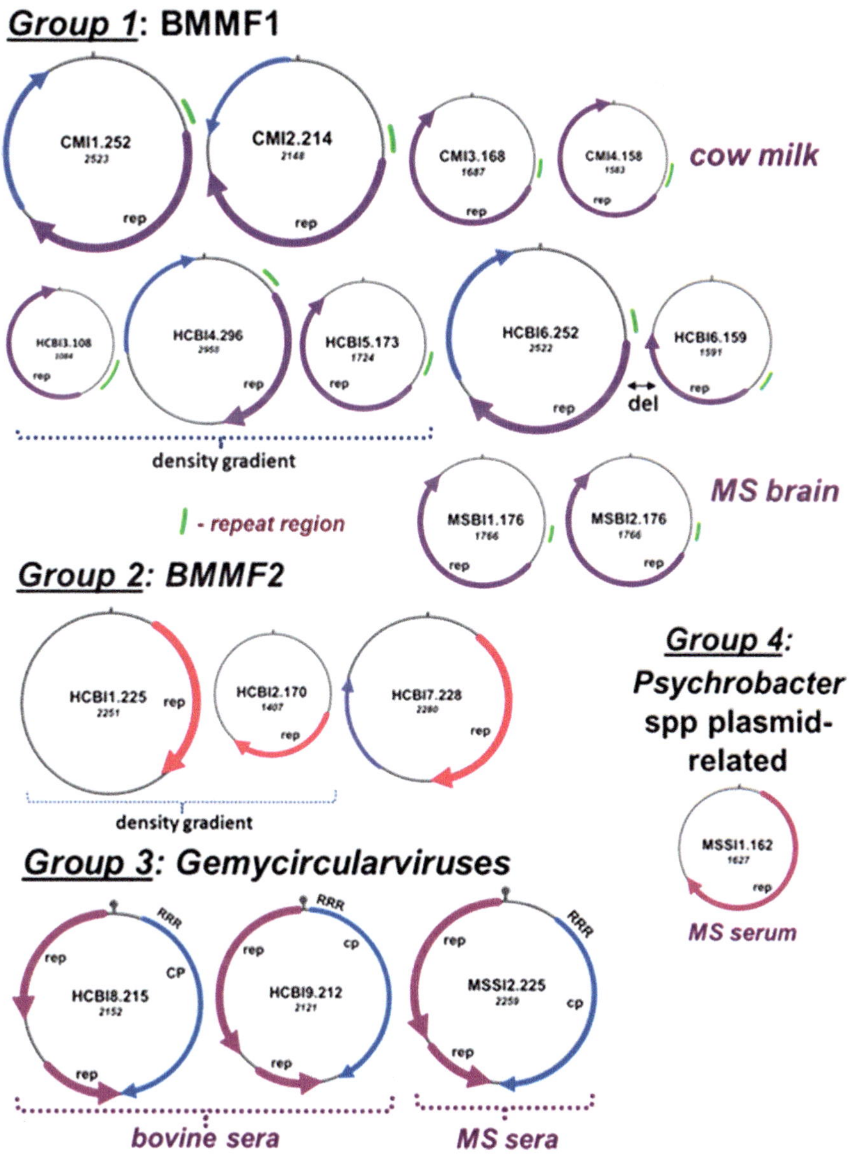

Fig. 3 Circular single-stranded DNA isolates

nucleotides 5′-TTTAAATGCTTTTAAA-3′ (16 nt) at 39–54 nt upstream, HCBI3.108 and HCBI4.296 each with 5′-TTAAATGCTTTTAA-3′ (14 nt) at 49–62 nt upstream and HCBI5.173 with 5′-TTAAATACTTTTAA-3′ (14 nt) at 46–59 nt upstream of its repeat region (Fig. 6).

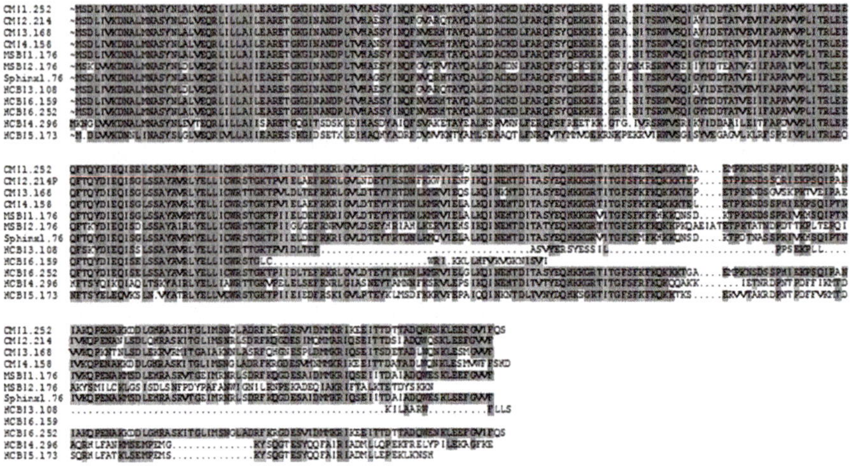

Fig. 4 MUSCLE alignment of the putative replication proteins of BMMF1

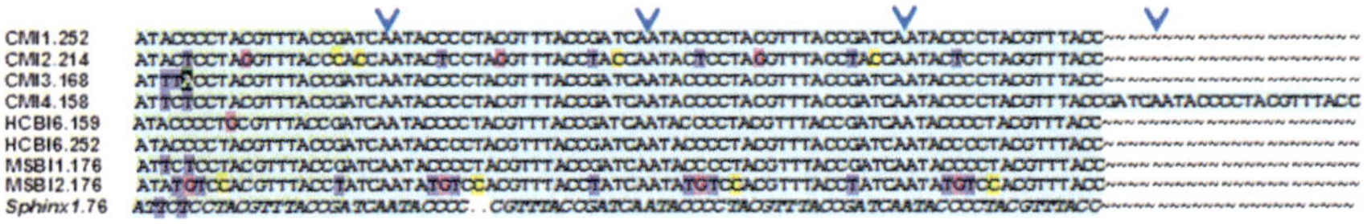

Fig. 5 CLUSTAL alignment of the repeat region of BMMF1. *Arrows* indicate the end of a single repeat

Fig. 6 Conserved palindromic structure as putative origin of replication

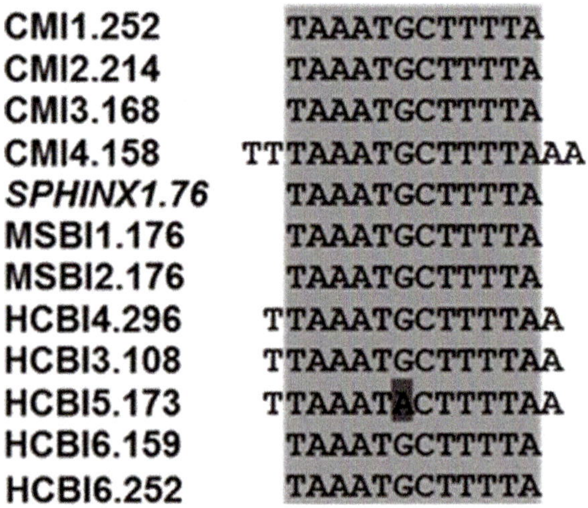

Manuelidis et al. (2011) described the paired activity between the “larger” Sphinx 2.36 and the “smaller” Sphinx 1.76. We demonstrated in vitro transcription of all isolates tested in human cells (293TT cell line) after transfection (Eilebrecht et al., in preparation). The combined transfection of the “larger” CMI1.252 with the “smaller” MSBI1.176 resulted in an augmented replication of the latter. This clearly shows a “helper” function provided by CMI1.252 for the replication of MSBI1.176. It is tempting to speculate that our isolates may act in vivo in a similar synergistic way as monopartite *begomoviruses* and their satellites of the family *Geminiviridae*, in that pairing/complementation of two genomes—each with own genomic information—is necessary for a full replication cycle of these organisms in vivo (Hanley-Bowdoin et al. 2013). Complementation is also suggested by alignments of the repeat regions, of the palindromic sequences, as well as of their replication proteins. The replication proteins of MSBI1 and CMI1 (Funk et al. 2014; Gunst et al. 2014; Whitley et al. 2014) and those of Geminiviruses (Rizvi et al. 2015) show similarity to the rolling-circle replication initiator proteins of bacterial plasmids. It is also clear from the differences between isolates that our isolates most likely represent only a fraction of a multitude of existing, unknown agents which may form part of other complementation groups/infectious agents, all under an umbrella of a large family of agents, each possibly playing its role in the pathogenesis of a specific disease.

Three isolates from cattle serum belong to a second group, BMMF2, based on their sequence relatedness to Sphinx 2.36 (Manuelidis et al. 2011). These are HCBI1.225, HCBI2.170, and HCBI7.228 (Funk et al. 2014). Interestingly, 2 isolates have “larger” genomes (2251 and 2280 nt, respectively), whereas one genome HCBI2.170 has a “smaller” genome (1407 nt)—in analogy to the BMMF1 group (see above). These isolates share a nucleotide sequence identity of 75–81% with those of Sphinx 2.36, whereas their sequence identity between each other varies between 80 and 89%. All 3 isolates have ORFs encoding for a putative replication protein which share between 80 and 97% amino acid identity to replication proteins of *Acinetobacter* spp., although their genome nucleotide sequence homology to these bacteria does not exceed 70%. No putative functions can be attributed to almost all other larger ORFs present. Only one of these in HCBI7.228 could encode for a putative capsid protein, but in vitro studies have not yet been performed to substantiate this in silico analysis.

Two isolates from cattle sera and one from MS serum and brain (Group 3) were characterized as novel *Gemycircularviruses* (Lamberto et al. 2014) belonging to the new family *Genomoviridae* of widespread single-stranded DNA viruses (Krupovic et al. 2016). The first known Gemycircularvirus was isolated from fungi, but has since been isolated from various animals including cattle, insects, and birds, as well as from humans (Krupovic et al. 2016). Our isolates (HCBI8.215 and HCBI9.212) were present in cattle serum (Lamberto et al. 2014), whereas additional cattle isolates were identified in their feces (Steel et al. 2016). Our third novel Gemycircularvirus MSSI2.225 was isolated from serum and brain of MS patients (Lamberto et al. 2014). Other Gemycircularviruses reported from humans originated from samples of cerebrospinal and pericardial fluid, plasma, and feces (Phan et al. 2015; Halary et al. 2016; Zhang et al. 2016b).

The nucleotide sequence of one isolate MSSI1.162 (1627 nt) from MS serum was distantly related to a *Psychrobacter* species plasmid (Group 4) (Gunst et al. 2014). Psychrobacter species are frequently transmitted as food contaminants and are regarded as opportunistic human pathogens. Although frequently present in human blood samples, it is interesting to note that *Psychrobacter* sp. were identified in cerebrospinal fluid in single reports of pediatric meningitis (Lloyd-Puyear et al. 1991; Le Guern et al. 2014; Ortiz-Alcántara et al. 2016).

Antibodies directed against the rep protein in a high percentage of humans (see Bund et al., to be published) with significantly elevated titers in MS further underline the infectivity of the BMMF group 1 for humans. At present, it remains an open question whether these antibodies result either from repeated exposures to the respective antigens (by re-infections or occasional reactivations), or from a long-lasting latency state with a limited protein expression.

8 Synergistic Model for Interactions of These Infections with Other Viral Infections and Genetic or Epigenetic Modifications

8.1 Multiple Sclerosis

The isolation of two BMMF isolates from a brain lesion of a patient with MS directed our interest to a possible involvement of BMMF infections in the development of this neurodegenerative disease. An analysis of the available epidemiological data revealed three main risk factors for MS: cow milk consumption, vitamin D deficiency, and herpesvirus reactivation (in particular Epstein–Barr virus, but also human herpesvirus type 6 and Varicella zoster virus) (reviewed in zur Hausen 2015; zur Hausen and de Villiers 2016). Dairy cow milk consumption in early childhood, vitamin D3 deficiency (probably explaining the north–south gradient in MS incidence) linked to low sun exposure, as well as serological, immunochemical evidence and excretion of reactivated virus (mainly EBV) under conditions of vitamin D deficiency have been discussed as major MS risk factors (reviewed in zur Hausen 2015; zur Hausen and de Villiers 2016).

In vitro studies performed in our laboratory during past decades had demonstrated that latent EBV infections can be reactivated by tumor growth factor-β (TGFβ) (Bauer et al. 1982, 1991). A negative regulation of TGFβ synthesis by vitamin D3 has been reported through activation of the vitamin D3 receptor by vitamin D3 (Tao et al. 2015; Beilfuss et al. 2015; Zerr et al. 2015). The fact that lack of sun exposure in antarctic expeditioners was linked to increased EBV reactivation and excretion in the saliva (Mehta et al. 2000) and that vitamin D supplementation mitigated this EBV reactivation (Zwart et al. 2011) permits us to draw a parallel between the in vitro data and EBV reactivation in humans. This would fit to a link of both vitamin D deficiency and EBV reactivation as risk factors for MS pathogenesis.

Other in vitro studies performed in our laboratory demonstrated that herpesvirus infection or herpesvirus reactivation in cells co-infected by small DNA viruses (adeno-associated viruses, polyoma-, papilloma-, or Torque teno-viruses) led to a substantial amplification of the latter, at the same time inhibiting the replication of the herpesvirus DNA. This was interpreted as the likely consequence of competition for DNA polymerases, favoring a more rapid amplification of small DNA molecules (Heilbronn and zur Hausen 1989; Heilbronn et al. 1990; Borkosky et al. 2012).

Taken together, these data represented the basis for a pathogenesis concept of multiple sclerosis (zur Hausen 2015; zur Hausen and de Villiers 2016). A slightly modified version of this concept is presented in Fig. 7:

This concept permits several predictions, part of which has been experimentally analyzed:

1. *Infections early in life by BMMF should result in immune reactivity. Repeated subsequent exposures to related agents during the following years should lead to their neutralization and to a simultaneous booster for a specific immune response. Due to exposure of the vast majority of humans to dairy products and bovine meat, a substantial percentage of the human population is expected to reveal sero-reactivity against BMMF antigens. This prediction is serologically confirmed* (Bund et al., in preparation).

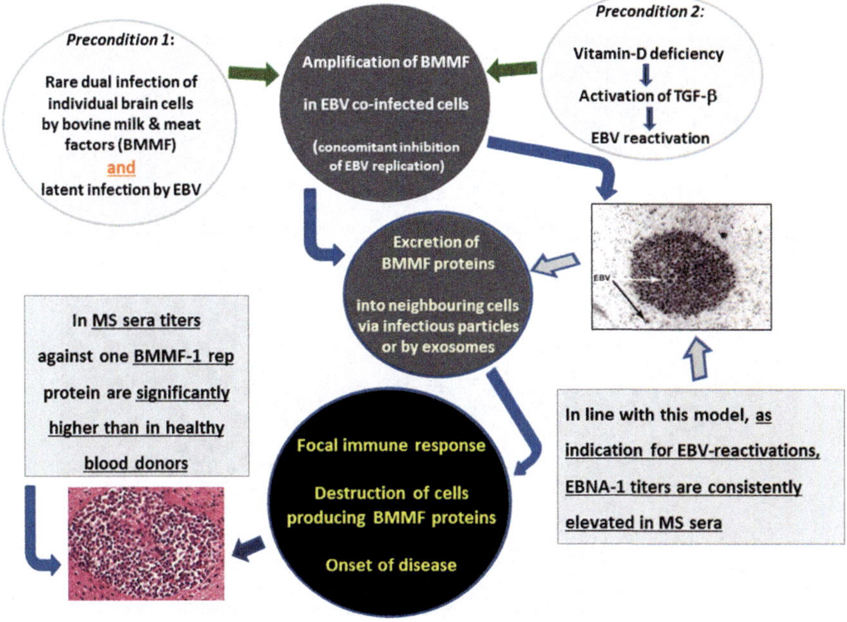

Fig. 7 Concept of multiple sclerosis pathogenesis (modified from zur Hausen 2015)

2. *Infections by herpes group viruses are widely spread in human populations and frequently acquired at a young age. They regularly result in viral genome latency in specific target cells. Reactivation of these genomes commonly results in local spreading and in a booster of an early immune response which leads to usually low levels of persisting antibodies. Frequent reactivations clustered in time, as anticipated in focal multiple MS lesions, however, should result in higher titer antibody responses. This is in line with previously published and our own preliminary data showing an increased EBV-EBNA-1 sero-reactivity in MS patients* (reviewed in Borkosky et al. 2012; Bund et al., in preparation).
3. *EBV reactivation has been reported in humans not exposed to sunlight for prolonged periods of time (supposedly vitamin D deficiency)* (Mehta et al. 2000; Zwart et al. 2011). *EBNA1 titers are elevated under conditions of vitamin D deficiency* (Wergeland et al. 2016). *High-dose oral vitamin D3 supplementation reduces humoral immune responses against the latent EBV antigen EBNA1* (Røsjø et al. 2016).
4. *Virus-like particles, suspected to represent TT viruses were discovered after transfection of TTV-DNA into lymphoblastoid cells with spontaneous reactivation of latent EBV* (Borkosky et al. 2012).
5. *Intensive inflammation was shown to accompany early MS lesions* (Fig. 8). *The latter frequently starts from a perivenous cuff containing EBV-positive memory cells* (Tan et al. 2000; Mistry et al. 2015).

None of these observations individually provides proof for the concept presented here. Taken together, however, they substantially support this hypothesis.

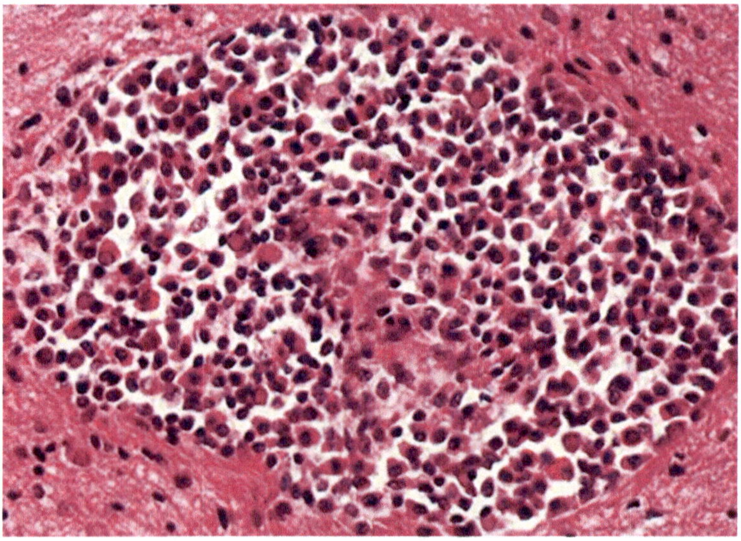

Fig. 8 Early lesion of multiple sclerosis revealing intense inflammation around an initial perivenular focus (*courtesy of Dr. Marta Margeta, University of San Francisco, Division of Neuropathology*)

8.2 Solid Cancers

Can these considerations be applied to an involvement of BMMF infections in malignant tumors?

Figure 9 depicts two possible outcomes, both of which would be compatible with previously discussed epidemiological analyses. Again, dual infections of the same cell are visualized as a prerequisite for those cancers with epidemiological links to the consumption of meat and dairy products. A relatively large number of publications described an increased risk for such tumors in case of vitamin D deficiency (e.g., Dou et al. 2016; reviewed in Meeker et al. 2016). Herpesvirus infections, in particular EBV, but also human herpesvirus type 6, cytomegalovirus, and others, have repeatedly been reported in these malignancies (Halme et al. 2013; Richardson et al. 2015; Hu et al. 2016, Review), occasionally with evidence for EBV reactivation in inflammatory reactive tissue (Fiorina et al. 2014).

The patchy and, most often, regional appearance of lesions in Crohn's disease and ulcerative colitis accompanied by intensive inflammatory reactions might point to an interesting parallel to the multiple sclerosis model, although affecting a very different tissue. Some support for this hypothesis originates from a number of

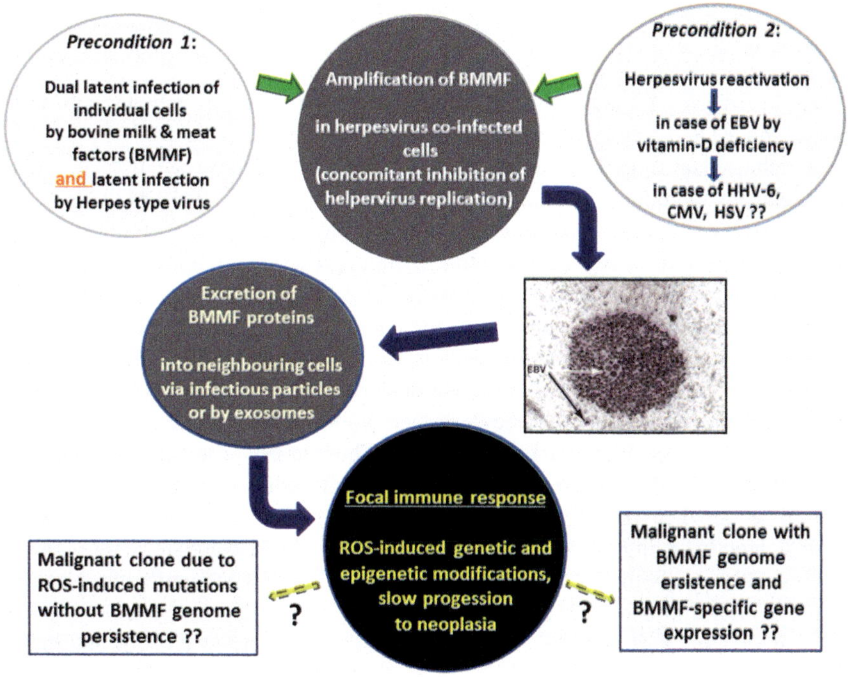

Fig. 9 Alternative models for a putative role of bovine milk and meat factors (BMMFs) in the development of specific solid tumors

reports on herpesvirus reactivations (Kangro et al. 1990; Sipponen et al. 2011; Silva et al. 2012) and vitamin D deficiency (de Bruyn et al. 2014; Torki et al. 2015; Sadeghian et al. 2016) predisposing for these conditions.

8.3 Cross-reactivity of BMMF Antibodies with Those of Neu5Gc?

A number of reports described an increased risk for multiple sclerosis after previous manifestation of infectious mononucleosis (IM) (reviewed in Haahr et al. 1995; Thacker et al. 2006). In MS, recurrence of acute attacks seemed to correlate with the development of heterophile antibodies (Jans et al. 1982). Prior to the identification of Epstein–Barr virus as causative agent of infectious mononucleosis and even subsequently, the development of heterophile antibodies, commonly demonstrated by the agglutination of bovine or sheep erythrocytes, served as a somewhat less specific marker for IM. Two test systems had been developed: the *Hanganutziu–Deicher* test (HD test) (Hanganutziu 1924; Deicher 1926) and the more specific Paul–Bunnell test (PB test) (Paul and Bunnell 1932). Heterophile Paul–Bunnell antibodies most frequently emerge as the consequence of Epstein–Barr virus infections, appearing slightly later than antibodies against EBV (Hurt and Tammaro 2007; reviewed in Dunmire et al. 2015), and are in about 25% falsely negative during the first week of illness (Womack and Jimenez 2015).

Besides in IM, HD antibodies can often be demonstrated in patients with melanomas, breast and colon cancers, and a few additional cancers, as well as in several other pathological disorders (Inoue et al. 2010). Among these are hematopoietic malignancies, including Hodgkin's disease (Masaki et al. 1981; Malykh et al. 2001). PB tests are more specific for IM after prior absorption of human sera, commonly with guinea pig kidney extracts, although some remaining reactivity has been noted in a low percentage of cancers, in particular in lymphomas and leukemias and in rheumatoid arthritis (Nishimaki et al. 1979).

HD, as well as PB reactive antigens, has been identified in membranes of bovine erythrocytes. Here, several different sialoglycoproteins (17–50 kDa) have been noted playing a decisive role in the described agglutination reaction. Desialylation completely abolishes HD and PB reactivity of bovine erythrocytes (Gołaszewska et al. 2003). According to these authors, the PB antibodies recognize the structure *NeuGcα2-3Galβ1-3(Neu5Gcα2-6)GalNAcα1-threonine/serine*, with Neu5Gc as the terminal reactive component. This structure differs from Neu5Gc serum glycoproteins, although both HD and PB heterophilic antibodies recognize the Neu5Gc as an essential immunogenic epitope.

The interesting part of these studies concerns the co-emergence of PB antibodies with infectious mononucleosis, although, as previously mentioned, with some delay after initiation of the immune response against Epstein–Barr virus. Since humans are unable to synthesize Neu5Gc, it remains an open question how the synthesis or uptake of this sialin is mediated by the EBV infection. Gołaszewska et al. (2003) proposed that the virus-infected or responsive cells concentrate Neu5Gc during the course of EBV infection. It is, however, unlikely that an enhanced uptake of Neu5Gc is occurring in a disease characterized by a loss of appetite (Simon 1998). Due to the failure of humans to convert Neu5Ac into Neu5Gc, potential alternative pathways for Neu5Gc synthesis have also been discussed (Malyhk et al. 2001). Yet, one additional possibility apparently has not been mentioned previously: the conversion of Neu5Ac to Neu5Gc by a Herpesvirus-specific hydroxylase. O-glycosylation for viral envelope proteins has been described for Herpes simplex, cytomegalo-, Varicella zoster, and Epstein–Barr viruses (Bagdonaite et al. 2016). It should be of interest that lymphomas arising after Marek's herpesvirus infection, in general Neu5Gc-deficient chicken, contain HD-reactive Neu5Gc (Higashi et al. 1984).

Why is it tempting to speculate on Neu5Gc glycoproteins as potential receptors for BMMF infections?

1. Neu5Gc glycoprotein acts as receptor for human polyomavirus type 9 (Khan et al. 2014), for canine and feline parvoviruses (Löfling et al. 2013), for specific types of influenza A and parainfluenza viruses (Song et al. 2011), and probably for other not specified agents. This receptor seems to determine the cell type specificity for certain infections.
2. The development of HD heterophile antibodies has been identified in a relatively broad spectrum of disorders (Sect. 8.1), although with some preference for infectious mononucleosis. The reactive components are terminally bound Neu5Gc sialoproteins or sialoganglosides (Gołaszewska et al. 2003). HD antibodies react with the membranes of bovine erythrocytes which contain the reactive antigen not as a single molecule, but in several glycoproteins, all terminating with N-glycolylneuraminic acid (Neu5Gc) (Gołaszewska et al. 2003). Although the more specific PB test for infectious mononucleosis also interacts with Neu5Gc, it lost the part of the broad reactivity of HD antibodies, suggesting that here a different epitope of the reactive sialin is recognized. If Neu5Gc-containing glycoproteins in bovine erythrocytes bind BMMFs (frequently present in sera of dairy cattle), it is tempting to speculate that receptor-bound BMMF particles modify the presentation of Neu5Gc reactive sites. The absorption of heterophile sera in guinea pig kidney extracts, as initially performed for the PB test, may remove to a large part HD reactivity against unbound Neu5Gc glycoproteins. PB antibodies would thus react more specifically with BMMF–sialin-receptor-bound epitopes.
3. Amplification of small DNA viral genomes by different types of reactivated or lytically infecting herpesviruses has been described before (see Sect. 8). If a number of cells are latently infected with BMMF and co-infected with EBV, the lytic replication or reactivation of the latter should stimulate BMMF amplification. This in turn should lead to excretion of BMMF proteins into neighboring cells (via exosomes or particles), resulting in a booster of anti-BMMF reactivity and, in case of the previously discussed conversion of Neu5Ac to Neu5Gc by a

Herpesvirus-specific O-glycosylating hydroxylase, in emergence of PB antibodies.
4. Alternatively, EBV infections may concomitantly occur with BMMF virus-like particles. This should result in the same BMMF amplification effect, provided co-infections occur within the same cells. Small virus-like particles (20–35 nm in size) have been documented repeatedly in EBV-producing clinical lesions of hairy leukoplakias (Greenspan et al. 1984) and in EBV-producing Burkitt's lymphoma cells in tissue culture (zur Hausen, unpublished). They are shown in Fig. 10. Their nature has not yet been determined.
5. An interesting observation was the development of Neu5Gc antibodies during the weaning period (Taylor et al. 2010). This corresponds to the period when the risk for uptake of BMMFs should be very high, correlated with increasing consumption of cow milk and dairy products.
6. The reported increased reactivity against Neu5Gc in several solid tumors, neurodegenerative diseases, and cardiovascular conditions also requires plausible explanations. Inflammatory reactions are a common feature, particularly in early stages of solid tumor development (Li et al. 2016; Castellon and Bogdanova 2016). The presence of different types of herpesviruses has repeatedly been observed—most prominently of Epstein–Barr virus, but also of cytomegalovirus, human herpesvirus type 6, and a few others (Roizman 1974;

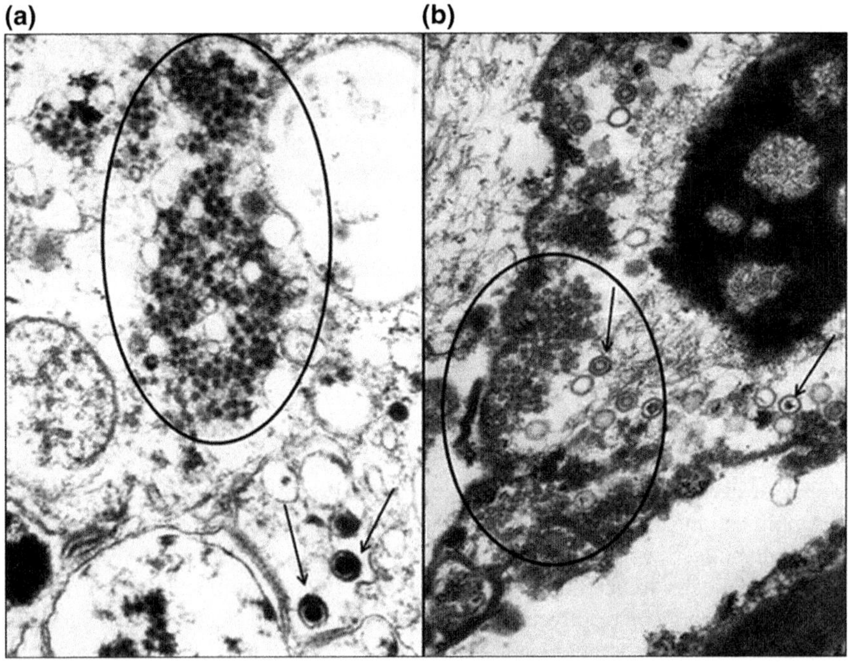

Fig. 10 Small virus-like particles in Epstein–Barr virus (*arrows*) producing Burkitt's lymphoma cells. **a** P3HR-1 cells after TPA induction (zur Hausen, unpublished) and **b** in an EBV-producing hairy leukoplakia of an HIV-infected person (*courtesy of Dr. John Greenspan*)

Söderberg-Nauclér 2006; Cao et al. 2016). As proposed in Sect. 8, interactions between a reactivated herpes-type virus and a persistent BMMF-like genome may result in strong inflammatory responses producing reactive oxygen species (ROS). Mutagenic or epigenetic modifications as a consequence of ROS activation are also common in neurodegenerative diseases (Limongi and Baldelli 2016). Under these circumstances, the actual manifestation of the disease would not require a persistent and continued genetic activity of the initial trigger of the inflammatory reaction. In this respect, it is interesting to note that long-time application of low-dose aspirin, as an anti-inflammatory drug, has been repeatedly reported to reduce the risk for colon, breast, and prostate cancers (Elwood et al. 2016).

9 Conclusions

Dietary habits, in particular consumption of beef and dairy products, substantially not only contribute as risk factors for a number of prevalent human cancers, but also pose as risks for several neurodegenerative and cardiovascular diseases. The concept of virus-like infections transmitted from specific ungulates to humans, most likely by dietary ingestion, is presently receiving experimental support by the isolation and characterization of small single-stranded DNA molecules from dairy cattle serum and cow milk and products derived therefrom. The genetic activity of several of these isolates in human cells has been documented.

Demonstration of antibodies against one of their major proteins (at this stage in MS sera) supports the view that they represent candidates for a link to some of these very common human diseases. Available surveys indicate that similar antibodies are prevalent in the human population. Yet, the titers in multiple sclerosis are significantly elevated in comparison with human control sera.

The previously outlined concept of a synergistic function of dual infections of the same cell by a herpes group virus (apparently frequently by EBV) and BMMF genomes (zur Hausen 2015) receives further support in MS patients by concomitant elevated antibody titers against BMMF antigens and EBNA1 (Bund et al., to be published).

The relationship between heterophile Paul–Bunnell antibodies, shown to be directed against the sialin Neu5Gc, and anti-BMMF reactivity, requires further studies. Neu5Gc has been shown to be an important terminal component of a set of cellular glycoproteins and gangliosides, both of them documented as determining the specificity of receptors for several viral infections.

The serological identification of the virus-like BMMF antigenicity is presently based on BMMF1 Rep proteins as antigens. Despite the high degree of conservation between these proteins, an accurate assignment of the sero-response to specific BMMF genotypes, except for the two isolates from MS brain autopsy material, remains to be determined. In view of the heterogeneity of available isolates, further studies are required, to determine the potential existence of "high"- and "low"-risk types and their pathogenicity for specific diseases. Thus, a number of important questions remain open.

Acknowledgements We thank Prof. Dr. Hermann Müller (Veterinary Faculty, University of Leipzig, Germany) for providing the serum samples from cattle.

An unrestricted grant was provided by Oryx, München.

References

Abbott RD, Ross GW, Petrovitch H, Masaki KH, Launer LJ, Nelson JS, White LR, Tanner CM (2016) Midlife milk consumption and substantia nigra neuron density at death. Neurology 86:512–519

Abou-Dakn M, Scheele M, Strecker JR (2003) Vorbeugung von Brustkrebs durch Stillen? Zentralbl Gynäkol 125:48–52 Review

Alisson-Silva F, Kawanishi K, Varki A (2016) Human risk of diseases associated with red meat intake: analysis of current theories and proposed role for metabolic incorporation of a non-human sialic acid. Mol Aspects Med pii: S0098-2997(16)30040-1. doi:10.1016/j.mam.2016.07.002

Amin M, Jeyaganth S, Fahmy N, Bégin LR, Aronson S, Jacobson S, Tanguay S, Kassouf W, Aprikian A (2008) Dietary habits and prostate cancer detection: a case-control study. Can Urol Assoc J 2:510–515

Andersen V, Vogel U (2015) Interactions between meat intake and genetic variation in relation to colorectal cancer. Genes Nutr 10:448. doi:10.1007/s12263-014-0448-9

Aykan NF (2015) Red meat and colorectal cancer. Oncol Rev 9:b288 E-collection

Bagdonaite I, Nordén R, Joshi HJ, King SL, Vakhrushev SY, Olofsson S, Wandall HH (2016) Global mapping of O-glycosylation of varicella zoster virus, human cytomegalovirus, and Epstein-Barr virus. J Biol Chem 291:12014–12028

Baron T, Biacabe AG, Arsac JN, Benestad S, Groschup MH (2007) Atypical transmissible spongiform encephalopathies (TSEs) in ruminants. Vaccine 25:5625–5630

Battaglia Richi E, Baumer B, Conrad B, Darioli R, Schmid A, Keller U (2015) Health risks associated with meat consumption: a review of epidemiological studies. Int J Vitam Nutr Res 85:70–78

Bauer G, Höfler P, zur Hausen H (1982) Epstein-Barr virus induction by a serum factor. I. Induction and cooperation with additional inducers. Virology 121:184–194

Bauer G, Götschl M, Höfler P (1991) Tumor-promoting activity of Epstein-Barr-virus-inducing factor transforming growth factor type beta (EIF/TGF-beta) is due to the induction of irreversible transformation. Int J Cancer 47:881–888

Beilfuss A, Sowa JP, Sydor S, Beste M, Bechmann LP, Schlattjan M, Syn WK, Wedemeyer I, Mathé Z, Jochum C, Gerken G, Gieseler RK, Canbay A (2015) Vitamin D counteracts fibrogenic TGF-β signalling in human hepatic stellate cells both receptor-dependently and independently. Gut 64:791–799

Bener A, Denic S, Galadari S (2001) Longer breast-feeding and protection against childhood leukaemia and lymphomas. Eur J Cancer 37:234–238

Bener A, Hoffmann GF, Afify Z, Rasul K, Tewfik I (2008) Does prolonged breastfeeding reduce the risk for childhood leukemia and lymphomas? Minerva Pediatr 60:155–161

Boleij A, van Gelder MM, Swinkels DW, Tjalsma H (2011) Clinical importance of Streptococcus gallolyticus infection among colorectal cancer patients: systematic review and meta-analysis. Clin Infect Dis 53:870–878 Review

Borkosky SS, Whitley C, Kopp-Schneider A, zur Hausen H, de Villiers EM (2012) Epstein-Barr virus stimulates torque teno virus replication: a possible relationship to multiple sclerosis. PLoS One 7(2):e32160. doi:10.1371/journal.pone.0032160

Bosetti C, Micelotta S, Dal Maso L, Talamini R, Montella M, Negri E, Conti E, Franceschi S, La Vecchia C (2004) Food groups and risk of prostate cancer in Italy. Int J Cancer 110:424–428

Botsios S, Manuelidis L (2016) CJD and Scrapie require agent-associated nucleic acids for infection. J Cell Biochem 117(8):1947–1958 doi:10.1002/jcb.25495

Bouvard V, Loomis D, Guyton KZ, Grosse Y, Ghissassi FE, Benbrahim-Tallaa L, Guha N, Mattock H, Straif K, International Agency for Research on Cancer Monograph Working Group (2015) Carcinogenicity of consumption of red and processed meat. Lancet Oncol 16:1599–1600

Bovalino S, Charleson G, Szoeke C (2016) The impact of red and processed meat consumption on cardiovascular disease risk in women. Nutrition 32:349–354

Butcher J (1976) The distribution of multiple sclerosis in relation to the dairy industry and milk consumption. N Z Med J 83:427–430

Caini S, Masala G, Gnagnarella P, Ermini I, Russell-Edu W, Palli D, Gandini S (2016) Food of animal origin and risk of non-Hodgkin lymphoma and multiple myeloma: a review of the literature and meta-analysis. Crit Rev Oncol Hematol 100:16–24

Cantorna MT, Mahon BD (2004) Mounting evidence for vitamin D as an environmental factor affecting autoimmune disease prevalence. Expo Biol Med (Maywood) 229:1136–1142

Cao S, Wendl MC, Wyczalkowski MA, Wylie K, Ye K, Jayasinghe R, Xie M, Wu S, Niu B, Grubb R 3rd, Johnson KJ, Gay H, Chen K, Rader JS, Dipersio JF, Chen F, Ding L (2016) Divergent viral presentation among human tumors and adjacent normal tissues. Sci Rep 6:28294. doi:10.1038/srep28294

Carr PR, Walter V, Brenner H, Hoffmeister M (2016) Meat subtypes and their association with colorectal cancer: systematic review and meta-analysis. Int J Cancer 138:293–302 Review

Castellon X, Bogdanova V (2016) Chronic inflammatory diseases and endothelial dysfunction. Aging Dis 7:81–89

Chattoraj DK (2000) Control of plasmid DNA replication by iterons: no longer paradoxical. Mol Microbiol 37:467–476

Chen H, O'Reilly E, McCullough ML, Rodriguez C, Schwarzschild MA, Calle EE, Thun MJ, Ascherio A (2007) Consumption of dairy products and risk of Parkinson's disease. Am J Epidemiol 165:998–1006

Chen WW, Zhang X, Huang WJ (2016) Role of physical exercise in Alzheimer's disease. Biomed Rep 4:403–407

Choi Y, Song S, Song Y, Lee JE (2013) Consumption of red and processed meat and esophageal cancer risk: meta-analysis. World J Gastroenterol 19:1020–1029 Review

Comoy EE, Mikol J, Luccantoni-Freire S, Correia E, Lescoutra-Etchegaray N, Durand V, Dehen C, Andreoletti O, Casalone C, Richt JA, Greenlee JJ, Baron T, Benestad SL, Brown P, Deslys JP (2015) Transmission of scrapie prions to primate after an extended silent incubation period. Sci Rep 5:11573. doi:10.1038/srep11573

Conradi S, Malzahn U, Paul F, Quill S, Harms L, Then Bergh F, Ditzenbach A, Georgi T, Heuschmann P, Rosche B (2013) Breastfeeding is associated with lower risk for multiple sclerosis. Mult Scler 19:553–558

Cross AJ, Lim U (2006) The role of dietary factors in the epidemiology of non-Hodgkin's lymphoma. Leuk Lymphoma 47:2477–2487 Review

Daniels JL, Olshan AF, Pollock BH, Shah NR, Stram DO (2002) Breast-feeding and neuroblastoma, USA and Canada. Cancer Causes Control 13:401–405

Davenport KA, Henderson DM, Bian J, Telling GC, Mathiason CK, Hoover EA (2015) Insights into chronic wasting disease and bovine spongiform encephalopathy species barriers by use of real-time conversion. J Virol 89:9524–9531

de Bruyn JR, van Heeckeren R, Ponsioen CY, van den Brink GR, Löwenberg M, Bredenoord AJ, Frijstein G, D'Haens GRJ (2014) Vitamin D deficiency in Crohn's disease and healthy controls: a prospective case-control study in the Netherlands. Crohns Colitis 8:1267–1273

Dechamma HJ, Natarajan C, Suryanarayanan VV, Rasool TJ (1998) Nucleotide sequence of DR-B exon of major histocompatibility complex from an Indian zebu cattle breed. Ind J Exp Biol 36:879–886
Deicher H (1926) Über die Erzeugung heterospezifischer Hämagglutinine durch Injektion artfremden Serums. Z Hyg 106:561–579
Diack AB, Head MW, McCutcheon S, Boyle A, Knight R, Ironside JW, Manson JC, Will RG (2014) Variant CJD. 18 years of research and surveillance. Prion 8:286–295 Review
Dou R, Ng K, Giovannucci EL, Manson JE, Qian ZR, Ogino S (2016) Vitamin D and colorectal cancer: molecular, epidemiological and clinical evidence. Br J Nutr 115:1643–1660
D'Souza S, Levy E, Mack D, Israel D, Lambrette P, Ghadirian P, Deslandres C, Morgan K, Seidman EG, Amre DK (2008) Dietary patterns and risk for Crohn's disease in children. Inflamm Bowel Dis 14:367–373
Dunmire SK, Hogquist KA, Balfour HH (2015) Infectious mononucleosis. Curr Top Microbiol Immunol 390:211–240 Review
Durusoy R, Boffetta P, Mannetje A, Zaridze D, Szeszenia-Dabrowska N, Rudnai P, Lissowska J, Fabiánová E, Cassidy A, Mates D, Bencko V, Salajka F, Janout V, Fevotte J, Fletcher T, Brennan P (2006) Lung cancer risk and occupational exposure to meat and live animals. Int J Cancer 118:2543–2547
Dziewit L, Cegielski A, Romaniuk K, Uhrynowski W, Szych A, Niesiobedzki P, Zmuda-Baranowska MJ, Zdanowski MK, Bartosik D (2013) Plasmid diversity in arctic strains of *Psychrobacter* spp. Extremophiles 17(3):433–444
Elwood PC, Morgan G, Pickering JE, Galante J, Weightman AL, Morris D, Kelson M, Dolwani S (2016) Aspirin in the treatment of cancer: reductions in metastatic spread and in mortality: a systematic review and meta-analyses of published studies. PLoS One 11(4):e0152402. doi:10.1371/journal.pone.0152402. eCollection 2016 Review
Epplein M, Zheng W, Li H, Peek RM Jr, Correa P, Gao J, Michel A, Pawlita M, Cai Q, Xiang YB, Shu XO (2014) Diet, Helicobacter pylori strain-specific infection, and gastric cancer risk among Chinese men. Nutr Cancer 66:550–557
Epstein MM, Chang ET, Zhang Y, Fung TT, Batista JL, Ambinder RF, Zheng T, Mueller NE, Birmann BM (2015) Dietary pattern and risk of Hodgkin lymphoma in a population-based case-control study. Am J Epidemiol 182:405–416
Etzold S, Bode L (2014) Glycan-dependent viral infection in infants and the role of human milk oligosaccharides. Curr Opin Virol 7:101–107 Review
Farvid MS, Cho E, Chen WY, Eliassen AH, Willett WC (2015) Adolescent meat intake and breast cancer risk. Int J Cancer 136:1909–1920
Felmus MT, Patten BM, Swanke L (1976) Antecedent events in amyotrophic lateral sclerosis. Neurology 26:167–172
Ferrís i Tortajada J, Ortega García JA, García i Castell J, López Andreu JA, Berbel Tornero O, Crehuá Gaudiza E (2005) Risk factors for neuroblastoma [Article in Spanish]. An Pediatr (Barc) 63:50–60
Fiorina L, Ricotti M, Vanoli A, Luinetti O, Dallera E, Riboni R, Paolucci S, Brugnatelli S, Paulli M, Pedrazzoli P, Baldanti F, Perfetti V (2014) Systematic analysis of human oncogenic viruses in colon cancer revealed EBV latency in lymphoid infiltrates. Infect Agent Cancer 9:18. doi:10.1186/1750-9378-9-18
Flynn KJ, Baxter NT, Schloss PD (2016) Metabolic and community synergy of oral bacteria in colorectal cancer. mSphere 1(3).pii:e00102-16. doi:10.1128/mSphere.00102-16. eCollection. Review
Fonseca-Nunes A, Jakszyn P, Agudo A (2014) Iron and cancer risk—a systematic review and meta-analysis of the epidemiological evidence. Cancer Epidemiol Biomarkers Prev 23:12–31
Funk M, Gunst K, Lucansky V, Müller H, zur Hausen H, de Villiers EM (2014) Isolation of protein-associated circular DNA from healthy cattle serum. Genome Announc 2(4):e00846-14
Gołaszewska E, Kurowska E, Duk M, Kościelak J (2003) Paul-Bunnell antigen and a possible mechanism of formation of heterophile antibodies in patients with infectious mononucleosis. Acta Biochim Pol 50:1205–1211

Gonia S, Tuepker M, Heisel T, Autran C, Bode L, Gale CA (2015) Human milk oligosaccharides inhibit *Candida albicans* invasion of human premature intestinal epithelial cells. J Nutr 145:1992–1998
González CA, López-Carrillo L (2010) Helicobacter pylori, nutrition and smoking interactions: their impact in gastric carcinogenesis. Scand J Gastroenterol 45:6–14 Review
Gouveri E, Papanas N, Hatzitolios AI, Maltezos E (2011) Breastfeeding and diabetes. Curr Diabetes Rev 7:135–142 Review
Graf R, Grüntzig K, Boo G, Hässig M, Axhausen KW, Fabrikant S, Welle M, Meier D, Guscetti F, Folkers G, Otto V, Pospischil A (2016) Swiss Feline Cancer Registry 1965–2008: the influence of sex, breed and age on tumour types and tumour locations. J Comp Pathol 154:195–210
Grant WB (2013) The role of milk protein in increasing risk of Parkinson's disease. Eur J Epidemiol 28:357
Graves J, Grandhe S, Weinfurtner K, Krupp L, Belman A, Chitnis T, Ness J, Weinstock-Guttman B, Gorman M, Patterson M, Rodriguez M, Lotze T, Aaen G, Mowry EM, Rose JW, Simmons T, Casper TC, James J, Waubant E, US Network of Pediatric Multiple Sclerosis Centers (2014) Protective environmental factors for neuromyelitis optica. Neurology 83:1923–1929
Greenop KR, Bailey HD, Miller M, Scott RJ, Attia J, Ashton LJ, Downie P, Armstrong BK, Milne E (2015) Breastfeeding and nutrition to 2 years of age and risk of childhood acute lymphoblastic leukemia and brain tumors. Nutr Cancer 67:431–441
Greenspan D, Greenspan JS, Conant M, Petersen V, Silverman S Jr, de Souza Y (1984) Oral "hairy" leucoplakia in male homosexuals: evidence of association with both papillomavirus and a herpes-group virus. Lancet 2(8407):831–834
Gu Y, Nieves JW, Stern Y, Luchsinger JA, Scarmeas N (2010) Food combination and Alzheimer disease risk: a protective diet. Arch Neurol 67:699–706
Guise JM, Austin D, Morris CD (2005) Review of case-control studies related to breastfeeding and reduced risk of childhood leukemia. Pediatrics 116:e724–e731
Gunst K, zur Hausen H, de Villiers EM (2014) Isolation of bacterial plasmid-related replication-associated circular DNA from serum sample of a multiple sclerosis patient. Genome Announc 2(4):e00847-14
Haahr S, Koch-Henriksen N, Møller-Larsen A, Eriksen LS, Andersen HM (1995) Increased risk of multiple sclerosis after late Epstein-Barr virus infection: a historical prospective study. Mult Scler 1:73–77
Halary S, Duraisamy R, Fancello L, Monteil-Bouchard S, Jardot P, Biagini P, Gouriet F, Raoult D, Desnues C (2016) Novel single-stranded DNA circular viruses in pericardial fluid of patient with recurrent pericarditis. Emerg Infect Dis 22:1839–1841
Halme L, Loginov R, Arola J, Turunen U, Lautenschlager I (2013) HHV-6 antigen and HHV-6 DNA expression in sporadic adenomatous polyps of the colon. Scand J Gastroenterol 1448:1423–1427
Han YW (2015) Fusobacterium nucleatum: a commensal-turned pathogen. Curr Opin Microbiol 23:141–147 Review
Hanganutziu M (1924) Hémagglutinines hétérogénétiques après injection de sérum de cheval. CR Séances Soc Biol 91:1457–1459
Hanley-Bowdoin L, Bejarano ER, Robertson D, Mansoor S (2013) Geminiviruses: masters in redirecting and reprogramming plant processes. Nature Rev Microbiol 11:777–788
Harrison PM, Chan HS, Prusiner SB, Cohen FE (2001) Conformational propagation with prion-like characteristics in a simple model of protein folding. Protein Sci 10:819–835
Heilbronn R, zur Hausen H (1989). A subset of herpes simplex virus replication genes induces DNA amplification within the host cell genome. J Virol 63:3683–3692
Heilbronn R, Bürkle A, Stephan S, zur Hausen H (1990) The adeno-associated virus rep gene suppresses herpes simplex virus-induced DNA amplification. J Virol 64:3012–3018
Hendrix DV, Newkirk KM (2014) Expression of epidermal growth factor receptor and human epidermal growth factor receptor 2 in periocular squamous cell carcinomas of horses. Am J Vet Res 75:912–917. doi:10.2460/ajvr.75.10.912

Hernández ÁR, Boada LD, Almeida-González M, Mendoza Z, Ruiz-Suárez N, Valeron PF, Camacho M, Zumbado M, Henríquez-Hernández LA, Luzardo OP (2015) An estimation of the carcinogenic risk associated with the intake of multiple relevant carcinogens found in meat and charcuterie products. Sci Total Environ 514:33–41

Higashi H, Ikuta K, Ueda S, Kato S, Hirabayashi Y, Matsumoto M, Naiki M (1984) Characterization of N-glycolyneuraminic acid-containing glycosphingolipids from a Marek's disease lymphoma-derived chicken cell line, MSB1, as tumor-associated heterophile Hanganutziu-Deicher antigens. J Biochem 95:785–794

Hou JK, Abraham B, El-Serag H (2011) Dietary intake and risk of developing inflammatory bowel disease: a systematic review of the literature. Am J Gastroenterol 106:563–573

Hu H, Luo ML, Desmedt C, Nabavi S, Yadegarynia S, Hong A, Konstantinopoulos PA, Gabrielson E, Hines-Boykin R, Pihan G, Yuan X, Sotirious C, Dittmer DP, Fingeroth JD, Wulf GM (2016) Epstein-Barr Virus infection of mammary epithelial cells promotes malignant transformation. EBioMedicine pii:S2352-3964(16)30209-2. doi:10.1016/j.ebiom.2016.05.025

Huisman MH, Seelen M, van Doormaal PT, de Jong SW, de Vries JH, van der Kooi AJ, de Visser M, Schelhaas HJ, van den Berg LH, Veldink JH (2015) Effect of presymptomatic body mass index and consumption of fat and alcohol on amyotrophic lateral sclerosis. JAMA Neurol 72:1155–1162

Hurt C, Tammaro D (2007) Diagnostic evaluation of mononucleosis-like illnesses. Am J Med 120:911.e1-8

Huxley RR, Ansary-Moghaddam A, Clifton P, Czernichow S, Parr CL, Woodward M (2009) The impact of dietary and lifestyle risk factors on risk of colorectal cancer: a quantitative overview of the epidemiological evidence. Int J Cancer 125:171–180

Inoue S, Sato C, Kitajima K (2010) Extensive enrichment of N-glycolylneuraminic acid in extracellular sialoglycoproteins abundantly synthesized and secreted by human cancer cells. Glycobiology 20:752–762

Ip S, Chung M, Raman G, Chew P, Magula N, DeVine D, Trikalinos T, Lau J (2007) Breastfeeding and maternal and infant health outcomes in developed countries. Evid Rep Technol Assess (Full Rep) 153:1–186

Jans H, Heltberg A, Raun NE (1982) Circulating immune complexes and complement in the course of multiple sclerosis. Acta Neurol Scand 66:488–496

Jantchou P, Morois S, Clavel-Chapelon F, Boutron-Ruault MC, Carbonnel F (2010) Animal protein intake and risk of inflammatory bowel disease: the E3 N prospective study. Am J Gastroenterol 105:2195–2201

Ji J, Sundquist J, Sundquist K (2015) Lactose intolerance and risk of lung, breast and ovarian cancers: aetiological clues from a population-based study in Sweden. Br J Cancer 112:149–152

Jiang W, Ju C, Jiang H, Zhang D (2014) Dairy foods intake and risk of Parkinson's disease: a dose-response meta-analysis of prospective cohort studies. Eur J Epidemiol 29:613–619

Juste RA (2010) Crohn's disease and ruminant farming. Got lactase? Med Hypotheses 75:7–13

Kabat GC, Cross AJ, Park Y, Schatzkin A, Hollenbeck AR, Rohan TE, Sinha R (2009) Meat intake and meat preparation in relation to risk of postmenopausal breast cancer in the NIH-AARP diet and health study. Int J Cancer 124:2430–2435

Kang H, Cha ES, Choi GJ, Lee WJ (2014) Amyotrophic lateral sclerosis and agricultural environments: a systematic review. J Korean Med Sci 29:1610–1617

Kangro HO, Chong SK, Hardiman A, Heath RB, Walker-Smith JA (1990) A prospective study of viral and mycoplasma infections in chronic inflammatory bowel disease. Gastroenterology 98:549–553

Kao TH, Chen S, Huang CW, Chen CJ, Chen BH (2014) Occurrence and exposure to polycyclic aromatic hydrocarbons in kindling-free-charcoal grilled meat products in Taiwan. Food Chem Toxicol 71:149–158

Khan ZM, Liu Y, Neu U, Gilbert M, Ehlers B, Feizi T, Stehle T (2014) Crystallographic and glycan microarray analysis of human polyomavirus 9 VP1 identifies N-glycolyl neuraminic acid as a receptor candidate. J Virol 88:6100–6111

Khuder SA, Mutgi AB, Schaub EA, Tano BD (1999) Meta-analysis of Hodgkin's disease among farmers. Scand J Work Environ Health 25(5):436–441
Knize MG, Salmon CP, Pais P, Felton JS (1999) Food heating and the formation of heterocyclic aromatic amine and polycyclic aromatic hydrocarbon mutagens/carcinogens. Adv Exp Med Biol 459:179–193 Review
Kolonel LN (2001) Fat, meat, and prostate cancer. Epidemiol Rev 23:72–81 Review
Krupovic M, Ghabrial SA, Jiang D, Varsani A (2016) Genomoviridae: a new family of widespread single-stranded DNA viruses. Arch Virol 161:2633–2643
Lamberto I, Gunst K, Müller H, zur Hausen H, de Villiers EM (2014) Mycovirus-like DNA virus sequences from cattle serum, human brain, and serum from multiple sclerosis patients. Genome Announc 2(4):e00848-14
Larsson SC, Orsini N (2014) Red meat and processed meat consumption and all-cause mortality: a meta-analysis. Am J Epidemiol 179:282–289
Larsson SC, Wolk A (2012) Red and processed meat consumption and risk of pancreatic cancer: meta-analysis of prospective studies. Br J Cancer 106:603–607
Le Guern R, Wallet F, Vega E, Courcol RJ, Loiez C (2014) Psychrobacter sanguinis: an unusual bacterium for nosocomial meningitis. J Clin Microbiol 52:3475–3477
Leunda A, Van Vaerenbergh B, Baldo A, Roels S, Herman P (2013) Laboratory activities involving transmissible spongiform encephalopathy causing agents: risk assessment and biosafety recommendations in Belgium. Prion 7:420–433 Review
Li Z, Zheng Z, Ruan J, Li Z, Tzeng CM (2016) Chronic Inflammation links cancer and Parkinson's disease. Front Aging Neurosci 8:126. doi:10.3389/fnagi.2016.00126. eCollection 2016 Review
Limongi D, Baldelli S (2016) Redox imbalance and viral infections in neuro-degenerative diseases. Oxid Med Cell Longev 2016:6547248. doi:10.1155/2016/6547248
Linos E, Willett WC, Cho E, Colditz G, Frazier LA (2008) Red meat consumption during adolescence among premenopausal women and risk of breast cancer. Cancer Epidemiol Biomarkers Prev 17:2146–2151
Lippi G, Mattiuzzi C, Cervellin G (2016) Meat consumption and cancer risk: a critical review of published meta-analyses. Crit Rev Oncol Hematol 97:1–14
Lloyd-Puryear M, Wallace D, Baldwin T, Hollis DG (1991) Meningitis caused by Psychrobacter immobilis in an infant. J Clin Microbiol 29:2014–2042
Löfling J, Lyi SM, Parrish CR, Varki A (2013) Canine and feline parvoviruses preferentially recognize the non-human cell surface sialic acid N-glycolylneuraminic acid. Virology 440:89–96
Lohman-Payne B, Slyker JA, Moore S, Maleche-Obimbo E, Wamalwa DC, Richardson BA, Rowland-Jones S, Mbori-Ngacha D, Farquhar C, Overbaugh J, John-Stewart G (2012) Breast milk cellular HIV-specific interferon γ responses are associated with protection from peripartum HIV transmission. AIDS 26:2007–2016
Luethy D, Habecker P, Murphy B, Nolen-Walston R (2016) Computed tomographic appearance of melanomas in the equine head: 13cases. Vet Radiol Ultrasound 57:246–252
Machado NO, Chopra PJ, Hamdani AA (2010) Crohn's disease of the appendix with enterocutaneous fistula post-appendicectomy: an approach to management. N Am J Med Sci 2:158–161
Małolepsza EB (2008) Modeling of protein misfolding in disease. Methods Mol Biol 443:297–330 Review
Malosse D, Perron H (1993) Correlation analysis between bovine populations, other farm animals, house pets, and multiple sclerosis prevalence. Neuroepidemiology 12:15–27
Malosse D, Perron H, Sasco A, Seigneurin JM (1992) Correlation between milk and dairy product consumption and multiple sclerosis prevalence: a worldwide study. Neuroepidemiology 11:304–312
Malykh YN, Schauer R, Shaw L (2001) N-Glycolylneuraminic acid in human tumours. Biochimie 83:623–634 Review
Manuelidis L (2011) Nuclease resistant circular DNAs copurify with infectivity in scrapie and CJD. J Neurovirol 17:131–145

Manuelidis L (2013) Infectious particles, stress, and induced prion amyloids: a unifying perspective. Virulence 4:373–383
Martin RM, Gunnell D, Owen CG, Smith GD (2005a) Breast-feeding and childhood cancer: a systematic review with meta-analysis. Int J Cancer 117:1020–1031 Review
Martin RM, Middleton N, Gunnell D, Owen CG, Smith GD (2005b) Breast-feeding and cancer: the Boyd Orr cohort and a systematic review with meta-analysis. J Natl Cancer Inst 97:1446–1457
Masaki M, Kano K, Merrick JM, Milgrom F (1981) Studies on Paul-Bunnell (P-B) antigen-antibody system. II. P-B antigens in extracts of lymphoma-leukemia spleens and pathologic sera. J Immunol 127:5–8
Matos E, Brandani A (2002) Review on meat consumption and cancer in South America. Mutat Res 506–507:243–249 Review
Maytsetseg B, Riichiro I (2006) Changes and actual state of Mongolian meat market and distribution system: a case study of Ulaanbaatar City's "Khuchit Shondor" Food Market. International Association of Agricultural Economists, Annual Meeting, August 12–18, 2006 Queensland, Australia. http://purl.umn.edu/25528
Mayuri M, Garg V, Mukherji C, Aggarwal D, Ganguly S (2012) Bovine milk usage and feeding practices for infants in India. Ind J Public Health 56:75–81
McElroy MC, Bassett HF (2010) Mammary carcinoma in a ewe. J Vet Diagn Invest 22:1006–1007
McNally RJ, Parker L (2006) Environmental factors and childhood acute leukemias and lymphomas. Leuk Lymphoma 47:583–598 Review
Meeker S, Seamons A, Maggio-Price L, Paik J (2016) Protective links between vitamin D, inflammatory bowel disease and colon cancer. World J Gastroenterol 22:933–948
Mehta SK, Pierson DL, Cooley H, Dubow R, Lugg D (2000) Epstein-Barr virus reactivation associated with diminished cell-mediated immunity in antarctic expeditioners. J Med Virol 61:235–240
Merlo DF, Rossi L, Pellegrino C, Ceppi M, Cardellino U, Capurro C, Ratto A, Sambucco PL, Sestito V, Tanara G, Bocchini V (2008) Cancer incidence in pet dogs: findings of the Animal Tumor Registry of Genoa, Italy. J Vet Intern Med 22:976–984
Meyer T, Brink U, Unterberg C, Stöhr S, Kreuzer H, Buchwald AB (1994) Expression of meta-vinculin in human coronary arteriosclerosis is related to the histological grade of plaque formation. Atherosclerosis 111:111–119
Micha R, Michas G, Mozaffarian D (2012) Unprocessed red and processed meats and risk of coronary artery disease and type 2 diabetes—an updated review of the evidence. Curr Atheroscler Rep 14:515–524
Michels KB, Trichopoulos D, Rosner BA, Hunter DJ, Colditz GA, Hankinson SE, Speizer FE, Willett WC (2001) Being breastfed in infancy and breast cancer incidence in adult life: results from the two nurses' health studies. Am J Epidemiol 153:275–283
Miller MA, Moore GE, Bertin FR, Kritchevsky JE (2016) What's new in old horses? Postmortem diagnoses in mature and aged equids. Vet Pathol 53:390–398
Mistry N, Abdel-Fahim R, Samaraweera A, Mougin O, Tallantyre E, Tench C, Jaspan T, Morris P, Morgan PS, Evangelou N (2015) Imaging central veins in brain lesions with 3-T T2*-weighted magnetic resonance imaging differentiates multiple sclerosis from microangiopathic brain lesions. Mult Scler 22:1289–1296. doi:10.1177/1352458515616700 Epub 2015
Monti L, Cattaneo TM, Orlandi M, Curadi MC (2015) Capillary electrophoresis of sialylated oligosaccharides in milk from different species. J Chromatogr A 1409:288–291
Munger KL, Chitnis T, Frazier AL, Giovannucci E, Spiegelman D, Ascherio A (2011) Dietary intake of vitamin D during adolescence and risk of multiple sclerosis. J Neurol 258:479–485
Munson L, Moresco A (2007) Comparative pathology of mammary gland cancers in domestic and wild animals. Breast Dis 28:7–21
Ng PS, Böhm R, Hartley-Tassell LE, Steen JA, Wang H, Lukowski SW, Hawthorne PL, Trezise AE, Coloe PJ, Grimmond SM, Haselhorst T, von Itzstein M, Paton AW, Paton JC, Jennings MP (2014) Ferrets exclusively synthesize Neu5Ac and express naturally humanized influenza A virus receptors. Nat Commun 5:5750

Nishimaki T, Kano K, Milgrom F (1979) Paul-Bunnell antigen in malignancies and rheumatoid arthritis. Int Arch Allergy Appl Immunol 60:115–119

Norat T, Bingham S, Ferrari P, Slimani N, Jenab M, Mazuir M, Overvad K, Olsen A, Tjønneland A, Clavel F, Boutron-Ruault MC, Kesse E, Boeing H, Bergmann MM, Nieters A, Linseisen J, Trichopoulou A, Trichopoulos D, Tountas Y, Berrino F, Palli D, Panico S, Tumino R, Vineis P, Bueno-de-Mesquita HB, Peeters PH, Engeset D, Lund E, Skeie G, Ardanaz E, González C, Navarro C, Quirós JR, Sanchez MJ, Berglund G, Mattisson I, Hallmans G, Palmqvist R, Day NE, Khaw KT, Key TJ, San Joaquin M, Hémon B, Saracci R, Kaaks R, Riboli E (2005) Meat, fish, and colorectal cancer risk: the European Prospective Investigation into cancer and nutrition. J Natl Cancer Inst 97:906–916

Ortiz-Alcántara JM, Segura-Candelas JM, Garcés-Ayala F, Gonzalez-Durán E, Rodríguez-Castillo A, Alcántara-Pérez P, Wong-Arámbula C, González-Villa M, León-Ávila G, García-Chéquer AJ, Diaz-Quiñonez JA, Méndez-Tenorio A, Ramírez-González JE (2016) Fatal *Psychrobacter* sp. infection in a pediatric patient with meningitis identified by metagenomic next-generation sequencing in cerebrospinal fluid. Arch Microbiol 198:129–135

Otaegui D, Mostafavi S, Bernard CC, Lopez de Munain A, Mousavi P, Oksenberg JR, Baranzini SE (2007) Increased transcriptional activity of milk-related genes following the active phase of experimental autoimmune encephalomyelitis and multiple sclerosis. J Immunol 179:4074–4082

Owston MA, Ramsay EC, Rotstein DS (2008) Neoplasia in felids at the Knoxville Zoological Gardens, 1979–2003. J Zoo Wildl Med 39:608–613

Pan A, Sun Q, Bernstein AM, Schulze MB, Manson JE, Willett WC, Hu FB (2011) Red meat consumption and risk of type 2 diabetes: 3 cohorts of US adults and an updated meta-analysis. Am J Clin Nutr 94:1088–1096

Pan A, Sun Q, Bernstein AM, Schulze MB, Manson JE, Stampfer MJ, Willett WC, Hu FB (2012) Red meat consumption and mortality: results from 2 prospective cohort studies. Arch Intern Med 172:555–563

Park M, Ross GW, Petrovitch H, White LR, Masaki KH, Nelson JS, Tanner CM, Curb JD, Blanchette PL, Abbott RD (2005) Consumption of milk and calcium in midlife and the future risk of Parkinson disease. Neurology 64:1047–1051

Paul JR, Bunnell WW (1932) The presence of heterophile antibodies in infectious mononucleosis. Am J Med Sci 183:91–104

Pedditizi E, Peters R, Beckett N (2016) The risk of overweight/obesity in mid-life and late life for the development of dementia: a systematic review and meta-analysis of longitudinal studies. Age Ageing 45:14–21

Peterson JA, Patton S, Hamosh M (1998) Glycoproteins of the human milk fat globule in the protection of the breast-fed infant against infections. Biol Neonate 74:143–162 Review

Peterson R, Cheah WY, Grinyer J, Packer N (2013) Glycoconjugates in human milk: protecting infants from disease. Glycobiology 23(12):1425–1438

Phan TG, Mori D, Deng X, Rajindrajith S, Ranawaka U, Fan Ng TF, Bucardo-Rivera F, Orlandi P, Ahmed K, Delwart E (2015) Small circular single stranded DNA viral genomes in unexplained cases of human encephalitis, diarrhea, and in untreated sewage. Virology 482:98–104

Pisacane A, Impagliazzo N, Russo M, Valiani R, Mandarini A, Florio C, Vivo P (1994) Breast feeding and multiple sclerosis. BMJ 308:1411–1412

Price AB, Morson BC (1975) Inflammatory bowel disease: the surgical pathology of Crohn's disease and ulcerative colitis. Hum Pathol 6:7–29 Review

Prusiner SB (1982) Novel proteinaceous infectious particles cause scrapie. Science 216:136–144

Ragnedda G, Leoni S, Parpinel M, Casetta I, Riise T, Myhr KM, Wolfson C, Pugliatti M (2015) Reduced duration of breastfeeding is associated with a higher risk of multiple sclerosis in both Italian and Norwegian adult males: the EnvIMS study. J Neurol 262:1271–1277

Richardson AK, Currie MJ, Robinson BA, Morrin H, Phung Y, Pearson JF, Anderson TP, Potter JD, Walker LC (2015) Cytomegalovirus and Epstein-Barr virus in breast cancer. PLoS One 10(2):e0118989. doi:10.1371/journal.pone.0118989. eCollection 2015 Review

Rizvi I, Choudhury NR, Tuteja N (2015) Insights into the functional characteristics of geminivirus rolling-circle replication initiator protein and its interaction with host factors affecting viral DNA replication. Arch Virol 160:375–387

Rohrmann S, Linseisen J, Nöthlings U, Overvad K, Egeberg R, Tjønneland A, Boutron-Ruault MC, Clavel-Chapelon F, Cottet V, Pala V, Tumino R, Palli D, Panico S, Vineis P, Boeing H, Pischon T, Grote V, Teucher B, Khaw KT, Wareham NJ, Crowe FL, Goufa I, Orfanos P, Trichopoulou A, Jeurnink SM, Siersema PD, Peeters PH, Brustad M, Engeset D, Skeie G, Duell EJ, Amiano P, Barricarte A, Molina-Montes E, Rodríguez L, Tormo MJ, Sund M, Ye W, Lindkvist B, Johansen D, Ferrari P, Jenab M, Slimani N, Ward H, Riboli E, Norat T, Bueno-de-Mesquita HB (2013) Meat and fish consumption and risk of pancreatic cancer: results from the European Prospective Investigation into Cancer and Nutrition. Int J Cancer 132:617–624

Roizman B (1974) Herpesviruses, latency and cancer: a biochemical approach. J Reticuloendothel Soc 15:312–321 Review

Røsjø E, Lossius A, Abdelmagid N, Lindstrøm JC, Kampman MT, Jørgensen L, Sundström P, Olsson T, Steffensen LH, Torkildsen Ø, Holmøy T (2016) Effect of high-dose vitamin D3 supplementation on antibody responses against Epstein-Barr virus in relapsing-remitting multiple sclerosis. Mult Scler pii:1352458516654310

Sadeghian M, Saneei P, Siassi F, Esmaillzadeh A (2016) Vitamin D status in relation to Crohn's disease: meta-analysis of observational studies. Nutrition 32:505–514 Review

Samraj AN, Pearce OM, Läubli H, Crittenden AN, Bergfeld AK, Banda K, Gregg CJ, Bingman AE, Secrest P, Diaz SL, Varki NM, Varki A (2015) A red meat-derived glycan promotes inflammation and cancer progression. Proc Natl Acad Sci U S A 112:542–547

Sepcić J, Mesaros E, Materljan E, Sepić-Grahovac D (1993) Nutritional factors and multiple sclerosis in Gorski Kotar, Croatia. Neuroepidemiology 12:234–240

Sharma RL, Godara R, Thilagar MB (2011) Epizootiology, pathogenesis and immunoprophylactic trends to control tropical bubaline fasciolosis: an overview. J Parasit Dis 35:1–9

Silva MA, Menezes J, Dionne S, Levy E, Amre DK, Seidman EG (2012) Herpes simplex virus-1 infection of colonic explants as a model of viral-induced activation of Crohn's disease. J Crohns Colitis 6:454–463

Simmons MM, Spiropoulos J, Hawkins SA, Bellworthy SJ, Tongue SC (2008) Approaches to investigating transmission of spongiform encephalopathies in domestic animals using BSE as an example. Vet Res 39:34

Simon MW (1998) Anorexia and failure to grow associated with Epstein-Barr virus infection. J Ky Med Assoc 96:13–15

Sipponen T, Turunen U, Lautenschlager I, Nieminen U, Arola J, Halme L (2011) Human herpesvirus 6 and cytomegalovirus in ileocolonic mucosa in inflammatory bowel disease. Scand J Gastroenterol 46:1324–1333

Söderberg-Nauclér C (2006) Does cytomegalovirus play a causative role in the development of various inflammatory diseases and cancer? J Intern Med 259:219–246 Review

Song X, Yu H, Chen X, Lasanajak Y, Tappert MM, Air GM, Tiwari VK, Cao H, Chokhawala HA, Zheng H, Cummings RD, Smith DF (2011) A sialylated glycan microarray reveals novel interactions of modified sialic acids with proteins and viruses. J Biol Chem 286:31610–31622

Song P, Lu M, Yin Q, Wu L, Zhang D, Fu B, Wang B, Zhao Q (2014) Red meat consumption and stomach cancer risk: a meta-analysis. J Cancer Res Clin Oncol 140:979–992

Springer SA, Diaz SL, Gagneux P (2014) Parallel evolution of a self-signal: humans and new world monkeys independently lost the cell surface sugar Neu5Gc. Immunogenetics 66:671–674

Steel O, Kraberger S, Sikorski A, Young LM, Catchpole RJ, Steven AJ, Ladley JJ, Coray DS, Stainton D, Dayaram A, Julian L, van Bysterveldt K, Varsani A (2016) Circular replication-associated protein encoding DNA viruses identified in the faecal matter of various animals in New Zealand. Infect Genet Evol 43:151–164

Stuebe A (2009) The risks of not breastfeeding for mothers and infants. Rev Obstet Gynecol 2:222–231

Sugimura T, Wakabayashi K, Nakagama H, Nagao M (2004) Heterocyclic amines: mutagens/carcinogens produced during cooking of meat and fish. Cancer Sci 95:290–299 Review

Takimori S, Shimaoka H, Furukawa J, Yamashita T, Amano M, Fujitani N, Takegawa Y, Hammarström L, Kacskovics I, Shinohara Y, Nishimura S (2011) Alteration of the N-glycome of bovine milk glycoproteins during early lactation. FEBS J 278:3769–3781

Tan IL, van Schijndel RA, Pouwels PJ, van Walderveen MA, Reichenbach JR, Manoliu RA, Barkhof F (2000) MR venography of multiple sclerosis. AJNR Am J Neuroradiol 21:1039–1042

Tao Q, Wang B, Zheng Y, Jiang X, Pan Z, Ren J (2015) Vitamin D prevents the intestinal fibrosis via induction of vitamin D receptor and inhibition of transforming growth factor-beta1/Smad3 pathway. Dig Dis Sci 60:868–875

Taylor EF, Burley VJ, Greenwood DC, Cade JE (2007) Meat consumption and risk of breast cancer in the UK Woman's Cohort Study. Br J Cancer 96:1139–1146

Taylor RE, Gregg CJ, Padler-Karavani V, Ghaderi D, Yu H, Huang S, Sorensen RU, Chen X, Inostroza J, Nizet V, Varki A (2010) Novel mechanism for the generation of human xeno-autoantibodies against the nonhuman sialic acid N-glycolylneuraminic acid. J Exp Med 207:1637–1646

Thacker EL, Mirzaei F, Ascherio A (2006) Infectious mononucleosis and risk for multiple sclerosis: a meta-analysis. Ann Neurol 59:499–503

Torki M, Gholamrezaei A, Mirbagher L, Danesh M, Kheiri S, Emami MH (2015) Vitamin D deficiency associated with disease activity in patients with inflammatory bowel diseases. Dig Dis Sci 60:3085–3091

Urashima T, Odaka G, Asakuma S, Uemura Y, Goto K, Senda A, Saito T, Fukuda K, Messer M, Oftedal OT (2009) Chemical characterization of oligosaccharides in chimpanzee, bonobo, gorilla, orangutan, and siamang milk or colostrum. Glycobiology 19:499–508

Weichert S, Koromyslova A, Singh BK, Hansman S, Jennewein S, Schroten H, Hansman GS (2016) Structural basis for norovirus inhibition by human milk oligosaccharides. J Virol 90:4843–4848

Wergeland S, Myhr KM, Løken-Amsrud KI, Beiske AG, Bjerve KS, Hovdal H, Midgard R, Kvistad SS, Holmøy T, Riise T, Torkildsen Ø (2016) Vitamin D, HLA-DRB1 and Epstein-Barr virus antibody levels in a prospective cohort of multiple sclerosis patients. Eur J Neurol 23:1064–1070

Whitley C, Gunst K, Müller H, Funk M, zur Hausen H, de Villiers EM (2014) Novel replication-competent circular DNA molecules from healthy cattle serum, milk, and multiple sclerosis-affected human brain tissue. Genome Announc 2(4):e00849-14

Winer S, Astsaturov I, Cheung RK, Schrade K, Gunaratnam L, Wood DD, Moscarello MA, O'Connor P, McKerlie C, Becker DJ, Dosch HM (2001) T cells of multiple sclerosis patients target a common environmental peptide that causes encephalitis in mice. J Immunol 166:4751–4756

Wise LA, Titus-Ernstoff L, Newcomb PA, Trentham-Dietz A, Trichopoulos D, Hampton JM, Egan KM (2009) Exposure to breast milk in infancy and risk of breast cancer. Cancer Causes Control 20:1083–1090

Womack J, Jimenez M (2015) Common questions about infectious mononucleosis. Am Fam Physician 91:372–376

Xue XJ, Gao Q, Qiao JH, Zhang J, Xu CP, Liu J (2014) Red and processed meat consumption and the risk of lung cancer: a dose-response meta-analysis of 33 published studies. Int J Clin Exp Med 7:1542–1553

Zerr P, Vollath S, Palumbo-Zerr K, Tomcik M, Huang J, Distler A, Beyer C, Dees C, Gela K, Distler O, Schett G, Distler JH (2015) Vitamin D receptor regulates TGF-β signalling in systemic sclerosis. Ann Rheum Dis 74(3):e20. doi:10.1136/annrheumdis-2013-204378. Epub 2014 Jan 21

Zhang Y (2013) Epidemiology of esophageal cancer. World J Gastroenterol 19:5598–5606 Review

Zhang L, de Waard M, Verheijen H, Boeren S, Hageman JA, van Hooijdonk T, Vervoort J, van Goudoever JB, Hettinga K (2016) Changes over lactation in breast milk serum proteins involved in the maturation of immune and digestive system of the infant. J Proteomics pii: S1874-391, 9(16) 30028-8

Zhang W, Li L, Deng X, Blümel J, Nübling CM, Hunfeld A, Baylis SA, Delwart E (2016b) Viral nucleic acids in human plasma pools. Transfusion 56:2248–2255

zur Hausen H (2001) Proliferation-inducing viruses in non-permissive systems as possible causes of human cancers. Lancet 357:381–384

zur Hausen H (2009) The search for infectious agents of human cancers: where and why. Nobel lecture. Virology 392:1–10

zur Hausen H (2012) Red meat consumption and cancer: reasons to suspect involvement of bovine infectious factors in colorectal cancer. Int J Cancer 130:2475–2483

zur Hausen H (2015) What do breast and CRC cancers and MS have in common? Nat Rev Clin Oncol 12:569–570

zur Hausen H, de Villiers EM (2014a) Cancer "causation" by infections—individual contributions and synergistic networks. Sem Oncol 41:868–883

zur Hausen H, de Villiers EM (2014b) Prenatal infections with subsequent immune tolerance could explain the epidemiology of common childhood cancers. World Cancer Report IARC Lyon, pp 261–265

zur Hausen H, de Villiers EM (2015) Dairy cattle serum and milk factors contributing to the risk of colon and breast cancers. Int J Cancer 137:959–967

zur Hausen H, de Villiers EM (2016) Risikofaktoren für Krebs und MS in Kuhmilch und Rindfleisch? Biol Unserer Zeit 46:26–31. doi:10.1002/bluz.201610583

Zwart SR, Mehta SK, Ploutz-Snyder R, Bourbeau Y, Locke JP, Pierson DL, Smith SM (2011) Response to vitamin D supplementation during antarctic winter is related to BMI, and supplementation can mitigate Epstein-Barr virus reactivation. J Nutr 141:692–697

MYC and RAF: Key Effectors in Cellular Signaling and Major Drivers in Human Cancer

Eduard Stefan and Klaus Bister

Abstract The prototypes of the human *MYC* and *RAF* gene families are orthologs of animal proto-oncogenes that were originally identified as transduced alleles in the genomes of highly oncogenic retroviruses. *MYC* and *RAF* genes are now established as key regulatory elements in normal cellular physiology, but also as major cancer driver genes. Although the predominantly nuclear MYC proteins and the cytoplasmic RAF proteins have different biochemical functions, they are functionally linked in pivotal signaling cascades and circuits. The MYC protein is a transcription factor and together with its dimerization partner MAX holds a central position in a regulatory network of bHLH-LZ proteins. MYC regulates transcription conducted by all RNA polymerases and controls virtually the entire transcriptome. Fundamental cellular processes including distinct catabolic and anabolic branches of metabolism, cell cycle regulation, cell growth and proliferation, differentiation, stem cell regulation, and apoptosis are under MYC control. Deregulation of *MYC* expression by rearrangement or amplification of the *MYC* locus or by defects in kinase-mediated upstream signaling, accompanied by loss of apoptotic checkpoints, leads to tumorigenesis and is a hallmark of most human cancers. The critically controlled serine/threonine RAF kinases are central nodes of the cytoplasmic MAPK signaling cascade transducing converted extracellular signals to the nucleus for reshaping transcription factor controlled gene expression profiles. Specific mutations of RAF kinases, such as the prevalent BRAF(V600E) mutation in melanoma, or defects in upstream signaling or feedback loops cause decoupled kinase activities which lead to tumorigenesis. Different strategies for pharmacological interference with MYC- or RAF-induced tumorigenesis are being developed and several RAF kinase inhibitors are already in clinical use.

Dedication: This article is dedicated to Peter K. Vogt. His pioneering work has profound and sustained impact on the fields of virology and cancer genetics.

E. Stefan · K. Bister (✉)
Institute of Biochemistry and Center for Molecular Biosciences Innsbruck,
University of Innsbruck, Innrain 80–82, 6020 Innsbruck, Austria
e-mail: klaus.bister@uibk.ac.at

Current Topics in Microbiology and Immunology (2017) 407:117–152
DOI 10.1007/82_2017_4

Published Online: 03 May 2017

Contents

1 Introduction

The original discoveries of several important cancer driver genes, including *MYC* and *RAF*, are closely linked to animal retrovirology (Vogt 2012). The first biochemical identification of a cancer gene (Duesberg and Vogt 1970)—the v-*src* oncogene of Rous sarcoma virus—and the striking proof of its origin from a normal cellular gene (c-*src* proto-oncogene) (Stehelin et al. 1976) were landmark discoveries in molecular cancer research (Bister 2015). They immediately stimulated the search for the transforming principles of other highly oncogenic retroviruses (Vogt 2012; Bister 2015). A specific class of retroviruses, the avian acute leukemia viruses, proved to be a rich source of novel cancer genes, including *myc*, *mil* (*raf*), *erb*B, *erb*A, *myb*, and *ets* (Bister and Jansen 1986). In 1977, specific protein products and nucleic acid sequences of the *myc* oncogene were discovered by biochemical analyses of acute leukemia virus MC29 (Bister et al. 1977; Duesberg et al. 1977). The viral genome was shown to be defective in all replicative genes (*gag*, *pol*, and *env*), to contain a contiguous novel insert unrelated to *src* and later termed v-*myc* (Coffin et al. 1981), and to encode a single protein product, a Gag-Myc hybrid protein (Fig. 1) (Bister et al. 1977; Duesberg et al. 1977; Bister and Vogt 1978; Mellon et al. 1978; Sheiness et al. 1978). Subsequently, v-*myc* alleles were identified in all other members of the MC29 subgroup of acute leukemia viruses, CMII, OK10, and MH2 (Bister and Jansen 1986). Following the *src* paradigm, the cellular origin of the v-*myc* alleles was soon proven and the chicken c-*myc* gene identified (Sheiness and Bishop 1979; Roussel et al. 1979; Robins et al. 1982; Vennström et al. 1982). The c-*myc* gene has been conserved throughout metazoan evolution and may even have pre-metazoan ancestors (Hartl et al. 2010; Young et al. 2011). The discovery of chromosomal translocations of the human *MYC* gene in Burkitt lymphoma cells provided the first evidence for the involvement of the cellular homolog of the

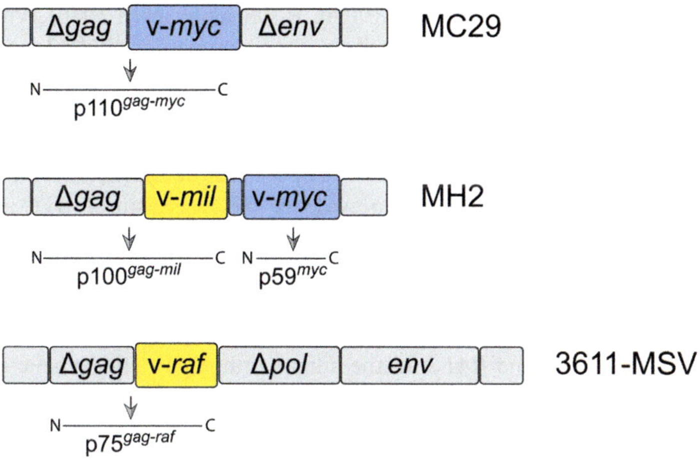

Fig. 1 Discoveries of *MYC* and *RAF*. The genome structures of avian acute leukemia viruses MC29 and MH2, and of murine sarcoma virus 3611 are depicted. The retroviral oncogenes v-*myc*, v-*mil*, and v-*raf* are transduced alleles derived from the chicken c-*myc* and c-*mil* genes, or the mouse c-*raf* gene, respectively. Chicken c-*mil* and mouse c-*raf* are orthologous genes. In the MH2 genome, v-*mil* and v-*myc* alleles are directly adjacent, and their coding regions are separated by c-*myc* intron derived sequences (*small blue box*). The protein products (p) of the retroviral oncogenes are expressed from genome-sized mRNAs as hybrid proteins containing N-terminal Gag sequences, or from a subgenomic mRNA (MH2 v-*myc*)

v-*myc* retroviral oncogene in human tumorigenesis (Dalla-Favera et al. 1982; Taub et al. 1982). Today, deregulated *MYC* expression is established as an important driving force in the majority of all human cancers (Vogt 2012; Dang 2012; Vogelstein et al. 2013; Stine et al. 2015; Tokheim et al. 2016).

In 1983, biochemical analyses of avian acute leukemia virus MH2 (Mill Hill virus 2) revealed that the viral genome—in addition to v-*myc*—contains a second unique insert termed v-*mil* (Fig. 1) (Jansen et al. 1983a, b; Coll et al. 1983; Kan et al. 1983). The cellular origin of v-*mil* was demonstrated by detailed structural comparisons of cloned MH2 v-*mil* and chicken c-*mil* genes (Jansen et al. 1983b). In the same year, a cell-derived insert (termed v-*raf*) was found in the genome of murine sarcoma virus 3611 (Fig. 1) isolated from mice that had been inoculated with a virus stock obtained by 5-iodo-2′-deoxyuridine induction in chemically transformed mouse cells (Rapp et al. 1983). It was subsequently shown that v-*mil* and v-*raf* were derived from orthologous genes in chicken and mice (Jansen et al. 1984). MH2 induces leukemia and carcinoma in infected fowl and is one of the oldest natural retroviral isolates (Begg 1927). It is an intriguing genetic curiosity that a highly oncogenic chicken virus arose long time ago by transduction of two cellular genes in tandem, both of which are recognized today as key drivers in human cancer. Indeed, elucidation of the genetic structure of MH2 provided a direct hint at possible oncogene cooperativity (Jansen et al. 1983b; Bister and Jansen 1986), which was supported by the observation of enhanced oncogenic properties of MH2 *in vitro* and *in vivo* (Graf et al. 1986) or by construction of a murine MH2 analog with a remarkably strong oncogenic potential in mice (Rapp et al. 1985). Today, mutated forms of the human ortholog of *mil/raf* and its paralogs are involved in a variety of human cancers (Wellbrock et al. 2004; Vogt 2012; Holderfield et al. 2014a; Tokheim et al. 2016), and there is also evidence for synergism of *MYC* and *RAF* genes in tumorigenesis (Tabor et al. 2014; Ratnikov et al. 2017).

In mammalian systems, including man, the gene designation *raf* coined for the murine oncogene is established and will be used henceforth in this article on human genes. Genes and proteins will be designated according to human nomenclature conventions, e.g., *MYC* and *RAF1* for the human orthologs of the avian c-*myc* and avian/murine c-*mil/raf* genes, respectively, and MYC and RAF1 for their protein products. Paralogous members of the human *MYC* gene family are *MYCN* and *MYCL*, encoding the N-MYC and L-MYC proteins. *ARAF* and *BRAF* are paralogs of *RAF1*, and specify the ARAF and BRAF proteins. Also, *MYC* or MYC and *RAF* or RAF will be used as generic terms when no specific member of the gene or protein families is addressed.

2 Protein Structure

2.1 MYC Proteins

The domain structure of the human MYC protein with a total length of 439 amino acids is shown in Fig. 2. MYC belongs to a family of proteins with a characteristic hallmark, a dimerization and DNA-binding domain (bHLH-LZ) encompassing a basic region (b) as DNA contact surface, a helix-loop-helix (HLH), and a leucine repeat/zipper (LZ) region as protein–protein interaction (PPI) domains. The preferred binding partner for MYC is another member of the bHLH-LZ protein family, MAX, and PPI between these proteins leads to formation of a stable MYC:MAX heterodimer. The discovery of MAX and the recognition of MYC:MAX as a sequence-specific DNA-binding complex enabled crucial leaps forward in the understanding of MYC biochemistry (Blackwood and Eisenman 1991; Eisenman 2001; Conacci-Sorrell et al. 2014). In the absence of MAX and at physiological concentrations, MYC is monomeric in solution and displays properties of an intrinsically disordered protein (IDP) with isolated regions of dynamic secondary structure elements and helical fraying (Fieber et al. 2001). MAX forms homodimers, albeit with lower stability than that of the MYC:MAX heterodimer. X-ray structures of the bHLH-LZ domains in MAX homodimers (Ferré-D'Amaré et al. 1993) or in MYC:MAX heterodimers (Nair and Burley 2003) revealed the structural details of the dimer-specific selective PPIs between the parallel protein chains

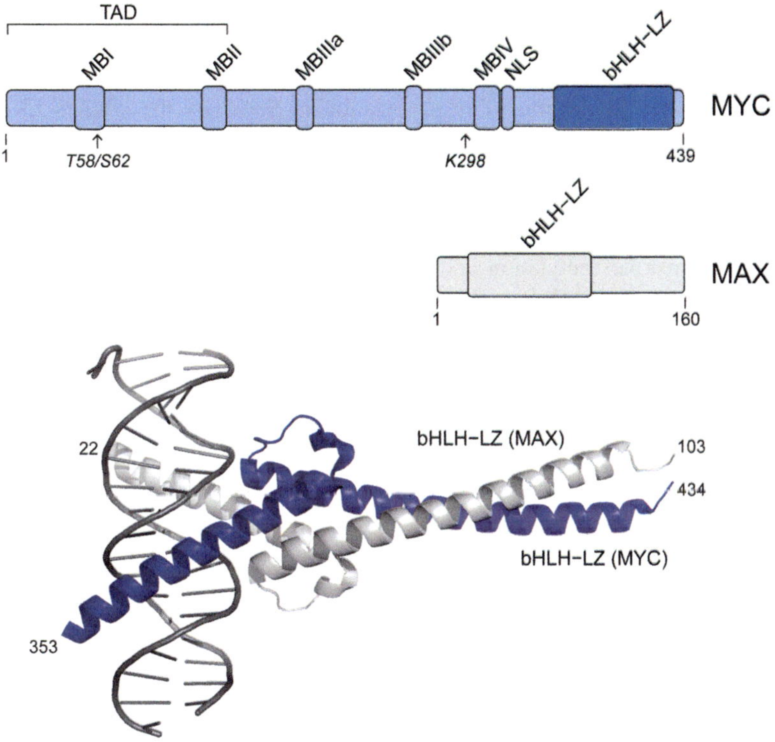

Fig. 2 Structure of the human MYC protein and its dimerization partner MAX. The dimerization and DNA-binding domains (bHLH-LZ) are indicated. On the MYC protein, conserved MYC boxes (MBI-IV), the transactivation domain (TAD), the nuclear localization signal (NLS), critical phosphorylation sites (Thr58, Ser62), and a calpain cleavage site (Lys298) are depicted. The X-ray structure of a dimer of the MYC and MAX bHLH-LZ domains bound to DNA is shown below (Nair and Burley 2003). The image was created from the PDB entry 1NKP using the PyMOL graphics system

(Fig. 2). The α-helical basic regions of MYC:MAX heterodimers bind to specific DNA sequence elements (E-boxes) with the preferred structure 5′-CACGTG-3′ by making specific base contacts in the major groove of DNA (Fig. 2) (Blackwood and Eisenman 1991; Nair and Burley 2003). MYC:MAX complexes induce transcription by binding to E-boxes in promoter or enhancer regions of target genes and form PPIs with a variety of other factors.

In addition to the distinctive bHLH-LZ domain and a proximate nuclear localization signal, MYC contains several regions (MYC boxes, MBI-IV) that are highly conserved among members of the MYC protein family (Fig. 2) (Conacci-Sorrell et al. 2014; Thomas et al. 2015). MBI and MBII are located within the transactivation (transcriptional activation) domain (TAD) and their precise biochemical functions are best defined. MBI contains a phospho-degron regulating ubiquitylation and proteolysis of the distinctively unstable MYC protein exhibiting half-lives of only 20–30 min. Phosphorylation at Ser62 by the kinases ERK or CDK increases MYC stability, whereas phosphorylation at Thr58 by GSK-3β initiates dephosphorylation at Ser62, ubiquitylation by Fbw7 E3 ligase, and proteasomal degradation (Farell and Sears 2014). MBII interacts with components of histone acetyltransferase (HAT) complexes including TRRAP and other cofactors to stimulate histone acetylation and gene activation. MBII is essential for most biological activities of MYC, and is also an interaction site for the E3 ligase Skp2 (Conacci-Sorrell et al. 2014). In contrast to MBI and MBII, the molecular functions of the other MYC boxes are not completely explored (Conacci-Sorrell et al. 2014; Thomas et al. 2015). However, MBIIIa, also named just MBIII (Conacci-Sorrell et al. 2014), interacts with histone deacetylase 3 (HDAC3) leading to transcriptional repression, the chromatin association of MYC depends in part on the interaction of MBIIIb with the WD40-repeat protein WDR5, and MBIV associates with the conserved transcriptional cofactor HCF–1 (Thomas et al. 2015, 2016). Specific cleavage of MYC at Lys298 by calcium-dependent calpain proteases yields MYC-nick, a cytoplasmic form of MYC that retains MYC boxes I through IIIb and is involved in α-tubulin acetylation and cell differentiation (Conacci-Sorrell et al. 2010; Anderson et al. 2016). Paralogs of *MYC* are *MYCN* and *MYCL* that were originally identified as amplified genes in human neuroblastoma or small cell lung cancer, respectively (Brodeur et al. 1984; Nau et al. 1985). The human N-MYC (464 amino acids) and L-MYC (364 amino acids) proteins share the principal domain topography with MYC (Fig. 2), with the exception that MBIIIa is missing from the shorter L-MYC protein (Conacci-Sorrell et al. 2014; Thomas et al. 2015). Notably, the transduced v-*myc* alleles of avian acute leukemia viruses contain the entire coding region of the chicken c-*myc* proto-oncogene, and the Myc amino acid sequences of the viral Gag-Myc or v-Myc protein products (cf. Fig. 1) differ from the 416-amino acid chicken c-Myc protein sequence only by individual amino acid substitutions and, in the case of MH2, a small internal deletion (Bister and Jansen 1986).

2.2 *RAF Proteins*

The domain structure of the human RAF1 protein (648 amino acids) is depicted in Fig. 3. The X-ray structure of the RAS binding domain (RBD) of RAF1 in complex with the G domain of HRAS, a member of the RAS protein family of small GTPases (Karnoub and Weinberg 2008; McCormick 2016), is shown below (Fetics et al. 2015). RAF proteins are cytoplasmic serine/threonine-specific protein kinases that share three conserved regions (CR). CR1 is located in the N-terminal auto-inhibitory domain of RAF and comprises the RBD and the cysteine-rich domain (CRD), which are both involved in a GTP-dependent interaction with RAS.

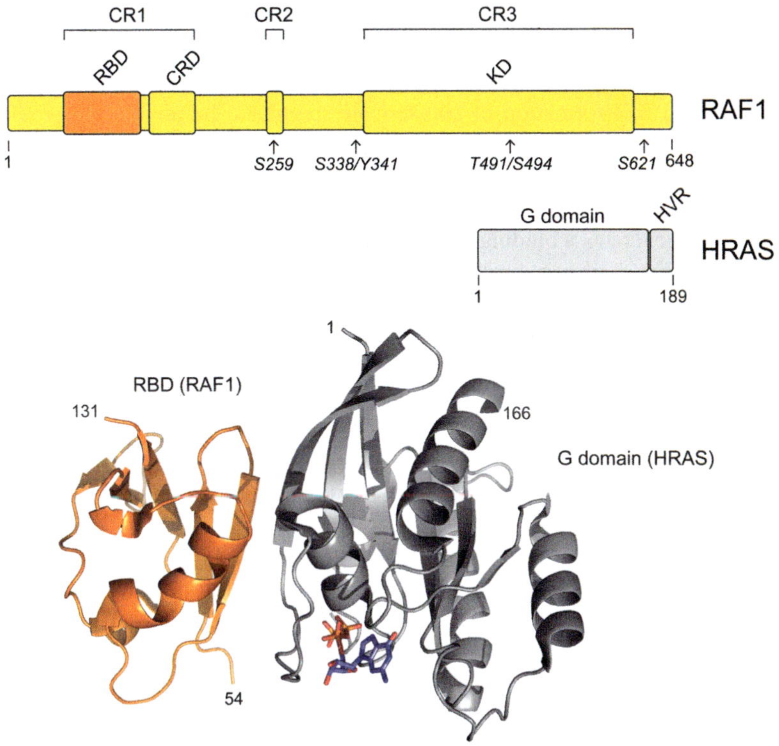

Fig. 3 Structure of the human RAF1 and HRAS proteins. The conserved regions (CR1-3), the RAS binding domain (RBD), the cysteine-rich domain (CRD), and the kinase domain (KD) of RAF1 are indicated. Critical phosphorylation sites in CR2 (Ser259), in the C-terminal region (Ser621), in the N-region upstream of CR3 (Ser338, Tyr341), and in the activation segment (Thr491, Ser494) are marked. On the HRAS protein, the G domain and the hypervariable region (HVR) are depicted. The X-ray structure of the RAF1 RBD bound to the G domain of HRAS loaded with the GTP analog GppNHp is shown below (Fetics et al. 2015). The image was created from the PDB entry 4G0N using the PyMOL graphics system

The central serine/threonine-rich CR2 is essential for phosphorylation-dependent regulation of RAF, and the C-terminal CR3 encompasses the catalytic kinase domain (Wellbrock et al. 2004; Baccarini 2005; Holderfield et al. 2014a; Lavoie and Therrien 2015; Desideri et al. 2015). Notably, the transduced v-*mil* and v-*raf* alleles in MH2 and 3611-MSV contain only the 3′ segment of the coding domains of c-*mil* and c-*raf*, respectively, and hence the Gag-Mil and Gag-Raf hybrid proteins lack the auto-inhibitory N-terminal domain (cf. Fig. 1). RAF proteins are essential effectors within the mitogen-activated protein kinase (MAPK) pathway and phosphorylate MEK (Dent et al. 1992; Howe et al. 1992; Kyriakis et al. 1992) which then activates ERK signaling (Lavoie and Therrien 2015; Desideri et al.

2015). RAF kinases are activated and relieved from intramolecular auto-inhibition by (i) interaction with GTP-loaded RAS proteins (Moodie et al. 1993; Van Aelst et al. 1993; Vojtek et al. 1993; Zhang et al. 1993), (ii) recruitment to the plasma membrane, and (iii) formation of allosterically regulated homo- and heterodimers (Wellbrock et al. 2004; Baccarini 2005; Lavoie and Therrien 2015; Desideri et al. 2015). These activating interactions are regulated by several phosphorylation and dephosphorylation events at crucial residues (Fig. 3). Phosphorylation of Ser259 in CR2 by PKA generates a binding site for 14-3-3 scaffold proteins and this interferes with RAS binding and membrane recruitment. For RAF activation, Ser259 has to be dephosphorylated by protein phosphatase 2A (Abraham et al. 2000; Desideri et al. 2015). In contrast, several other phosphorylation sites have positive effects on RAF activation. Phosphorylation of the C-terminal 14-3-3 binding site Ser621 facilitates RAF dimerization (Lavoie and Therrien 2015; Desideri et al. 2015). Other positive phosphorylation events involve Ser338 and Tyr341 in the N-region (*Negative* charge required for RAF activation) adjacent to CR3, and Thr491 and Ser494 within the activation segment of the kinase domain (Fig. 3) (Wellbrock et al. 2004; Lavoie and Therrien 2015).

The paralogs of *RAF1* are *ARAF* isolated by human and mouse cDNA screening (Huebner et al. 1986) and *BRAF* that was originally identified in human Ewing sarcoma DNA (Ikawa et al. 1988) and at the same time as a transduced *mil*-related oncogene (v-R*mil*) in avian retrovirus IC10 (Marx et al. 1988). The human ARAF (606 amino acids) and BRAF (766 amino acids) proteins exhibit the principal domain architecture of RAF1 (Fig. 3), with BRAF containing an additional N-terminal segment (Wellbrock et al. 2004). Phylogenetic sequence comparisons revealed that the *BRAF* gene is the human paralog that is most similar to the single *Raf* genes in non-vertebrate organisms and that BRAF is probably the prototypic RAF kinase (Desideri et al. 2015). The regulation of human BRAF shares common features with the other paralogs, but also displays important distinctions. The N-region of BRAF is negatively charged independent of regulatory events, with Asp449 corresponding to the position of Tyr341 in RAF1 and with Ser446 constitutively phosphorylated unlike the equivalent residue Ser338 in RAF1. While the N-regions of RAF1 and ARAF have to be negatively charged for activation, the BRAF N-region carries a constant negative charge requiring fewer steps in the activation of BRAF by RAS (Wellbrock et al. 2004; Baccarini 2005).

3 Cellular Signaling

Mitogenic signaling pathways connect membrane localized receptor activities with cytoplasmic and nuclear effector systems, as shown in Fig. 4 for the MAPK pathway. MYC and RAF proteins occupy apex positions in such signal transduction cascades. Remarkably, several protein effectors in these pathways are encoded by

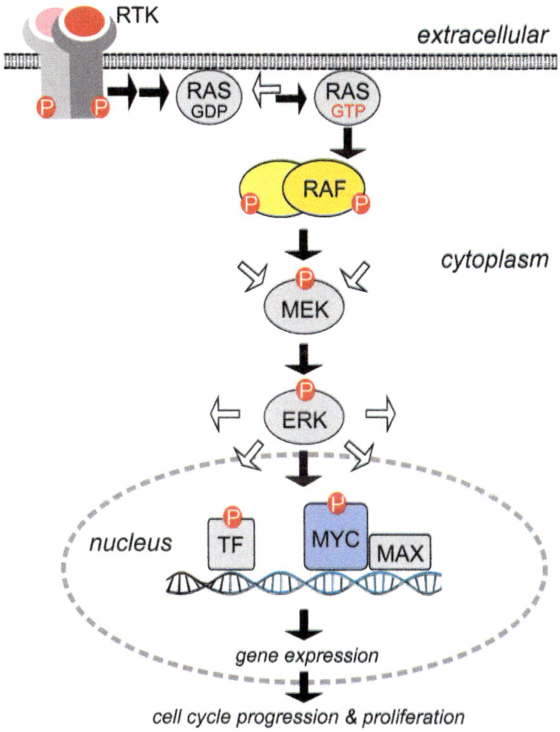

Fig. 4 Schematic diagram of the mitogen-activated protein kinase (MAPK) signaling pathway. Exposure to extracellular ligands leads to dimerization, autophosphorylation, and activation of receptor tyrosine kinases (RTK). This triggers recruitment of adaptor proteins and nucleotide exchange factors (not shown) that enhance GDP/GTP exchange leading to RAS activation (RAS-GTP). RAF, the first MAPK of the cascade, is activated through a multistep process including RAS binding, recruitment to the plasma membrane, conformational change, phosphorylation (P), and dimerization. Activated RAF phosphorylates and activates the gatekeeper kinase MEK which in turn phosphorylates ERK. Activated ERK directly or indirectly regulates a plethora of cytoplasmic and nuclear substrates including transcription factors (TF), such as transcriptional regulators of *MYC* expression. In addition, ERK directly phosphorylates the transcription factor MYC which enhances its stability. TF-mediated alterations of gene expression profiles are relevant for cell cycle progression and proliferation. The different modes of negative feedback regulation of RAS-RAF-MEK-ERK signaling are not shown

proto-oncogenes originally identified in transduced form as transforming principles in retroviral genomes. This includes the EGF receptor (EGFR), a receptor tyrosine kinase (RTK), and the GTPase RAS (Vogt 2012; Bister 2015), in addition to the cytoplasmic serine/threonine-specific protein kinase RAF and the transcription factor MYC (cf. Fig. 1). The most important regulatory functions of MYC will be addressed here, followed by an outline of the upstream cytoplasmic signaling involving RAF kinases.

3.1 The MYC Master Regulator

The broad physiological functions of the master regulator MYC reach to virtually all important cellular activities and compartments: transcriptional activation or repression of target genes, general transcriptional amplification, control of the entire non-coding transcriptome including microRNAs (miRNAs) and long non-coding RNAs (lncRNAs), DNA replication, energy metabolism, cell cycle progression, cell proliferation, apoptosis, differentiation, cell migration, and stem cell biology (Soucek and Evan 2010; Conacci-Sorrell et al. 2014; Stine et al. 2015). This huge and still expanding regulatory network of MYC represents an enormous challenge for exhaustive functional investigations. The canonical function of MYC is that of a transcriptional regulator in a dimeric DNA-binding complex with MAX (cf. Fig. 2). Indeed, most physiological and tumorigenic functions of MYC require dimerization with MAX, although MAX-independent and even cytoplasmic functions of MYC have also been reported. MYC:MAX dimers generally stimulate transcription, but a significant number of transcriptionally repressed target genes have also been identified (Eilers and Eisenman 2008; Conacci-Sorrell et al. 2014). A specific mechanism of transcriptional repression involves association with the zinc-finger transcription factor Miz–1 (Wanzel et al. 2003). The number of transcriptionally activated target genes is huge, and it has been estimated that over 15% of all human genomic loci are bound and possibly regulated by MYC (Conacci-Sorrell et al. 2014). In a global mapping analysis, more than 4000 MYC binding sites were identified in human B cells, and by gene expression profiling, more than 660 direct MYC-regulated target genes were identified (Zeller et al. 2006). A large number of the upregulated genes are involved in processes related to cell growth and proliferation, such as ribosomal and mitochondrial biogenesis, translation, biosynthetic pathways, glucose and glutamine metabolism, and cell cycle progression (Dang 2012, 2013; Stine et al. 2015). In agreement, MYC transcriptional regulation involves not only protein-coding or small RNA specifying genes transcribed by RNA polymerase II, but also genes transcribed by RNA polymerases I and III (Gomez-Roman et al. 2003; Grandori et al. 2005). In contrast to the common view of a gene-specific MYC transcriptional signature, it was proposed that MYC functions non-specifically as a global nonlinear amplifier of gene transcription, mainly enhancing the ongoing gene expression program in the cell by stimulating transcriptional elongation (Lin et al. 2012; Nie et al. 2012; Rahl and Young 2014). It has been argued that this *general amplification* model would not explain specific transcriptional repression by MYC or the MYC-induced specific changes in gene expression profiles in the transitions from resting to proliferating cells or from normal to transformed cells (Dang 2013; Stine et al. 2015). In the *selective amplification* model, the specific selection of MYC targets is largely determined by chromatin accessibility and interaction with other transcription factors to activate or repress gene transcription specifically (Dang 2014; Stine et al. 2015). However, the two models are not mutually exclusive, and MYC may function along both routes, depending on cell and chromatin status (Dang 2014; Wolf et al. 2015). Also, the net outcome of MYC-induced general amplification, including indirect effects on gene expression, can in the end still generate unique MYC transcriptional signatures (Rahl and Young 2014). It is important to note that the MYC:MAX interaction is inseparably embedded in a large network of selectively interacting bHLH-LZ proteins (Conacci-Sorrell et al. 2014; Wilde and Ayer 2015). MYC does not homodimerize under physiological conditions and forms heterodimers exclusively with MAX. In contrast, MAX forms homodimers and also heterodimers with other bHLH-LZ proteins, such as the MXD (formerly MAD) family of proteins (MXD1–4) or MNT and MGA (Ayer et al. 1993; Hurlin et al. 1997, 1999; Conacci-Sorrell et al. 2014). Furthermore, MXD proteins and MNT form heterodimers with MLX, which in turn can undergo complex formation with transcriptional activators of the Mondo protein family, such as MondoA or ChREBP (Conacci-Sorrell et al. 2014; Wilde and Ayer 2015). This extensive network of interacting bHLH-LZ proteins facilitates broad adjustments of transcriptional programs in response to mitogenic or metabolic signaling. MYC and the antagonistic MXD proteins compete for heterodimerization with MAX, and MYC:MAX and MXD:MAX compete for E-box binding sites. MYC:MAX heterodimers generally induce transcription, whereas MXD:MAX complexes repress transcription. Transcriptional repression by MXD:MAX involves recruitment of the mSIN3 and histone deacetylase corepressor complex (Conacci-Sorrell et al. 2014; Wilde and Ayer 2015), in contrast to transcriptional activation by MYC:MAX that involves recruitment of TRRAP and histone acetyltransferase GCN5 (McMahon et al. 2000). Heterodimer formation by the various members of the MYC/MAX/MXD network is dependent on the relative abundance of the bHLH-LZ proteins and also on the specific binding affinities. For instance, MYC:MAX heterodimers have very low dissociation constants and are significantly more stable than MAX:MAX homodimers (Fieber et al. 2001; Nair and Burley 2003). The relative abundance of MYC proteins is tightly controlled at all levels of gene expression, including transcription, translation, and protein degradation (Sears 2004; Liu and Levens 2006; Farell and Sears 2014). Transcriptional activation of *MYC* is strongly dependent on the cell cycle and differentiation status of the cell. While *MYC* expression is nearly shut off in quiescent cells, it is rapidly induced by mitogenic signals transduced by pathways like RTK-MAPK (Fig. 4) or WNT/β-catenin (Kelly et al. 1983; He et al. 1998; Liu and Levens 2006; Soucek and Evan 2010). The nuclear effectors of these pathways are members of the E-twenty-six (ETS) or T cell factor/lymphoid enhancer binding factor (TCF/Lef) families of transcription factors and regulate *MYC* expression, among many other factors (Wasylyk et al. 1998; Yochum et al. 2010). Transcriptional regulation of *MYC* is followed by a multitude of post-transcriptional and post-translational control elements including miRNAs, lncRNAs, protein modification and degradation (Lal et al. 2009; Hung et al. 2014; Tseng et al. 2014; Farell and Sears 2014; Stine et al. 2015).

Some of the earliest transcriptional targets of MYC to be identified encode metabolic enzymes, such as lactate dehydrogenase (LDHA) involved in the glycolytic pathway (Shim et al. 1997; Stine et al. 2015). Today, MYC is established as a master regulator of genes involved in ribosomal and mitochondrial biogenesis, glucose and glutamine metabolism, protein synthesis, nucleotide and lipid biosynthesis (Dang 2013; Stine et al. 2015). MYC-induced metabolism in normal cells is under control by mitogenic signaling and nutrient supply, whereas in cancer cells, deregulated MYC expression drives metabolic genes constitutively and leads to nutrient addiction, such as glucose and glutamine dependency (Dang 2013; Shroff et al. 2015; Stine et al. 2015; Altman et al. 2016). MYC directly regulates specific genes encoding essential factors of metabolic pathways, including glycolytic and glutaminolytic enzymes and transporters. The recognition of MYC as a master regulator of energy metabolism also links current molecular cancer research to one of the first observations of specific metabolic features of cancer cells, the increased conversion of glucose to lactate even in non-hypoxic conditions, a phenomenon known as aerobic glycolysis or Warburg effect (Warburg et al. 1924; Koppenol et al. 2011). Under hypoxic stress, normal cells activate expression of hypoxia inducible factors (HIFs) that induce genes for adaptation to hypoxia, including nearly all glycolytic genes. There are extensive interactions between the MYC/MAX/MXD network and HIFs, with antagonistic mechanisms in the response of normal cells to hypoxia, in contrast to cooperative functions in hypoxic cancer cells (Dang et al. 2008). In addition to the profound effect of MYC on the regulation of intermediary and energy metabolism, MYC also directly induces cell cycle progression by transcriptional activation of genes such as CDK4 or cyclin D2 (Hermeking et al. 2000; Bouchard et al. 2001). A non-transcriptional function of MYC involves direct interaction with the pre-replicative complex (pre-RC) controlling the initiation of DNA replication (Dominguez-Sola et al. 2007). In addition, the *CDT1* gene encoding a key component of the pre-RC, the origin licensing factor Cdt1, is transcriptionally activated by MYC (Valovka et al. 2013). Hence, DNA replication is both under transcriptional and non-transcriptional control by Myc. As a huge expansion of its regulatory sphere, MYC also controls virtually the entire non-coding transcriptome (Weinberg et al. 2015). MYC regulates the expression of miRNAs, short RNA molecules that regulate the half-life or translational efficiency of target mRNAs (Mendell 2005, 2008; Croce 2009; Di Leva et al. 2014). One of the best characterized miRNA clusters involved in tumorigenesis, miR-17-92, is activated by MYC, attenuates the function of E2F transcription factors and also has anti-apoptotic effects (He et al. 2005; O'Donnell et al. 2005; Mendell 2005, 2008; Croce 2009; Di Leva et al. 2014). MYC-regulated miRNAs are also involved in embryonic stem cell differentiation (Lin et al. 2009). In addition to miRNAs, MYC also regulates the transcription of lncRNAs (Hart et al. 2014a; Kim et al. 2015). Most of the lncRNAs are not well characterized yet, but some are involved in cell cycle regulation and tumorigenesis (Kim et al. 2015).

In apparent contrast to its pivotal role in cell proliferation, MYC is also critically involved in cell death. In the absence of survival factors, overexpression of MYC leads to apoptosis (Evan et al. 1992; Shortt and Johnstone 2012). Dependent on cellular context, MYC induces apoptosis by the ARF/MDM2/p53 axis, but also by p53-independent interactions of ARF and MYC (Zindy et al. 1998; Qi et al. 2004; Gregory et al. 2005; Boone et al. 2011). Disturbances of the fail-safe balance between the pro-proliferative and pro-apoptotic functions of MYC are critical in tumorigenesis and can be caused by perturbations of the ARF/MDM2/p53 pathway or by cooperative effects of MYC and anti-apoptotic proteins like BCL2 (Beverly and Varmus 2009; Shortt and Johnstone 2012). MYC has also crucial roles in differentiation and development. Targeted disruption of c-*myc* or N-*myc* genes in mice leads to lethality at around day 10 of gestation (Davis et al. 1993; Grandori et al. 2000; Laurenti et al. 2009), and MYC is an important regulator of stem cell self-renewal and differentiation (Lin et al. 2009; Scognamiglio et al. 2016; Kanatsu-Shinohara et al. 2016). Although all functions discussed so far relate to the canonical full-length nuclear MYC protein, cytoplasmic functions have also been described. Proteolytic cleavage of MYC by calpains, a family of Ca^{2+}-dependent cysteine proteases, generates a cytoplasmic form of MYC (MYC-nick) that lacks the C-terminal region of MYC beyond residue 298 (cf. Fig. 2) and promotes α-tubulin acetylation and cell migration (Conacci-Sorrell et al. 2010; Anderson et al. 2016). Interestingly, synergistic functions of MYC and Ca^{2+}-signaling in B cell differentiation, and direct Ca^{2+}-dependent interactions of MYC with the Ca^{2+}-sensor calmodulin have been reported (Habib et al. 2007; Raffeiner et al. 2017).

3.2 *The RAF Signaling Node*

Interconnected kinase activities are at the heart of signaling networks which transmit and propagate the extracellularly sensed input signal to the nuclear destination for dictating the cell fate (Fig. 4). In response to growth factor, cytokine, or hormone mediated activation of membrane receptors, kinases of the evolutionary conserved MAPK signaling pathway (RAS-RAF-MEK-ERK) take the center stage for spatiotemporally controlled signal transmission (Wellbrock et al. 2004). The simplified illustration in Fig. 4 emphasizes a linear architecture of the transient and adaptive RAS-RAF-MEK-ERK signal response leading to transmission of the signal to the nucleus. The integrity of consecutive phosphorylation events is mandatory for physiological processes such as cellular growth, proliferation, and differentiation (Robinson and Cobb 1997). At the plasma membrane, signal transmission is initiated by receptor stimulation which results in activation of the RAS-GTPase family members HRAS, NRAS, or KRAS linking the receptor sensed input signals with downstream kinases. RAS functions as molecular switch and cycles between GDP-bound inactive and GTP-bound active states. Many external but also internal signals converge on the RAS isoforms to trigger a collection of non-redundant biological functions (Karnoub and Weinberg 2008; Pylayeva-Gupta et al. 2011; McCormick 2016). The critical GDP/GTP exchange during this activation cycle is regulated through guanine nucleotide exchange factors (GEF), while inactivation results from GTP hydrolysis through the intrinsic GTPase activity of the small G-proteins which can be strongly enhanced by GTPase activating proteins (GAP). RAS-GTP signaling involves binary PPIs with central kinases such as RAF but also phosphatidylinositol-3 kinase (PI3K) (Cully et al. 2006; Desideri et al. 2015). The RAF enzymes (ARAF, BRAF, or RAF1) act as initiating kinases for

MAPK signaling. Physiological RAF activation depends on PPI with GTP-bound RAS via the RBD of RAF (Desideri et al. 2015; Lavoie and Therrien 2015). The consequences are membrane recruitment, dimerization, and the subsequent release of the auto-inhibitory RAF configuration leading to a shift to the open and active conformation of the full-length protein. This involves sequential phosphorylation and dephosphorylation events contributing to the intricate mode of RAF regulation by imposing either positive or negative constraints onto the RAF kinase function (Lavoie and Therrien 2015). Moreover, allosteric activation through side-to-side interaction of kinase domains of RAF homodimers or heterodimers, or interactions with the pseudokinase KSR (kinase suppressor of RAS) are functionally integrated (Brennan et al. 2011; Cseh et al. 2014). Activated RAF phosphorylates the downstream located gatekeeper enzymes MEK1 or MEK2 (Roskoski 2012a; Caunt et al. 2015). MEK1/2, acting as dual specific kinases, phosphorylate ERK1 and ERK2 which in turn phosphorylate a plethora of nuclear and cytoplasmic substrates such as kinases and transcription factors (Fig. 4) (Roskoski 2012b).

Cytoplasmic substrates of ERK are upstream located kinases such as EGFR, RSK2, BRAF, MEK1, and the pseudokinase KSR1 (Lake et al. 2016). In the nucleus, ERK regulates transcription factors including members of the ternary complex factor (TCF) and ETS families (Wasylyk et al. 1998; Buchwalter et al. 2004). TCF and ETS induce the transcriptional activation of immediate early genes such as *MYC* and *FOS*. Their gene products MYC and FOS but also cyclin D1 are key factors for cell cycle progression, cell survival, cell division, and cell motility (Lito et al. 2013; Murphy and Blenis 2006; Roskoski 2012b). Another mechanism is the direct phosphorylation of transcription factors such as ELK1, FOS, or MYC by ERK (Fig. 4). ELK1 is a member of ETS transcription factors and the most thoroughly studied ERK target (Roskoski 2012b). ELK1 contains a C-terminal transcriptional activation domain which is the target for ERK phosphorylation leading to increased transcriptional activity (Hollenhorst et al. 2011). FOS forms a heterodimer with JUN for the assembly of the activator protein 1 (AP1) transcription factor complex. FOS phosphorylation in the nucleus extends the half-life of the protein to several hours (Okazaki and Sagata 1995). Double phosphorylation of FOS by ERK and Jun kinases (JNK) increases its transcriptional activity (Morton et al. 2003). Another critical transcription factor which is regulated by phosphorylation is MYC. In response to mitogenic signaling, the evolutionary conserved MYC residues such as Ser62 and Thr58 (cf. Fig. 2) are targets for phosphorylation (Sears 2004; Farell and Sears 2014). Phosphorylation of Ser62 not only by ERK but also by kinases of the CDK family transiently increases MYC stability. This is antagonized by Thr58 phosphorylation by GSK-3β which triggers PP2A mediated Ser62 dephosphorylation targeting MYC for the ubiquitin-proteasome system through E3-ligase SCF-Fbw7 (Welcker et al. 2004; Yada et al. 2004). Furthermore, ERK kinases activate additional kinases such as p90 ribosomal S6 kinase or MAPK interacting kinases (MNKs) which are also relevant for the transcriptional control of the RAS-RAF-MEK-ERK cascade transmitted signal to the nucleus. Dependent on the cellular context, diverse biological functions related to cell cycle progression, differentiation, or metabolic cell signaling are initiated. For the precise biological output, the MAPK-initiated adjustments of transcription factor controlled nuclear gene expression programs are crucial. The scheme in Fig. 4 is a very basic diagram of the feedforward regulation of the RAS-RAF-MEK-ERK signaling network, and schematics for the different modes of feedback regulation of the MAPK cascade have not been integrated. The complexity of MAPK dynamics is reflected by the activities of the signaling hubs RAS and ERK which control a large number of effectors. Alterations of functional interactions emanating from these two signaling nodes dramatically affect spatiotemporal MAPK dynamics (Rauch et al. 2016). In many cells, activation of the RAS-RAF-MEK-ERK pathway is transient through a fast negative feedback directly impacting components of the cascade through ERK-mediated phosphorylation. On a longer time scale, ERK activities enhance the transcriptional expression of DUSP family phosphatases to counterbalance kinase activities (Santos et al. 2007; Nagashima et al. 2015). These negative feedback loops have been implemented to ensure robustness and precise spatiotemporal control of transient or oscillating MAPK dynamics. The physiological response is determined by the integration and processing of the incoming signal with negative and positive feedback regulation. In particular, positive feedback loops are relevant for switch-like and bistable MAPK signaling responses (Santos et al. 2007; Rauch et al. 2016). An important strategy to achieve MAPK specificity is the organization of subsets of signaling units in space and time. For example, regulation of MAPK activities is ensured through molecular interactions with scaffolding proteins such as the pseudokinase KSR. The RAS-dependent compartmentalization of several MAPK through the KSR1 scaffold increases both the signaling strength and the fidelity of signal transmission (Yu et al. 1998; Brennan et al. 2011; Matallanas et al. 2011).

4 Tumorigenesis

4.1 Cancer Driver MYC

Deregulated *MYC* genes are a driving force in the majority of all human cancers (Vogt 2012; Dang 2012; Vogelstein et al. 2013; Stine et al. 2015; Tokheim et al. 2016), and *MYC* was dubbed the emperor of oncogenes (Weinberg et al. 2015). *MYC*, *MYCN*, and *MYCL* were listed in the census of human cancer genes (Futreal et al. 2004) and classified as driver genes in cancer genome landscapes (Vogelstein et al. 2013). Most importantly, *MYC* and *MYCN* are included in a consensus list of 401 cancer drivers predicted by at least one of the three top performing methods in a stringent reevaluation of cancer driver identification (Tokheim et al. 2016). In contrast to many other prominent cancer genes, including *RAS* and *RAF* (see below), oncogenic activation of *MYC* genes typically does not require mutations in the protein-coding region. Hence, deregulation of *MYC* gene expression and accompanying uncontrolled levels of MYC protein are the hallmarks of *MYC* oncogenicity. There are several genetic mechanisms that can cause *MYC* deregulation. Retroviral transduction uncouples c-*myc* expression from cellular controls and is the mechanism that led to the original discovery of *MYC* in the form of avian retroviral v-*myc* alleles (cf. Fig. 1). Retroviral insertion in the vicinity of the c-*myc* locus is another mechanism disturbing normal cellular control of c-*myc* expression and was identified in B cell lymphomas induced by avian leukosis viruses that lack oncogenic (v-*onc*) inserts in their genomes (Hayward et al. 1981; Payne et al. 1982). However, the important mechanisms relevant for human tumorigenesis do not require retroviral involvement and are characterized by major rearrangements or duplications of the *MYC* locus such as chromosomal translocation or gene amplification. Mapping of the human *MYC* locus to chromosome 8q24 and the identification of reciprocal t(8;14) translocations in Burkitt lymphoma promoting *MYC* transcription by juxtaposed immunoglobulin enhancers was the first demonstration of the involvement of *MYC* in human cancer (Dalla-Favera et al. 1982; Taub et al. 1982). Gene amplification of *MYCN* in neuroblastoma strongly correlates with advanced stages of the disease (Brodeur et al. 1984), and *MYCL* but also *MYC* and *MYCN* were found to be amplified in small cell lung cancer (Nau et al. 1985; Vita and Henriksson 2006). In all Burkitt lymphoma cases analyzed, the frequency of *MYC* translocations is 100%, and of *MYC* overexpression 91% (Vita and Henriksson 2006). *MYC* family gene amplification and/or overexpression has been observed at significant frequencies in a broad range of virtually all human cancer types, including myeloma, melanoma, ovarian, cervical, prostate, colon, breast, bladder, lung, and gastric cancers (Vita and Henriksson 2006). Although amplifications and less common translocations are the major genetic alterations leading to deregulation of *MYC* family genes in human tumors, it is important to bear in mind that aberrant *MYC* expression can also be caused by any defects in upstream signaling, like in the Wnt/β-catenin, Notch, Sonic hedgehog, or MAPK pathways (He et al. 1998; Weng et al. 2006; Knoepfler and Kenney 2006), or by disturbance of post-transcriptional and post-translational control elements, including loss of negative feedback control of signaling pathways.

Oncogenic activation manifested by constitutive elevated *MYC* expression is essential both for cancer initiation and maintenance, and established *MYC*-driven tumors become addictive to constitutive *MYC* expression (Dang 2012; Gabay et al. 2014; Hsieh et al. 2015). Furthermore, based on the broad influence of *MYC* on metabolic regulation, cancer cells with deregulated *MYC* expression are also addicted to continuous nutrient supply, such as glucose or glutamine, whereas *MYC* expression and *MYC*-controlled energy metabolism in normal cells can be shut down by withdrawal of mitogenic signals or insufficient nutrient supply. In cancer cells with runaway *MYC* expression, frequent loss of p53 or ARF tumor suppressor checkpoints impedes the apoptotic safeguard mechanism and allows uncontrolled cellular growth and proliferation (Stine et al. 2015). In addition to deactivation of tumor suppressors, *MYC*-induced tumorigenesis frequently involves cooperative activation of other oncogenes such as *BCL2* or *RAS* (Beverly and Varmus 2009; McFadden et al. 2016). There is also evidence that cooperative interactions between

MYC and effectors of the PI3K/AKT/mTOR signaling pathway contribute to oncogenesis and that resistance to therapeutic PI3K inhibition may be mediated by *MYC* amplification (Liu et al. 2011; Hsieh et al. 2015). Furthermore, there are mutual control mechanisms of *MYC* and miRNA expression that adopt important roles in tumorigenesis (Croce 2009; Di Leva et al. 2014). Loss of *MYC* repressive miRNAs such as *let-7*, or MYC-mediated transactivation of oncogenic miRNA clusters such as miR-17-92 are involved in many cancer types (He et al. 2005; O'Donnell et al. 2005; Kumar et al. 2007; Cairo et al. 2010). Deregulated *MYC* expression has also been implicated in the emergence of chromosomal instability (CIN) (Felsher and Bishop 1999; Louis et al. 2005; Dang 2012). CIN causes aneuploidy, a hallmark of most human cancers, and is possibly involved in tumorigenesis and metastasis (Rajagopalan et al. 2003; Gao et al. 2016; Turajlic and Swanton 2016).

4.2 Oncogenic RAF Signaling

Hyperactivation of many components of the RAS-RAF-MEK-ERK pathway contributes to the initiation or progression of cancer. *HRAS*, *KRAS*, *NRAS*, *BRAF*, and *RAF1* are all included in the stringent consensus list of cancer drivers (Tokheim et al. 2016). In human cancers, recurrent oncogenic point mutations at the RAS codon positions Gly12, Gly13, or Gln61 eliminate the precisely controlled GDP/GTP exchange cycle and cause constitutive RAS activation. This leads to uncontrolled RAS signaling to a multitude of effector molecules and pathways including kinases such as RAF and PI3K (Cully et al. 2006; Karnoub and Weinberg 2008; Stephen et al. 2014; Lavoie and Therrien 2015). Under physiological conditions, the formation of homo- and heterodimers of RAF is a central element for phosphotransferase activation which is attenuated and disrupted by ERK-mediated feedback phosphorylation (Freeman et al. 2013; Lake et al. 2016). Oncogenic mutations in RAF kinases lead to decoupled phosphorylation events. Although the RAF isoforms share substantial sequence similarity, their mode of regulation differs in crucial details and this is important for the predominant role of BRAF in human cancers (Wellbrock et al. 2004; Desideri et al. 2015). In the entire superfamily of kinase genes, *BRAF* has been identified as the most frequently mutated proto-oncogene (Davies et al. 2002; Fleuren et al. 2016). An important structural basis for its paramount cancerogenic role is the unique presence of a phospho-mimetic Asp449 and a constitutively phosphorylated Ser446 confering a constant negative charge to the functionally crucial N-region in the BRAF sequence (Lito et al. 2013). Hence, BRAF is poised for activation, and single point mutations are sufficient to convert BRAF into a constitutively active kinase. BRAF mutations have been identified in many cancers including malignant melanomas (50–60%), thyroid (30–50%), colorectal (10%) and lung (3%) cancers, hairy cell leukemia (100%), and others (Tiacci et al. 2011; Lito et al. 2013). The predominant oncogenic gain-of-function mutation in BRAF is the substitution of glutamic acid for valine at position 600 (V600E) which is found in 60% of all malignant melanomas (Lito et al. 2013). This mutation is located in the activation segment of the BRAF kinase domain between two regulatory phosphorylation sites (Thr599, Ser602) and converts it into the active state. Based on molecular modeling and crystallographic evidence, the V600E mutation promotes formation of a salt bridge between Lys507 and Glu600, thereby stabilizing the active conformation of the activation segment. Either phosphorylation of the activation segment (Thr599, Ser602) or substitution of charged amino acids (Glu, Lys, Asp, or Arg have been found in tumors) for Val600 converts the kinase into the active conformation. Furthermore, additional phosphorylation events throughout the RAF activation cycle control the activity status of the kinase. Differential RAF activation involves post-translational modifications of the activation segment, the negatively charged N-region, and the C-terminal 14-3-3 binding site (Lavoie and Therrien 2015). However, the critical role of the phosphorylated N-region is not fully understood. Recent data suggest that it could determine the transactivation direction in RAF dimers (Hu et al. 2013).

So far, nearly 300 distinct missense mutations of BRAF have been identified in tumor samples and cancer cell lines (Forbes et al. 2011; Holderfield et al. 2014a). Most of the mutations occur in the activation (A) loop or in the phosphate-binding (P) loop of the kinase domain. However, the biochemistry of altered BRAF proteins varies substantially. In addition to kinase activating mutations, some amino acid substitutions lead to intermediate or low kinase activity. A subgroup of these mutations endorses MEK-ERK signaling through heterodimerization-mediated activation of RAF1 (Wan et al. 2004; Lito et al. 2013; Desideri et al. 2015). This suggests that independent of the BRAF catalytic function, the dimerization with RAF monomers releases the intrinsic activity of BRAF-bound RAF1. Interestingly, it has been shown that kinase-impaired activities of BRAF mutants coexist with RAS mutations (Andreadi et al. 2012). Analyses of the crystal structures of RAF kinase domains revealed a conserved side-to-side interface (Rajakulendran et al. 2009). Amino acid substitutions in the interface such as R509H impeded RAF dimerization and activity underscoring that side-to-side homo- and heterodimer formation leads to protomer transactivation (Lavoie and Therrien 2015). In addition, the dimerization-dependent RAF activation involves the allosteric reorganization of the N- and C-lobe of the kinase domain to obtain an active-like conformation. The alignment of hydrophobic residues spanning the N- and C-lobes are critical for the catalysis (Taylor and Kornev 2011; Hu et al. 2015; Lavoie and Therrien 2015). However, in contrast to wild-type and kinase-impaired BRAF mutants, tumors harboring the BRAF(V600E) mutation show hyperactivated ERK signaling resulting in increased proliferation and evasion of apoptosis, independent of RAF dimerization (Poulikakos et al. 2011; Samatar and Poulikakos 2014). Since BRAF(V600E) is not dependent on dimerization and upstream regulation, it bypasses the inhibitory effect of negative feedback regulation of ERK (Tsavachidou et al. 2004; Lake et al. 2016). ERK-dependent negative feedback regulation results in low RAS-GTP levels in cancer cells with BRAF mutations (Lito et al. 2013). However, the RTK/RAS-uncoupled BRAF(V600E) mutant acts as a functional monomer, constitutively activating MEK and cytoplasmic and nuclear ERK signaling. MEK1/2 are activated as a result of RAF-catalyzed phosphorylation in the kinase activation loop, in the case of MEK1 at Ser218 and Ser222 (Roskoski 2012a; Caunt et al. 2015). MEK1/2 in turn are the only activators of ERK1/2 and phosphorylate specific residues, such as Thr202 and Tyr204 in ERK1. MEK1/2 serve as critical gatekeeper kinases for ERK1/2 which regulate multiple cellular events through phosphorylation of over 150 substrates in the cytoplasm and nucleus, many involved in cell proliferation and survival (Yoon and Seger 2006; Roskoski 2012b; Caunt et al. 2015; Lake et al. 2016). In contrast to BRAF, activating mutations in MEK1 or MEK2 are only found at low frequency in human tumors (Marks et al. 2008; Murugan et al. 2009).

5 Pharmacological Interference

5.1 MYC Inhibition

In view of the strikingly broad involvement of *MYC* family genes in virtually all human cancers, strategies to interfere with *MYC*-induced oncogenesis, directly or indirectly, are an obviously urgent goal in cancer research and clinical treatment (Soucek et al. 2008; Prochownik and Vogt 2010; Dang 2012; McKeown and Bradner 2014; Stefan et al. 2015). A genetic approach using an engineered dominant negative MYC mutant with altered dimerization specificity provided principal proof of guided MYC inhibition as a means to interfere with cell transformation. The engineered mutant, termed Omomyc, competes with MAX for dimerization with MYC and blocks MYC-dependent transcriptional activation (Soucek et al. 1998, 2008). In tissue culture and in preclinical mouse models, systemic Omomyc expression interfered with cell transformation and tumorigenesis, with surprisingly tolerable and fully reversible effects on normal tissues (Soucek et al. 2008). In another genetic approach, ectopic expression of the repressed MYC target gene *BASP1* was shown to inhibit MYC-induced cell transformation (Hartl et al. 2009). Obviously, for pharmacological interference and therapeutic applications, the development of small-molecule inhibitors is of highest priority. In view of the striking success in the development of small-molecule inhibitors for the therapy of cancers driven by oncogenic kinases (Holderfield et al. 2014a), MYC also became a target for possible drug development. However, in contrast to well-structured enzymes with defined catalytic clefts, MYC is intrinsically disordered in free form (Fieber et al. 2001) and its biological activities are mainly based on macromolecular interactions, such as PPIs involving large flat surface areas that are difficult to target with small molecules. Nevertheless, several small-molecule inhibitors of MYC:MAX dimerization and MYC-induced cell transformation were isolated from combinatorial chemical libraries using a fluorescence resonance energy transfer (FRET) assay or a yeast two-hybrid approach (Berg et al. 2002; Yin et al. 2003). The compounds displayed diverse specificity and efficacy, but represented an important proof of concept for targeting the MYC:MAX interface. Small-molecule inhibitors of MYC:MAX dimerization with pharmacokinetic properties suitable for possible drug development were isolated from a Kröhnke pyridine library in a fluorescence polarization screen (Hart et al. 2014b; Raffeiner et al. 2014). The compounds KJ-Pyr-9 (Fig. 5) and KJ-Pyr-10 proved to be highly effective in the nanomolar range and interfered specifically with MYC:MAX complex formation, the MYC-driven transcriptional signature, MYC-induced cell transformation, and the growth of breast cancer MDA-MB-231 xenografts carrying *MYC* amplifications (Hart et al. 2014b; Raffeiner et al. 2014).

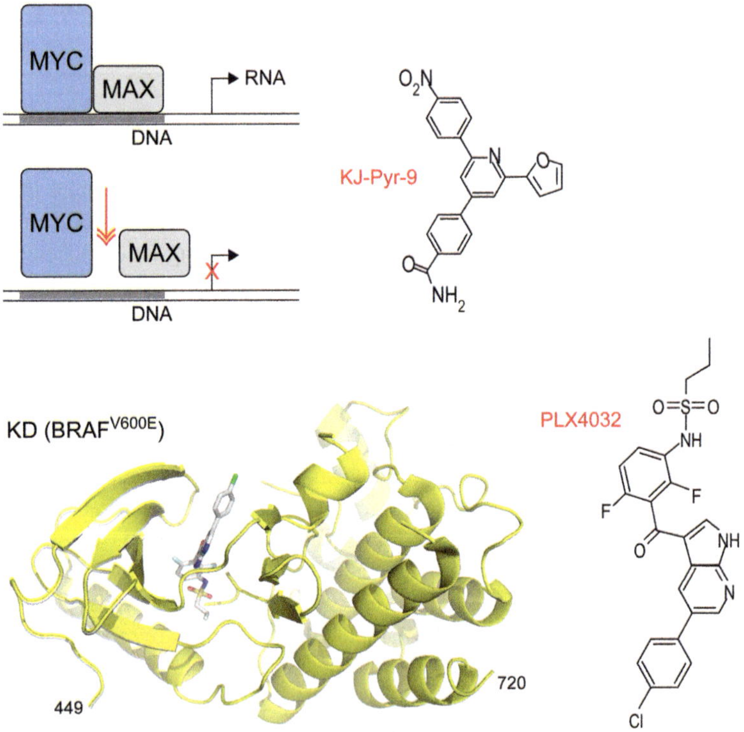

Fig. 5 Strategies for MYC and RAF inhibition. *Top*: The MYC:MAX heterodimer is shown, bound to a DNA E-box sequence (*shaded*) and activating transcription. The inhibitor KJ-Pyr-9 (Hart et al. 2014b) interferes with MYC:MAX dimerization, preventing DNA binding and transcriptional activation. KJ-Pyr-9 is effective both *in vitro* and *in vivo* in experimental settings. *Bottom*: The X-ray structure of the BRAF(V600E) kinase domain (KD) in complex with the PLX4032 (Vemurafenib) inhibitor is shown (Bollag et al. 2010). Vemurafenib is in clinical use for treatment of melanoma patients. The image was created from the PDB entry 3OG7 using the PyMOL graphics system

The intrinsic difficulties of targeting MYC proteins directly have stimulated various efforts to develop indirect inhibitory strategies, such as attenuating *MYC* transcription. G-quadruplex (G4) regulatory regions in chromatin are implicated in transcriptional regulation and have been identified in several cancer associated genes, including *MYC*. Endogenous G4 structures in promoters were linked to elevated transcriptional activity, whereas G4 stabilization by small molecules was shown to repress *MYC* expression (Balasubramanian et al. 2011; Hänsel-Hertsch et al. 2016). An important approach to target *MYC* transcription involves small-molecule inhibition of members of the bromodomain and extraterminal (BET) subfamily of bromodomain proteins, such as BRD4 (McKeown and Bradner 2014; Shu and Polyak 2017). BET bromodomain proteins are histone modification readers that recognize acetylated lysine residues and facilitate transcriptional activation by recruiting transcriptional cofactors to chromatin. Small-molecule inhibitors (BETi) were developed, such as the very potent thienodiazepine-based JQ1 compound, that compete for the acetyl binding pocket of BET bromodomains causing displacement of BET proteins from chromatin and repression of gene expression (McKeown and Bradner 2014; Shu and Polyak 2017). BETi were shown to efficiently downregulate *MYC* transcription and the MYC-driven transcriptional program (Delmore et al. 2011; Mertz et al. 2011). Several BETi have entered clinical trials, and clinical responses have been linked to *MYC* repression (Shu and Polyak 2017). A dual-activity small-molecule compound has been developed that simultaneously inhibits BRD4 and PI3K leading to concomitant decrease of *MYC* transcription and increase of MYC degradation, respectively. In mouse models, the compound effectively blocked tumor growth and metastasis (Andrews et al. 2017). In view of the pivotal role of the MYC master regulator in nearly all fundamental processes in normal cells (see above), the success of inhibitory strategies for clinical treatment of MYC-driven cancers will largely depend on the specificity of targeting the uncontrolled MYC expression and subsequent changes in cancer cells. The addiction of many tumors to deregulated *MYC* expression and metabolic reprogramming may offer additional therapeutic opportunities such as targeting key enzymes of *MYC*-controlled glucose or glutamine metabolism (Dang 2012; Stine et al. 2015; Altman et al. 2016). In principle, the pharmacological inhibition of genes whose function is specifically required for the survival of *MYC*-driven cancer cells is a promising strategy. Screens for *MYC* synthetic lethality, including genome-wide RNA interference, have identified several potential *MYC* synthetic lethal genes encoding aurora-B kinase, CSNK1e kinase, or the SUMO-activating enzyme SAE1/2 (Yang et al. 2010; Kessler et al. 2012; Toyoshima et al. 2012; Cermelli et al. 2014).

5.2 RAF Inhibition

Kinases have become one of the most intensively pursued classes of drug targets. They are involved in virtually every signal transduction process, and the human genome encodes over 500 different protein kinases (Manning et al. 2002). Despite the high degree of conservation of the ATP-binding pocket in the catalytic kinase domain, a collection of highly selective ATP-competitive small-molecule inhibitors have been developed. So far, more than 25 oncology drugs that target kinases have been approved for cancer therapy (Gross et al. 2015). In the early years of this millennium, the identification of BRAF mutations in a wide variety of human tumor types, most notably in melanoma, colorectal cancer, glioma, and lung adenocarcinoma (Davies et al. 2002; Fukushima et al. 2003), initiated extensive research efforts to develop RAF inhibitors (RAFi). The first general RAFi was Sorafenib acting as a multi-targeted kinase inhibitor. The incomplete specificity of Sorafenib promoted the development of mutation-selective BRAFi such as Vemurafenib (Fig. 5) and Dabrafenib which block the phosphotransferase activity of RAF mutants such as BRAF(V600E) (Samatar and Poulikakos 2014). Both BRAFi were approved for clinical use and brought unprecedented clinical benefits to patients with melanomas containing the BRAF(V600E) or BRAF(V600K) mutations (Flaherty et al. 2010; Chapman et al. 2011; Hauschild et al. 2012; Lito et al. 2013). These inhibitors selectively block monomeric BRAF(V600E/K) activities without impacting the activities of the wild-type BRAF dimer. The selective and direct blocking of mutant BRAF-initiated downstream signaling to MEK-ERK is a remarkable feature of BRAFi in a subset of tumors harboring the specific mutations (Solit and Rosen 2011; Holderfield et al. 2014a). The discrimination for mutant BRAF inhibition is partially responsible for the broad therapeutic index of BRAFi which are either FDA-approved (Vemurafenib, Dabrafenib) or currently in clinical trials, for example PLX8394 or Encorafenib (Samatar and Poulikakos 2014; Zhang et al. 2015; Adelmann et al. 2016). Of the more than 300 different missense mutations identified in BRAF, BRAF(V600E) is clearly the most common cancer driver mutation in melanoma (Forbes et al. 2011; Holderfield et al. 2014a). However, oncogenic mutations are not restricted to BRAF, and RAF1 mutations have also been found but at a much lower rate (Lavoie and Therrien 2015). Furthermore, BRAF mutations do not always predict favorable responsiveness to BRAFi. In addition to the type of mutation, the cellular context in which BRAFi are used is an important determinant for drug effectiveness. BRAFi are ineffective in BRAF(V600E)-dependent metastatic colorectal carcinoma and thyroid cancer (Holderfield et al. 2014a; Desideri et al. 2015). In these cancer types, a crucial negative feedback inhibition of EGFR signaling is lost upon BRAFi exposure, leading to increased EGFR signaling and intrinsic resistance to the drug (Prahallad et al. 2012). In contrast, melanoma cells express low basal levels of EGFR, evade this mechanism, and show favorable initial response to BRAFi treatment (Lito et al. 2012; Sun et al. 2014; Holderfield et al. 2014a; Desideri et al. 2015). However, the therapeutic benefits of BRAFi in melanoma patients are temporary and limited due to the development of drug resistance. A collection of mechanisms causing primary or acquired drug resistance have been described that limit the application of this approach (Hatzivassiliou et al. 2010; Heidorn et al. 2010; Poulikakos et al. 2010; Lito et al. 2013; Samatar and Poulikakos 2014; Yao et al. 2015). BRAFi resistance involves RTK and ERK reactivation through multiple mechanisms such as increased expression of RTKs, BRAF(V600E) or RAF1 amplification, occurrence of BRAF splice variants, mutational activation of NRAS, MEK mutations, or COT kinase overexpression (Lito et al. 2013; Desideri et al. 2015). Another unwanted consequence of BRAFi treatment is the so-called *paradoxical kinase activation* of ERK signaling in BRAF wild-type cells with activated RAS (Hatzivassiliou et al. 2010; Heidorn et al. 2010; Poulikakos et al. 2010; Holderfield et al. 2014b). The mechanism is based on formation of RAF dimers and BRAFi-induced conformational changes of a protomer leading to transactivation in the dimer of the other non-drug bound RAF isoform (Hatzivassiliou et al. 2010; Heidorn et al. 2010; Poulikakos et al. 2010; Lito et al. 2013; Lavoie et al. 2013; Holderfield et al. 2014a; Lavoie and Therrien 2015). Hence, the drug binding allosterically activates signaling by the drug-free RAF kinase to MEK-ERK thereby enhancing cellular proliferation. This BRAFi-driven mechanism can promote skin cancers such as cutaneous squamous-cell carcinoma and keratoacanthomas that often harbor RAS mutations (Chapman et al. 2011; Su et al. 2012). The multiple mechanisms of acquired BRAFi resistance and off-target effects underscore the need to develop improved drugs for cancer therapy. There is rising interest in identifying RAFi with different mechanistic functions, such as paradox breakers, with the ultimate goal to develop inhibitors with improved efficacy and selectivity (Zhang et al. 2015; Okimoto et al. 2016). Moreover, the implementation of poly-pharmacology approaches using a collection of kinase inhibitors may improve clinical drug efficacies by delaying or preventing the onset of drug resistance.

6 Conclusions

By now, four or almost four decades, respectively, have passed since the prototypes of the *MYC* and *RAF* gene families were discovered, initially as avian or murine orthologs transduced in retroviral genomes (cf. Fig. 1). In these past decades, a spectacular amount of detailed knowledge of the structure and function of these genes and their protein products has accumulated. It is one of the most stunning outcomes that MYC is apparently involved directly or indirectly in all major signaling and regulatory pathways and hence a true master regulator of cell biology and biochemistry. Furthermore, it is also one of the most prominent cancer driver genes, apparently deregulated and causally involved in the majority of all human cancers. In a way, it is not surprising that unleashing a master regulator such as MYC profoundly abrogates the normal cellular regime. Accordingly, MYC has also broad influence in tumorigenesis, especially when crucial checkpoints designed to guard against uncontrolled cell proliferation are lost simultaneously with *MYC* deregulation. Although the mere complexity of the MYC regulatory sphere poses an enormous challenge for biochemical and biomedical research, it may also provide multiple opportunities to interfere with *MYC*-driven cancer growth. Efficient direct targeting of MYC PPIs, attenuating the expression of *MYC*, exploiting the unique oncogene and nutrient addiction of *MYC*-driven cancers, or systematic screening for synthetic lethality could all be feasible strategies for therapeutic intervention. As for the RAF kinases, pharmacological inhibition initially appeared to be more straightforward, since competitive or allosteric inhibition of enzymes is standard biochemical technology, and the stunning clinical success of several BRAFi is compelling proof. However, the inevitable development of drug resistance and the occurrence of even adverse effects of clinically approved BRAFi caused by perturbance of feedback mechanisms or activation of redundant signaling pathways are severe drawbacks. Here, the development of more specific inhibitors, combinations of inhibitors, or specific targeting of PPIs appear to be promising strategies. Obviously, the paramount medical relevance of deregulated or mutated *MYC* or *RAF* genes is the impulse for intensive research into their role as major cancer driver genes. Similarly, the physiological roles of MYC and RAF as master regulator of transcription or key cytoplasmic signal transducer, respectively, are an important source for further research into fundamental principles of regulatory networks in normal cells. Also, the transition from physiological to oncogenic function, particularly in the case of *MYC*, may be gradual and dependent on cellular context and critical threshold levels. Hence, the oncogenic and proto-oncogenic facets of *MYC* and *RAF* will remain in the focus of biochemical and biomedical research.

Acknowledgements We thank Gabriele Reiter for help with the preparation of the manuscript. Work by the authors has been supported by grants (P23652, P27606) from the Austrian Science Fund (FWF).

References

Abraham D, Podar K, Pacher M, Kubicek M, Welzel N, Hemmings BA, Dilworth SM, Mischak H, Kolch W, Baccarini M (2000) Raf-1-associated protein phosphatase 2A as a positive regulator of kinase activation. J Biol Chem 275:22300–22304

Adelmann CH, Ching G, Du L, Saporito RC, Bansal V, Pence LJ, Liang R, Lee W, Tsai KY (2016) Comparative profiles of BRAF inhibitors: the paradox index as a predictor of clinical toxicity. Oncotarget 7:30453–30460

Altman BJ, Stine ZE, Dang CV (2016) From Krebs to clinic: glutamine metabolism to cancer therapy. Nat Rev Cancer 16:619–634

Anderson S, Poudel KR, Roh-Johnson M, Brabletz T, Yu M, Borenstein-Auerbach N, Grady WN, Bai J, Moens CB, Eisenman RN, Conacci-Sorrell M (2016) MYC-nick promotes cell migration by inducing fascin expression and Cdc42 activation. Proc Natl Acad Sci USA 113:E5481–E5490

Andreadi C, Cheung LK, Giblett S, Patel B, Jin H, Mercer K, Kamata T, Lee P, Williams A, McMahon M, Marais R, Pritchard C (2012) The intermediate-activity (L597V)BRAF mutant acts as an epistatic modifier of oncogenic RAS by enhancing signaling through the RAF/MEK/ERK pathway. Genes Dev 26:1945–1958

Andrews FH, Singh AR, Joshi S, Smith CA, Morales GA, Garlich JR, Durden DL, Kutateladze TG (2017) Dual-activity PI3K-BRD4 inhibitor for the orthogonal inhibition of MYC to block tumor growth and metastasis. Proc Natl Acad Sci USA 114: E1072–1080

Ayer DE, Kretzner L, Eisenman RN (1993) Mad: a heterodimeric partner for Max that antagonizes Myc transcriptional activity. Cell 72:211–222

Baccarini M (2005) Second nature: biological functions of the Raf-1 "kinase". FEBS Lett 579:3271–3277

Balasubramanian S, Hurley LH, Neidle S (2011) Targeting G-quadruplexes in gene promoters: a novel anticancer strategy? Nat Rev Drug Discov 10:261–275

Begg AM (1927) A filtrable endothelioma of the fowl. Lancet 209:912–915

Berg T, Cohen SB, Desharnais J, Sonderegger C, Maslyar DJ, Goldberg J, Boger DL, Vogt PK (2002) Small-molecule antagonists of Myc/Max dimerization inhibit Myc-induced transformation of chicken embryo fibroblasts. Proc Natl Acad Sci USA 99:3830–3835

Beverly LJ, Varmus HE (2009) MYC-induced myeloid leukemogenesis is accelerated by all six members of the antiapoptotic BCL family. Oncogene 28:1274–1279

Bister K (2015) Discovery of oncogenes: the advent of molecular cancer research. Proc Natl Acad Sci USA 112:15259–15260

Bister K, Jansen HW (1986) Oncogenes in retroviruses and cells: biochemistry and molecular genetics. Adv Cancer Res 47:99–188

Bister K, Vogt PK (1978) Genetic analysis of the defectiveness in strain MC29 avian leukosis virus. Virology 88:213–221

Bister K, Hayman MJ, Vogt PK (1977) Defectiveness of avian myelocytomatosis virus MC29: isolation of long-term nonproducer cultures and analysis of virus-specific polypeptide synthesis. Virology 82:431–448

Blackwood EM, Eisenman RN (1991) Max: a helix-loop-helix zipper protein that forms a sequence-specific DNA-binding complex with Myc. Science 251:1211–1217

Bollag G, Hirth P, Tsai J, Zhang J, Ibrahim PN, Cho H, Spevak W, Zhang C, Zhang Y, Habets G, Burton EA, Wong B, Tsang G, West BL, Powell B, Shellooe R, Marimuthu A, Nguyen H, Zhang KY, Artis DR, Schlessinger J, Su F, Higgins B, Iyer R, D'Andrea K, Koehler A, Stumm M, Lin PS, Lee RJ, Grippo J, Puzanov I, Kim KB, Ribas A, McArthur GA, Sosman JA, Chapman PB, Flaherty KT, Xu X, Nathanson KL, Nolop K (2010) Clinical efficacy of a RAF inhibitor needs broad target blockade in *BRAF*-mutant melanoma. Nature 467:596–599

Boone DN, Qi Y, Li Z, Hann SR (2011) Egr1 mediates p53-independent c-Myc-induced apoptosis via a noncanonical ARF-dependent transcriptional mechanism. Proc Natl Acad Sci USA 108:632–637

Bouchard C, Dittrich O, Kiermaier A, Dohmann K, Menkel A, Eilers M, Lüscher B (2001) Regulation of cyclin D2 gene expression by the Myc/Max/Mad network: Myc-dependent TRRAP recruitment and histone acetylation at the cyclin D2 promoter. Genes Dev 15:2042–2047

Brennan DF, Dar AC, Hertz NT, Chao WC, Burlingame AL, Shokat KM, Barford D (2011) A Raf-induced allosteric transition of KSR stimulates phosphorylation of MEK. Nature 472:366–369

Brodeur GM, Seeger RC, Schwab M, Varmus HE, Bishop JM (1984) Amplification of N-myc in untreated human neuroblastomas correlates with advanced disease stage. Science 224: 1121–1124

Buchwalter G, Gross C, Wasylyk B (2004) Ets ternary complex transcription factors. Gene 324:1–14

Cairo S, Wang Y, de Reyniès A, Duroure K, Dahan J, Redon MJ, Fabre M, McClelland M, Wang XW, Croce CM, Buendia MA (2010) Stem cell-like micro-RNA signature driven by Myc in aggressive liver cancer. Proc Natl Acad Sci USA 107:20471–20476

Caunt CJ, Sale MJ, Smith PD, Cook SJ (2015) MEK1 and MEK2 inhibitors and cancer therapy: the long and winding road. Nat Rev Cancer 15:577–592

Cermelli S, Jang IS, Bernard B, Grandori C (2014) Synthetic lethal screens as a means to understand and treat MYC-driven cancers. Cold Spring Harb Perspect Med 4:a014209

Chapman PB, Hauschild A, Robert C, Haanen JB, Ascierto P, Larkin J, Dummer R, Garbe C, Testori A, Maio M, Hogg D, Lorigan P, Lebbe C, Jouary T, Schadendorf D, Ribas A, O'Day SJ, Sosman JA, Kirkwood JM, Eggermont AM, Dreno B, Nolop K, Li J, Nelson B, Hou J, Lee RJ, Flaherty KT, McArthur GA (2011) Improved survival with vemurafenib in melanoma with BRAF V600E mutation. N Engl J Med 364:2507–2516

Coffin JM, Varmus HE, Bishop JM, Essex M, Hardy WD Jr, Martin GS, Rosenberg NE, Scolnick EM, Weinberg RA, Vogt PK (1981) Proposal for naming host cell-derived inserts in retrovirus genomes. J Virol 40:953–957

Coll J, Righi M, Taisne C, Dissous C, Gegonne A, Stehelin D (1983) Molecular cloning of the avian acute transforming retrovirus MH2 reveals a novel cell-derived sequence (v-*mil*) in addition to the *myc* oncogene. EMBO J 2:2189–2194

Conacci-Sorrell M, Ngouenet C, Eisenman RN (2010) Myc-nick: a cytoplasmic cleavage product of Myc that promotes α-tubulin acetylation and cell differentiation. Cell 142:480–493

Conacci-Sorrell M, McFerrin L, Eisenman RN (2014) An overview of MYC and its interactome. Cold Spring Harb Perspect Med 4:a014357

Croce CM (2009) Causes and consequences of microRNA dysregulation in cancer. Nat Rev Genet 10:704–714

Cseh B, Doma E, Baccarini M (2014) "RAF" neighborhood: protein-protein interaction in the Raf/Mek/Erk pathway. FEBS Lett 588:2398–2406

Cully M, You H, Levine AJ, Mak TW (2006) Beyond PTEN mutations: the PI3K pathway as an integrator of multiple inputs during tumorigenesis. Nat Rev Cancer 6:184–192

Dalla-Favera R, Bregni M, Erikson J, Patterson D, Gallo RC, Croce CM (1982) Human *c-myc onc* gene is located on the region of chromosome 8 that is translocated in Burkitt lymphoma cells. Proc Natl Acad Sci USA 79:7824–7827

Dang CV (2012) *MYC* on the path to cancer. Cell 149:22–35

Dang CV (2013) MYC, metabolism, cell growth, and tumorigenesis. Cold Spring Harb Perspect Med 3:a014217

Dang CV (2014) Gene regulation: fine-tuned amplification in cells. Nature 511:417–418

Dang CV, Kim JW, Gao P, Yustein J (2008) The interplay between MYC and HIF in cancer. Nat Rev Cancer 8:51–56

Davies H, Bignell GR, Cox C, Stephens P, Edkins S, Clegg S, Teague J, Woffendin H, Garnett MJ, Bottomley W, Davis N, Dicks E, Ewing R, Floyd Y, Gray K, Hall S, Hawes R, Hughes J, Kosmidou V, Menzies A, Mould C, Parker A, Stevens C, Watt S, Hooper S, Wilson R, Jayatilake H, Gusterson BA, Cooper C, Shipley J, Hargrave D, Pritchard-Jones K, Maitland N, Chenevix-Trench G, Riggins GJ, Bigner DD, Palmieri G, Cossu A, Flanagan A, Nicholson A, Ho JW, Leung SY, Yuen ST, Weber BL, Seigler HF, Darrow TL, Paterson H, Marais R, Marshall CJ, Wooster R, Stratton MR, Futreal PA (2002) Mutations of the *BRAF* gene in human cancer. Nature 417:949–954

Davis AC, Wims M, Spotts GD, Hann SR, Bradley A (1993) A null c-*myc* mutation causes lethality before 10.5 days of gestation in homozygotes and reduced fertility in heterozygous female mice. Genes Dev 7:671–682

Delmore JE, Issa GC, Lemieux ME, Rahl PB, Shi J, Jacobs HM, Kastritis E, Gilpatrick T, Paranal RM, Qi J, Chesi M, Schinzel AC, McKeown MR, Heffernan TP, Vakoc CR, Bergsagel PL, Ghobrial IM, Richardson PG, Young RA, Hahn WC, Anderson KC, Kung AL, Bradner JE, Mitsiades CS (2011) BET bromodomain inhibition as a therapeutic strategy to target c-Myc. Cell 146:904–917

Dent P, Haser W, Haystead TA, Vincent LA, Roberts TM, Sturgill TW (1992) Activation of mitogen-activated protein kinase kinase by v-Raf in NIH 3T3 cells and in vitro. Science 257:1404–1407

Desideri E, Cavallo AL, Baccarini M (2015) Alike but different: RAF paralogs and their signaling outputs. Cell 161:967–970

Di Leva G, Garofalo M, Croce CM (2014) MicroRNAs in cancer. Annu Rev Pathol 9:287–314

Dominguez-Sola D, Ying CY, Grandori C, Ruggiero L, Chen B, Li M, Galloway DA, Gu W, Gautier J, Dalla-Favera R (2007) Non-transcriptional control of DNA replication by c-Myc. Nature 448:445–451

Duesberg PH, Vogt PK (1970) Differences between the ribonucleic acids of transforming and nontransforming avian tumor viruses. Proc Natl Acad Sci USA 67:1673–1680

Duesberg PH, Bister K, Vogt PK (1977) The RNA of avian acute leukemia virus MC29. Proc Natl Acad Sci USA 74:4320–4324

Eilers M, Eisenman RN (2008) Myc's broad reach. Genes Dev 22:2755–2766

Eisenman RN (2001) Deconstructing Myc. Genes Dev 15:2023–2030

Evan GI, Wyllie AH, Gilbert CS, Littlewood TD, Land H, Brooks M, Waters CM, Penn LZ, Hancock DC (1992) Induction of apoptosis in fibroblasts by c-myc protein. Cell 69:119–128

Farrell AS, Sears RC (2014) MYC degradation. Cold Spring Harb Perspect Med 4:a014365

Felsher DW, Bishop JM (1999) Transient excess of MYC activity can elicit genomic instability and tumorigenesis. Proc Natl Acad Sci USA 96:3940–3944

Ferré-D'Amaré AR, Prendergast GC, Ziff EB, Burley SK (1993) Recognition by Max of its cognate DNA through a dimeric b/HLH/Z domain. Nature 363:38–45

Fetics SK, Guterres H, Kearney BM, Buhrman G, Ma B, Nussinov R, Mattos C (2015) Allosteric effects of the oncogenic RasQ61L mutant on Raf-RBD. Structure 23:505–516

Fieber W, Schneider ML, Matt T, Kräutler B, Konrat R, Bister K (2001) Structure, function, and dynamics of the dimerization and DNA binding domain of oncogenic transcription factor v-Myc. J Mol Biol 307:1395–1410

Flaherty KT, Puzanov I, Kim KB, Ribas A, McArthur GA, Sosman JA, O'Dwyer PJ, Lee RJ, Grippo JF, Nolop K, Chapman PB (2010) Inhibition of mutated, activated BRAF in metastatic melanoma. N Engl J Med 363:809–819

Fleuren ED, Zhang L, Wu J, Daly RJ (2016) The kinome 'at large' in cancer. Nat Rev Cancer 16:83–98

Forbes SA, Bindal N, Bamford S, Cole C, Kok CY, Beare D, Jia M, Shepherd R, Leung K, Menzies A, Teague JW, Campbell PJ, Stratton MR, Futreal PA (2011) COSMIC: mining complete cancer genomes in the Catalogue of Somatic Mutations in Cancer. Nucleic Acids Res 39:D945–D950

Freeman AK, Ritt DA, Morrison DK (2013) Effects of Raf dimerization and its inhibition on normal and disease-associated Raf signaling. Mol Cell 49:751–758

Fukushima T, Suzuki S, Mashiko M, Ohtake T, Endo Y, Takebayashi Y, Sekikawa K, Hagiwara K, Takenoshita S (2003) BRAF mutations in papillary carcinomas of the thyroid. Oncogene 22:6455–6457

Futreal PA, Coin L, Marshall M, Down T, Hubbard T, Wooster R, Rahman N, Stratton MR (2004) A census of human cancer genes. Nat Rev Cancer 4:177–183

Gabay M, Li Y, Felsher DW (2014) MYC activation is a hallmark of cancer initiation and maintenance. Cold Spring Harb Perspect Med 4:a014241

Gao C, Su Y, Koeman J, Haak E, Dykema K, Essenberg C, Hudson E, Petillo D, Khoo SK, Vande Woude GF (2016) Chromosome instability drives phenotypic switching to metastasis. Proc Natl Acad Sci USA 113:14793–14798

Gomez-Roman N, Grandori C, Eisenman RN, White RJ (2003) Direct activation of RNA polymerase III transcription by c-Myc. Nature 421:290–294

Graf T, von Weizsaecker F, Grieser S, Coll J, Stehelin D, Patschinsky T, Bister K, Bechade C, Calothy G, Leutz A (1986) v-*mil* induces autocrine growth and enhanced tumorigenicity in v-*myc*-transformed avian macrophages. Cell 45:357–364

Grandori C, Cowley SM, James LP, Eisenman RN (2000) The Myc/Max/Mad network and the transcriptional control of cell behavior. Annu Rev Cell Dev Biol 16:653–699

Grandori C, Gomez-Roman N, Felton-Edkins ZA, Ngouenet C, Galloway DA, Eisenman RN, White RJ (2005) c-Myc binds to human ribosomal DNA and stimulates transcription of rRNA genes by RNA polymerase I. Nat Cell Biol 7:311–318

Gregory MA, Qi Y, Hann SR (2005) The ARF tumor suppressor: keeping Myc on a leash. Cell Cycle 4:249–252

Gross S, Rahal R, Stransky N, Lengauer C, Hoeflich KP (2015) Targeting cancer with kinase inhibitors. J Clin Invest 125:1780–1789

Habib T, Park H, Tsang M, de Alborán IM, Nicks A, Wilson L, Knoepfler PS, Andrews S, Rawlings DJ, Eisenman RN, Iritani BM (2007) Myc stimulates B lymphocyte differentiation and amplifies calcium signaling. J Cell Biol 179:717–731

Hänsel-Hertsch R, Beraldi D, Lensing SV, Marsico G, Zyner K, Parry A, Di Antonio M, Pike J, Kimura H, Narita M, Tannahill D, Balasubramanian S (2016) G-quadruplex structures mark human regulatory chromatin. Nat Genet 48:1267–1272

Hart JR, Roberts TC, Weinberg MS, Morris KV, Vogt PK (2014a) MYC regulates the non-coding transcriptome. Oncotarget 5:12543–12554

Hart JR, Garner AL, Yu J, Ito Y, Sun M, Ueno L, Rhee JK, Baksh MM, Stefan E, Hartl M, Bister K, Vogt PK, Janda KD (2014b) Inhibitor of MYC identified in a Kröhnke pyridine library. Proc Natl Acad Sci USA 111:12556–12561

Hartl M, Nist A, Khan MI, Valovka T, Bister K (2009) Inhibition of Myc-induced cell transformation by brain acid-soluble protein 1 (BASP1). Proc Natl Acad Sci USA 106:5604–5609

Hartl M, Mitterstiller AM, Valovka T, Breuker K, Hobmayer B, Bister K (2010) Stem cell-specific activation of an ancestral *myc* protooncogene with conserved basic functions in the early metazoan *Hydra*. Proc Natl Acad Sci USA 107:4051–4056

Hatzivassiliou G, Song K, Yen I, Brandhuber BJ, Anderson DJ, Alvarado R, Ludlam MJ, Stokoe D, Gloor SL, Vigers G, Morales T, Aliagas I, Liu B, Sideris S, Hoeflich KP, Jaiswal BS, Seshagiri S, Koeppen H, Belvin M, Friedman LS, Malek S (2010) RAF inhibitors prime wild-type RAF to activate the MAPK pathway and enhance growth. Nature 464:431–435

Hauschild A, Grob JJ, Demidov LV, Jouary T, Gutzmer R, Millward M, Rutkowski P, Blank CU, Miller WH Jr, Kaempgen E, Martín-Algarra S, Karaszewska B, Mauch C, Chiarion-Sileni V, Martin AM, Swann S, Haney P, Mirakhur B, Guckert ME, Goodman V, Chapman PB (2012) Dabrafenib in BRAF-mutated metastatic melanoma: a multicentre, open-label, phase 3 randomised controlled trial. Lancet 380:358–365

Hayward WS, Neel BG, Astrin SM (1981) Activation of a cellular *onc* gene by promoter insertion in ALV-induced lymphoid leukosis. Nature 290:475–480

He L, Thomson JM, Hemann MT, Hernando-Monge E, Mu D, Goodson S, Powers S, Cordon-Cardo C, Lowe SW, Hannon GJ, Hammond SM (2005) A microRNA polycistron as a potential human oncogene. Nature 435:828–833

He TC, Sparks AB, Rago C, Hermeking H, Zawel L, da Costa LT, Morin PJ, Vogelstein B, Kinzler KW (1998) Identification of c-MYC as a target of the APC pathway. Science 281:1509–1512

Heidorn SJ, Milagre C, Whittaker S, Nourry A, Niculescu-Duvas I, Dhomen N, Hussain J, Reis-Filho JS, Springer CJ, Pritchard C, Marais R (2010) Kinase-dead BRAF and oncogenic RAS cooperate to drive tumor progression through CRAF. Cell 140:209–221

Hermeking H, Rago C, Schuhmacher M, Li Q, Barrett JF, Obaya AJ, O'Connell BC, Mateyak MK, Tam W, Kohlhuber F, Dang CV, Sedivy JM, Eick D, Vogelstein B, Kinzler KW (2000) Identification of CDK4 as a target of c-MYC. Proc Natl Acad Sci USA 97:2229–2234

Holderfield M, Deuker MM, McCormick F, McMahon M (2014a) Targeting RAF kinases for cancer therapy: BRAF-mutated melanoma and beyond. Nat Rev Cancer 14:455–467

Holderfield M, Nagel TE, Stuart DD (2014b) Mechanism and consequences of RAF kinase activation by small-molecule inhibitors. Br J Cancer 111:640–645

Hollenhorst PC, McIntosh LP, Graves BJ (2011) Genomic and biochemical insights into the specificity of ETS transcription factors. Annu Rev Biochem 80:437–471

Howe LR, Leevers SJ, Gómez N, Nakielny S, Cohen P, Marshall CJ (1992) Activation of the MAP kinase pathway by the protein kinase raf. Cell 71:335–342

Hsieh AL, Walton ZE, Altman BJ, Stine ZE, Dang CV (2015) MYC and metabolism on the path to cancer. Semin Cell Dev Biol 43:11–21

Hu J, Stites EC, Yu H, Germino EA, Meharena HS, Stork PJ, Kornev AP, Taylor SS, Shaw AS (2013) Allosteric activation of functionally asymmetric RAF kinase dimers. Cell 154:1036–1046

Hu J, Ahuja LG, Meharena HS, Kannan N, Kornev AP, Taylor SS, Shaw AS (2015) Kinase regulation by hydrophobic spine assembly in cancer. Mol Cell Biol 35:264–276

Huebner K, ar-Rushdi A, Griffin CA, Isobe M, Kozak C, Emanuel BS, Nagarajan L, Cleveland JL, Bonner TI, Goldsborough MD, Croce CM, Rapp U (1986) Actively transcribed genes in the raf oncogene group, located on the X chromosome in mouse and human. Proc Natl Acad Sci USA 83:3934–3938

Hung CL, Wang LY, Yu YL, Chen HW, Srivastava S, Petrovics G, Kung HJ (2014) A long noncoding RNA connects c-Myc to tumor metabolism. Proc Natl Acad Sci USA 111:18697–18702

Hurlin PJ, Quéva C, Eisenman RN (1997) Mnt, a novel Max-interacting protein is coexpressed with Myc in proliferating cells and mediates repression at Myc binding sites. Genes Dev 11:44–58

Hurlin PJ, Steingrìmsson E, Copeland NG, Jenkins NA, Eisenman RN (1999) Mga, a dual-specificity transcription factor that interacts with Max and contains a T-domain DNA-binding motif. EMBO J 18:7019–7028

Ikawa S, Fukui M, Ueyama Y, Tamaoki N, Yamamoto T, Toyoshima K (1988) B raf, a new member of the raf family, is activated by DNA rearrangement. Mol Cell Biol 8:2651–2654

Jansen HW, Patschinsky T, Bister K (1983a) Avian oncovirus MH2: molecular cloning of proviral DNA and structural analysis of viral RNA and protein. J Virol 48:61–73

Jansen HW, Rueckert B, Lurz R, Bister K (1983b) Two unrelated cell–derived sequences in the genome of avian leukemia and carcinoma inducing retrovirus MH2. EMBO J 2:1969–1975

Jansen HW, Lurz R, Bister K, Bonner TI, Mark GE, Rapp UR (1984) Homologous cell–derived oncogenes in avian carcinoma virus MH2 and murine sarcoma virus 3611. Nature 307:281–284

Kan NC, Flordellis CS, Garon CF, Duesberg PH, Papas TS (1983) Avian carcinoma virus MH2 contains a transformation-specific sequence, *mht*, and shares the *myc* sequence with MC29, CMII, and OK10 viruses. Proc Natl Acad Sci USA 80:6566–6570

Kanatsu-Shinohara M, Tanaka T, Ogonuki N, Ogura A, Morimoto H, Cheng PF, Eisenman RN, Trumpp A, Shinohara T (2016) Myc/Mycn-mediated glycolysis enhances mouse spermatogonial stem cell self-renewal. Genes Dev 30:2637–2648

Karnoub AE, Weinberg RA (2008) Ras oncogenes: split personalities. Nat Rev Mol Cell Biol 9:517–531

Kelly K, Cochran BH, Stiles CD, Leder P (1983) Cell-specific regulation of the c-myc gene by lymphocyte mitogens and platelet-derived growth factor. Cell 35:603–610

Kessler JD, Kahle KT, Sun T, Meerbrey KL, Schlabach MR, Schmitt EM, Skinner SO, Xu Q, Li MZ, Hartman ZC, Rao M, Yu P, Dominguez-Vidana R, Liang AC, Solimini NL, Bernardi RJ, Yu B, Hsu T, Golding I, Luo J, Osborne CK, Creighton CJ, Hilsenbeck SG, Schiff R, Shaw CA, Elledge SJ, Westbrook TF (2012) A SUMOylation-dependent transcriptional subprogram is required for Myc-driven tumorigenesis. Science 335:348–353

Kim T, Jeon YJ, Cui R, Lee JH, Peng Y, Kim SH, Tili E, Alder H, Croce CM (2015) Role of MYC-regulated long noncoding RNAs in cell cycle regulation and tumorigenesis. J Natl Cancer Inst 107:dju505

Knoepfler PS, Kenney AM (2006) Neural precursor cycling at sonic speed: N-Myc pedals, GSK-3 brakes. Cell Cycle 5:47–52

Koppenol WH, Bounds PL, Dang CV (2011) Otto Warburg's contributions to current concepts of cancer metabolism. Nat Rev Cancer 11:325–337

Kumar MS, Lu J, Mercer KL, Golub TR, Jacks T (2007) Impaired microRNA processing enhances cellular transformation and tumorigenesis. Nat Genet 39:673–677

Kyriakis JM, App H, Zhang XF, Banerjee P, Brautigan DL, Rapp UR, Avruch J (1992) Raf-1 activates MAP kinase-kinase. Nature 358:417–421

Lake D, Correa SA, Müller J (2016) Negative feedback regulation of the ERK1/2 MAPK pathway. Cell Mol Life Sci 73:4397–4413

Lal A, Navarro F, Maher CA, Maliszewski LE, Yan N, O'Day E, Chowdhury D, Dykxhoorn DM, Tsai P, Hofmann O, Becker KG, Gorospe M, Hide W, Lieberman J (2009) miR-24 inhibits cell proliferation by targeting E2F2, MYC, and other cell-cycle genes via binding to "seedless" 3′ UTR microRNA recognition elements. Mol Cell 35:610–625

Laurenti E, Wilson A, Trumpp A (2009) Myc's other life: stem cells and beyond. Curr Opin Cell Biol 21:844–854

Lavoie H, Therrien M (2015) Regulation of RAF protein kinases in ERK signalling. Nat Rev Mol Cell Biol 16:281–298

Lavoie H, Thevakumaran N, Gavory G, Li JJ, Padeganeh A, Guiral S, Duchaine J, Mao DY, Bouvier M, Sicheri F, Therrien M (2013) Inhibitors that stabilize a closed RAF kinase domain conformation induce dimerization. Nat Chem Biol 9:428–436

Lin CH, Jackson AL, Guo J, Linsley PS, Eisenman RN (2009) Myc-regulated microRNAs attenuate embryonic stem cell differentiation. EMBO J 28:3157–3170

Lin CY, Lovén J, Rahl PB, Paranal RM, Burge CB, Bradner JE, Lee TI, Young RA (2012) Transcriptional amplification in tumor cells with elevated c-Myc. Cell 151:56–67

Lito P, Pratilas CA, Joseph EW, Tadi M, Halilovic E, Zubrowski M, Huang A, Wong WL, Callahan MK, Merghoub T, Wolchok JD, de Stanchina E, Chandarlapaty S, Poulikakos PI, Fagin JA, Rosen N (2012) Relief of profound feedback inhibition of mitogenic signaling by RAF inhibitors attenuates their activity in BRAFV600E melanomas. Cancer Cell 22:668–682

Lito P, Rosen N, Solit DB (2013) Tumor adaptation and resistance to RAF inhibitors. Nat Med 19:1401–1409

Liu J, Levens D (2006) Making myc. Curr Top Microbiol Immunol 302:1–32

Liu P, Cheng H, Santiago S, Raeder M, Zhang F, Isabella A, Yang J, Semaan DJ, Chen C, Fox EA, Gray NS, Monahan J, Schlegel R, Beroukhim R, Mills GB, Zhao JJ (2011) Oncogenic PIK3CA-driven mammary tumors frequently recur via PI3K pathway-dependent and PI3K pathway-independent mechanisms. Nat Med 17:1116–1120

Louis SF, Vermolen BJ, Garini Y, Young IT, Guffei A, Lichtensztejn Z, Kuttler F, Chuang TC, Moshir S, Mougey V, Chuang AY, Kerr PD, Fest T, Boukamp P, Mai S (2005) c-Myc induces chromosomal rearrangements through telomere and chromosome remodeling in the interphase nucleus. Proc Natl Acad Sci USA 102:9613–9618

Manning G, Whyte DB, Martinez R, Hunter T, Sudarsanam S (2002) The protein kinase complement of the human genome. Science 298:1912–1934

Marks JL, Gong Y, Chitale D, Golas B, McLellan MD, Kasai Y, Ding L, Mardis ER, Wilson RK, Solit D, Levine R, Michel K, Thomas RK, Rusch VW, Ladanyi M, Pao W (2008) Novel MEK1 mutation identified by mutational analysis of epidermal growth factor receptor signaling pathway genes in lung adenocarcinoma. Cancer Res 68:5524–5528

Marx M, Eychène A, Laugier D, Béchade C, Crisanti P, Dezélée P, Pessac B, Calothy G (1988) A novel oncogene related to c-mil is transduced in chicken neuroretina cells induced to proliferate by infection with an avian lymphomatosis virus. EMBO J 7:3369–3373

Matallanas D, Birtwistle M, Romano D, Zebisch A, Rauch J, von Kriegsheim A, Kolch W (2011) Raf family kinases: old dogs have learned new tricks. Genes Cancer 2:232–260

McCormick F (2016) K-Ras protein as a drug target. J Mol Med 94:253–258

McFadden DG, Politi K, Bhutkar A, Chen FK, Song X, Pirun M, Santiago PM, Kim-Kiselak C, Platt JT, Lee E, Hodges E, Rosebrock AP, Bronson RT, Socci ND, Hannon GJ, Jacks T, Varmus H (2016) Mutational landscape of *EGFR-*, *MYC-*, and *Kras*-driven genetically engineered mouse models of lung adenocarcinoma. Proc Natl Acad Sci USA 113:E6409–E6417

McKeown MR, Bradner JE (2014) Therapeutic strategies to inhibit MYC. Cold Spring Harb Perspect Med 4:a014266

McMahon SB, Wood MA, Cole MD (2000) The essential cofactor TRRAP recruits the histone acetyltransferase hGCN5 to c-Myc. Mol Cell Biol 20:556–562

Mellon P, Pawson A, Bister K, Martin GS, Duesberg PH (1978) Specific RNA sequences and gene products of MC29 avian acute leukemia virus. Proc Natl Acad Sci USA 75:5874–5878
Mendell JT (2005) MicroRNAs: critical regulators of development, cellular physiology and malignancy. Cell Cycle 4:1179–1184
Mendell JT (2008) miRiad roles for the miR-17-92 cluster in development and disease. Cell 133:217–222
Mertz JA, Conery AR, Bryant BM, Sandy P, Balasubramanian S, Mele DA, Bergeron L, Sims RJ III (2011) Targeting MYC dependence in cancer by inhibiting BET bromodomains. Proc Natl Acad Sci USA 108:16669–16674
Moodie SA, Willumsen BM, Weber MJ, Wolfman A (1993) Complexes of Ras.GTP with Raf-1 and mitogen-activated protein kinase kinase. Science 260:1658–1661
Morton S, Davis RJ, McLaren A, Cohen P (2003) A reinvestigation of the multisite phosphorylation of the transcription factor c-Jun. EMBO J 22:3876–3886
Murphy LO, Blenis J (2006) MAPK signal specificity: the right place at the right time. Trends Biochem Sci 31:268–275
Murugan AK, Dong J, Xie J, Xing M (2009) MEK1 mutations, but not ERK2 mutations, occur in melanomas and colon carcinomas, but none in thyroid carcinomas. Cell Cycle 8:2122–2124
Nagashima T, Inoue N, Yumoto N, Saeki Y, Magi S, Volinsky N, Sorkin A, Kholodenko BN, Okada-Hatakeyama M (2015) Feedforward regulation of mRNA stability by prolonged extracellular signal-regulated kinase activity. FEBS J 282:613–629
Nair SK, Burley SK (2003) X-ray structures of Myc-Max and Mad-Max recognizing DNA. Molecular bases of regulation by proto-oncogenic transcription factors. Cell 112:193–205
Nau MM, Brooks BJ, Battey J, Sausville E, Gazdar AF, Kirsch IR, McBride OW, Bertness V, Hollis GF, Minna JD (1985) L-myc, a new myc-related gene amplified and expressed in human small cell lung cancer. Nature 318:69–73
Nie Z, Hu G, Wei G, Cui K, Yamane A, Resch W, Wang R, Green DR, Tessarollo L, Casellas R, Zhao K, Levens D (2012) c-Myc is a universal amplifier of expressed genes in lymphocytes and embryonic stem cells. Cell 151:68–79
O'Donnell KA, Wentzel EA, Zeller KI, Dang CV, Mendell JT (2005) c-Myc-regulated microRNAs modulate E2F1 expression. Nature 435:839–843
Okazaki K, Sagata N (1995) The Mos/MAP kinase pathway stabilizes c-Fos by phosphorylation and augments its transforming activity in NIH 3T3 cells. EMBO J 14:5048–5059
Okimoto RA, Lin L, Olivas V, Chan E, Markegard E, Rymar A, Neel D, Chen X, Hemmati G, Bollag G, Bivona TG (2016) Preclinical efficacy of a RAF inhibitor that evades paradoxical MAPK pathway activation in protein kinase BRAF-mutant lung cancer. Proc Natl Acad Sci USA 113:13456–13461
Payne GS, Bishop JM, Varmus HE (1982) Multiple arrangements of viral DNA and an activated host oncogene in bursal lymphomas. Nature 295:209–214
Poulikakos PI, Zhang C, Bollag G, Shokat KM, Rosen N (2010) RAF inhibitors transactivate RAF dimers and ERK signalling in cells with wild-type BRAF. Nature 464:427–430
Poulikakos PI, Persaud Y, Janakiraman M, Kong X, Ng C, Moriceau G, Shi H, Atefi M, Titz B, Gabay MT, Salton M, Dahlman KB, Tadi M, Wargo JA, Flaherty KT, Kelley MC, Misteli T, Chapman PB, Sosman JA, Graeber TG, Ribas A, Lo RS, Rosen N, Solit DB (2011) RAF inhibitor resistance is mediated by dimerization of aberrantly spliced BRAF(V600E). Nature 480:387–390
Prahallad A, Sun C, Huang S, Di Nicolantonio F, Salazar R, Zecchin D, Beijersbergen RL, Bardelli A, Bernards R (2012) Unresponsiveness of colon cancer to BRAF(V600E) inhibition through feedback activation of EGFR. Nature 483:100–103
Prochownik EV, Vogt PK (2010) Therapeutic targeting of Myc. Genes. Cancer 1:650–659
Pylayeva-Gupta Y, Grabocka E, Bar-Sagi D (2011) RAS oncogenes: weaving a tumorigenic web. Nat Rev Cancer 11:761–774
Qi Y, Gregory MA, Li Z, Brousal JP, West K, Hann SR (2004) p19ARF directly and differentially controls the functions of c-Myc independently of p53. Nature 431:712–717

Raffeiner P, Röck R, Schraffl A, Hartl M, Hart JR, Janda KD, Vogt PK, Stefan E, Bister K (2014) In vivo quantification and perturbation of Myc-Max interactions and the impact on oncogenic potential. Oncotarget 5:8869–8878

Raffeiner P, Schraffl A, Schwarz T, Röck R, Ledolter K, Hartl M, Konrat R, Stefan E, Bister K (2017) Calcium-dependent binding of Myc to calmodulin. Oncotarget 8:3327–3343

Rahl PB, Young RA (2014) MYC and transcription elongation. Cold Spring Harb Perspect Med 4: a020990

Rajagopalan H, Nowak MA, Vogelstein B, Lengauer C (2003) The significance of unstable chromosomes in colorectal cancer. Nat Rev Cancer 3:695–701

Rajakulendran T1, Sahmi M, Lefrançois M, Sicheri F, Therrien M (2009) A dimerization-dependent mechanism drives RAF catalytic activation. Nature 461:542–545

Rapp UR, Goldsborough MD, Mark GE, Bonner TI, Groffen J, Reynolds FH Jr, Stephenson JR (1983) Structure and biological activity of v-*raf*, a unique oncogene transduced by a retrovirus. Proc Natl Acad Sci USA 80:4218–4222

Rapp UR, Cleveland JL, Fredrickson TN, Holmes KL, Morse HC III, Jansen HW, Patschinsky T, Bister K (1985) Rapid induction of hemopoietic neoplasms in newborn mice by a *raf(mil)/myc* recombinant murine retrovirus. J Virol 55:23–33

Ratnikov BI, Scott DA, Osterman AL, Smith JW, Ronai ZA (2017) Metabolic rewiring in melanoma. Oncogene 36:147–157

Rauch N, Rukhlenko OS, Kolch W, Kholodenko BN (2016) MAPK kinase signalling dynamics regulate cell fate decisions and drug resistance. Curr Opin Struct Biol 41:151–158

Robins T, Bister K, Garon C, Papas T, Duesberg PH (1982) Structural relationship between a normal chicken DNA locus and the transforming gene of the avian acute leukemia virus MC29. J Virol 41:635–642

Robinson MJ, Cobb MH (1997) Mitogen-activated protein kinase pathways. Curr Opin Cell Biol 9:180–186

Roskoski R Jr (2012a) MEK1/2 dual-specificity protein kinases: structure and regulation. Biochem Biophys Res Commun 417:5–10

Roskoski R Jr (2012b) ERK1/2 MAP kinases: structure, function, and regulation. Pharmacol Res 66:105–143

Roussel M, Saule S, Lagrou C, Rommens C, Beug H, Graf T, Stehelin D (1979) Three new types of viral oncogene of cellular origin specific for haematopoietic cell transformation. Nature 281:452–455

Samatar AA, Poulikakos PI (2014) Targeting RAS-ERK signalling in cancer: promises and challenges. Nat Rev Drug Discov 13:928–942

Santos SD, Verveer PJ, Bastiaens PI (2007) Growth factor-induced MAPK network topology shapes Erk response determining PC-12 cell fate. Nat Cell Biol 9:324–330

Scognamiglio R, Cabezas-Wallscheid N, Thier MC, Altamura S, Reyes A, Prendergast ÁM, Baumgärtner D, Carnevalli LS, Atzberger A, Haas S, von Paleske L, Boroviak T, Wörsdörfer P, Essers MA, Kloz U, Eisenman RN, Edenhofer F, Bertone P, Huber W, van der Hoeven F, Smith A, Trumpp A (2016) Myc depletion induces a pluripotent dormant state mimicking diapause. Cell 164:668–680

Sears RC (2004) The life cycle of c-Myc: from synthesis to degradation. Cell Cycle 3:1133–1137

Sheiness D, Bishop JM (1979) DNA and RNA from uninfected vertebrate cells contain nucleotide sequences related to the putative transforming gene of avian myelocytomatosis virus. J Virol 31:514–521

Sheiness D, Fanshier L, Bishop JM (1978) Identification of nucleotide sequences which may encode the oncogenic capacity of avian retrovirus MC29. J Virol 28:600–610

Shim H, Dolde C, Lewis BC, Wu CS, Dang G, Jungmann RA, Dalla-Favera R, Dang CV (1997) c-Myc transactivation of LDH-A: implications for tumor metabolism and growth. Proc Natl Acad Sci USA 94:6658–6663

Shortt J, Johnstone RW (2012) Oncogenes in cell survival and cell death. Cold Spring Harb Perspect Biol 4:a009829

Shroff EH, Eberlin LS, Dang VM, Gouw AM, Gabay M, Adam SJ, Bellovin DI, Tran PT, Philbrick WM, Garcia-Ocana A, Casey SC, Li Y, Dang CV, Zare RN, Felsher DW (2015) MYC oncogene overexpression drives renal cell carcinoma in a mouse model through glutamine metabolism. Proc Natl Acad Sci USA 112:6539–6544

Shu S, Polyak K (2017) BET bromodomain proteins as cancer therapeutic targets. Cold Spring Harb Symp Quant Biol 81:030908

Solit DB, Rosen N (2011) Resistance to BRAF inhibition in melanomas. N Engl J Med 364:772–774

Soucek L, Evan GI (2010) The ups and downs of Myc biology. Curr Opin Genet Dev 20:91–95

Soucek L, Helmer-Citterich M, Sacco A, Jucker R, Cesareni G, Nasi S (1998) Design and properties of a Myc derivative that efficiently homodimerizes. Oncogene 17:2463–2472

Soucek L, Whitfield J, Martins CP, Finch AJ, Murphy DJ, Sodir NM, Karnezis AN, Swigart LB, Nasi S, Evan GI (2008) Modelling Myc inhibition as a cancer therapy. Nature 455:679–683

Stefan E, Hart JR, Bister K (2015) Stopping MYC in its tracks. Aging 7:463–464

Stehelin D, Varmus HE, Bishop JM, Vogt PK (1976) DNA related to the transforming gene(s) of avian sarcoma viruses is present in normal avian DNA. Nature 260:170–173

Stephen AG, Esposito D, Bagni RK, McCormick F (2014) Dragging ras back in the ring. Cancer Cell 25:272–281

Stine ZE, Walton ZE, Altman BJ, Hsieh AL, Dang CV (2015) MYC, metabolism, and cancer. Cancer Discov 5:1024–1039

Su F, Viros A, Milagre C, Trunzer K, Bollag G, Spleiss O, Reis-Filho JS, Kong X, Koya RC, Flaherty KT, Chapman PB, Kim MJ, Hayward R, Martin M, Yang H, Wang Q, Hilton H, Hang JS, Noe J, Lambros M, Geyer F, Dhomen N, Niculescu-Duvaz I, Zambon A, Niculescu-Duvaz D, Preece N, Robert L, Otte NJ, Mok S, Kee D, Ma Y, Zhang C, Habets G, Burton EA, Wong B, Nguyen H, Kockx M, Andries L, Lestini B, Nolop KB, Lee RJ, Joe AK, Troy JL, Gonzalez R, Hutson TE, Puzanov I, Chmielowski B, Springer CJ, McArthur GA, Sosman JA, Lo RS, Ribas A, Marais R (2012) RAS mutations in cutaneous squamous-cell carcinomas in patients treated with BRAF inhibitors. N Engl J Med 366:207–215

Sun C, Wang L, Huang S, Heynen GJ, Prahallad A, Robert C, Haanen J, Blank C, Wesseling J, Willems SM, Zecchin D, Hobor S, Bajpe PK, Lieftink C, Mateus C, Vagner S, Grernrum W, Hofland I, Schlicker A, Wessels LF, Beijersbergen RL, Bardelli A, Di Nicolantonio F, Eggermont AM, Bernards R (2014) Reversible and adaptive resistance to BRAF(V600E) inhibition in melanoma. Nature 508:118–122

Tabor V, Bocci M, Alikhani N, Kuiper R, Larsson LG (2014) MYC synergizes with activated BRAFV600E in mouse lung tumor development by suppressing senescence. Cancer Res 74:4222–4229

Taub R, Kirsch I, Morton C, Lenoir G, Swan D, Tronick S, Aaronson S, Leder P (1982) Translocation of the c-*myc* gene into the immunoglobulin heavy chain locus in human Burkitt lymphoma and murine plasmacytoma cells. Proc Natl Acad Sci USA 79:7837–7841

Taylor SS, Kornev AP (2011) Protein kinases: evolution of dynamic regulatory proteins. Trends Biochem Sci 36:65–77

Thomas LR, Wang Q, Grieb BC, Phan J, Foshage AM, Sun Q, Olejniczak ET, Clark T, Dey S, Lorey S, Alicie B, Howard GC, Cawthon B, Ess KC, Eischen CM, Zhao Z, Fesik SW, Tansey WP (2015) Interaction with WDR5 promotes target gene recognition and tumorigenesis by MYC. Mol Cell 58:440–452

Thomas LR, Foshage AM, Weissmiller AM, Popay TM, Grieb BC, Qualls SJ, Ng V, Carboneau B, Lorey S, Eischen CM, Tansey WP (2016) Interaction of MYC with host cell factor-1 is mediated by the evolutionarily conserved Myc box IV motif. Oncogene 35:3613–3618

Tiacci E, Trifonov V, Schiavoni G, Holmes A, Kern W, Martelli MP, Pucciarini A, Bigerna B, Pacini R, Wells VA, Sportoletti P, Pettirossi V, Mannucci R, Elliott O, Liso A, Ambrosetti A, Pulsoni A, Forconi F, Trentin L, Semenzato G, Inghirami G, Capponi M, Di Raimondo F, Patti C, Arcaini L, Musto P, Pileri S, Haferlach C, Schnittger S, Pizzolo G, Foà R, Farinelli L,

Haferlach T, Pasqualucci L, Rabadan R, Falini B (2011) BRAF mutations in hairy-cell leukemia. N Engl J Med 364:2305–2315
Tokheim CJ, Papadopoulos N, Kinzler KW, Vogelstein B, Karchin R (2016) Evaluating the evaluation of cancer driver genes. Proc Natl Acad Sci USA 113:14330–14335
Toyoshima M, Howie HL, Imakura M, Walsh RM, Annis JE, Chang AN, Frazier J, Chau BN, Loboda A, Linsley PS, Cleary MA, Park JR, Grandori C (2012) Functional genomics identifies therapeutic targets for MYC-driven cancer. Proc Natl Acad Sci USA 109:9545–9550
Tsavachidou D, Coleman ML, Athanasiadis G, Li S, Licht JD, Olson MF, Weber BL (2004) SPRY2 is an inhibitor of the ras/extracellular signal-regulated kinase pathway in melanocytes and melanoma cells with wild-type BRAF but not with the V599E mutant. Cancer Res 64:5556–5559
Tseng YY, Moriarity BS, Gong W, Akiyama R, Tiwari A, Kawakami H, Ronning P, Reuland B, Guenther K, Beadnell TC, Essig J, Otto GM, O'Sullivan MG, Largaespada DA, Schwertfeger KL, Marahrens Y, Kawakami Y, Bagchi A (2014) PVT1 dependence in cancer with MYC copy-number increase. Nature 512:82–86
Turajlic S, Swanton C (2016) Metastasis as an evolutionary process. Science 352:169–175
Valovka T, Schönfeld M, Raffeiner P, Breuker K, Dunzendorfer-Matt T, Hartl M, Bister K (2013) Transcriptional control of DNA replication licensing by Myc. Sci Rep 3:3444
Van Aelst L, Barr M, Marcus S, Polverino A, Wigler M (1993) Complex formation between RAS and RAF and other protein kinases. Proc Natl Acad Sci USA 90:6213–6217
Vennstrom B, Sheiness D, Zabielski J, Bishop JM (1982) Isolation and characterization of c-*myc*, a cellular homolog of the oncogene (v-*myc*) of avian myelocytomatosis virus strain 29. J Virol 42:773–779
Vita M, Henriksson M (2006) The Myc oncoprotein as a therapeutic target for human cancer. Semin Cancer Biol 16:318–330
Vogelstein B, Papadopoulos N, Velculescu VE, Zhou S, Diaz LA Jr, Kinzler KW (2013) Cancer genome landscapes. Science 29:1546–1558
Vogt PK (2012) Retroviral oncogenes: a historical primer. Nat Rev Cancer 12:639–648
Vojtek AB, Hollenberg SM, Cooper JA (1993) Mammalian Ras interacts directly with the serine/threonine kinase Raf. Cell 74:205–214
Wan PTC, Garnett MJ, Roe SM, Lee S, Niculescu-Duvaz D, Good VM, Cancer Genome Project, Jones CM, Marshall CJ, Springer CJ, Barford D, Marais R (2004) Mechanism of activation of the RAF-ERK signaling pathway by oncogenic mutations of B-RAF. Cell 116:855–867
Wanzel M, Herold S, Eilers M (2003) Transcriptional repression by Myc. Trends Cell Biol 13:146–150
Warburg O, Posener K, Negelein E (1924) Über den Stoffwechsel der Carcinomzelle. Biochem Z 152:309–344
Wasylyk B, Hagman J, Gutierrez-Hartmann A (1998) Ets transcription factors: nuclear effectors of the Ras-MAP-kinase signaling pathway. Trends Biochem Sci 23:213–216
Weinberg MS, Hart JR, Vogt PK (2015) A brave new MYC-amplified world. Aging 7:459–460
Welcker M, Orian A, Jin J, Grim JE, Harper JW, Eisenman RN, Clurman BE (2004) The Fbw7 tumor suppressor regulates glycogen synthase kinase 3 phosphorylation-dependent c-Myc protein degradation. Proc Natl Acad Sci USA 101:9085–9090
Wellbrock C, Karasarides M, Marais R (2004) The RAF proteins take centre stage. Nat Rev Mol Cell Biol 5:875–885
Weng AP, Millholland JM, Yashiro-Ohtani Y, Arcangeli ML, Lau A, Wai C, Del Bianco C, Rodriguez CG, Sai H, Tobias J, Li Y, Wolfe MS, Shachaf C, Felsher D, Blacklow SC, Pear WS, Aster JC (2006) c-Myc is an important direct target of Notch1 in T-cell acute lymphoblastic leukemia/lymphoma. Genes Dev 20:2096–2109
Wilde BR, Ayer DE (2015) Interactions between Myc and MondoA transcription factors in metabolism and tumourigenesis. Br J Cancer 113:1529–1533
Wolf E, Lin CY, Eilers M, Levens DL (2015) Taming of the beast: shaping Myc-dependent amplification. Trends Cell Biol 25:241–248

Yada M, Hatakeyama S, Kamura T, Nishiyama M, Tsunematsu R, Imaki H, Ishida N, Okumura F, Nakayama K, Nakayama KI (2004) Phosphorylation-dependent degradation of c-Myc is mediated by the F-box protein Fbw7. EMBO J 23:2116–2125

Yang D, Liu H, Goga A, Kim S, Yuneva M, Bishop JM (2010) Therapeutic potential of a synthetic lethal interaction between the MYC proto-oncogene and inhibition of aurora-B kinase. Proc Natl Acad Sci USA 107:13836–13841

Yao Z, Torres NM, Tao A, Gao Y, Luo L, Li Q, de Stanchina E, Abdel-Wahab O, Solit DB, Poulikakos PI, Rosen N (2015) BRAF mutants evade ERK-dependent feedback by different mechanisms that determine their sensitivity to pharmacologic inhibition. Cancer Cell 28:370–383

Yin X, Giap C, Lazo JS, Prochownik EV (2003) Low molecular weight inhibitors of Myc-Max interaction and function. Oncogene 22:6151–6159

Yochum GS, Sherrick CM, Macpartlin M, Goodman RH (2010) A β-catenin/TCF-coordinated chromatin loop at MYC integrates 5′ and 3′ Wnt responsive enhancers. Proc Natl Acad Sci USA 107:145–150

Yoon S, Seger R (2006) The extracellular signal-regulated kinase: multiple substrates regulate diverse cellular functions. Growth Factors 24:21–44

Young SL, Diolaiti D, Conacci-Sorrell M, Ruiz-Trillo I, Eisenman RN, King N (2011) Premetazoan ancestry of the Myc-Max network. Mol Biol Evol 28:2961–2971

Yu W, Fantl WJ, Harrowe G, Williams LT (1998) Regulation of the MAP kinase pathway by mammalian Ksr through direct interaction with MEK and ERK. Curr Biol 8:56–64

Zeller KI, Zhao X, Lee CW, Chiu KP, Yao F, Yustein JT, Ooi HS, Orlov YL, Shahab A, Yong HC, Fu Y, Weng Z, Kuznetsov VA, Sung WK, Ruan Y, Dang CV, Wei CL (2006) Global mapping of c-Myc binding sites and target gene networks in human B cells. Proc Natl Acad Sci USA 103:17834–17839

Zhang XF, Settleman J, Kyriakis JM, Takeuchi-Suzuki E, Elledge SJ, Marshall MS, Bruder JT, Rapp UR, Avruch J (1993) Normal and oncogenic p21ras proteins bind to the amino-terminal regulatory domain of c-Raf-1. Nature 364:308–313

Zhang C, Spevak W, Zhang Y, Burton EA, Ma Y, Habets G, Zhang J, Lin J, Ewing T, Matusow B, Tsang G, Marimuthu A, Cho H, Wu G, Wang W, Fong D, Nguyen H, Shi S, Womack P, Nespi M, Shellooe R, Carias H, Powell B, Light E, Sanftner L, Walters J, Tsai J, West BL, Visor G, Rezaei H, Lin PS, Nolop K, Ibrahim PN, Hirth P, Bollag G (2015) RAF inhibitors that evade paradoxical MAPK pathway activation. Nature 526:583–586

Zindy F, Eischen CM, Randle DH, Kamijo T, Cleveland JL, Sherr CJ, Roussel MF (1998) Myc signaling via the ARF tumor suppressor regulates p53-dependent apoptosis and immortalization. Genes Dev 12:2424–2433

Oncogenic Roles of the PI3K/AKT/mTOR Axis

Masahiro Aoki and Teruaki Fujishita

Abstract The PI3K/AKT/mTOR pathway is frequently activated in various human cancers and has been considered a promising therapeutic target. Many of the positive regulators of the PI3K/AKT/mTOR axis, including the catalytic (p110α) and regulatory (p85α), of class IA PI3K, AKT, RHEB, mTOR, and eIF4E, possess oncogenic potentials, as demonstrated by transformation assays in vitro and by genetically engineered mouse models in vivo. Genetic evidences also indicate their roles in malignancies induced by activation of the upstream oncoproteins including receptor tyrosine kinases and RAS and those induced by the loss of the negative regulators of the PI3K/AKT/mTOR pathway such as PTEN, TSC1/2, LKB1, and PIPP. Possible mechanisms by which the PI3K/AKT/mTOR axis contributes to oncogenic transformation include stimulation of proliferation, survival, metabolic reprogramming, and invasion/metastasis, as well as suppression of autophagy and senescence. These phenotypic changes are mediated by eIF4E-induced translation of a subset of mRNAs and by other downstream effectors of mTORC1 including S6K, HIF-1α, PGC-1α, SREBP, and ULK1 complex.

Contents

M. Aoki (✉) · T. Fujishita
Division of Molecular Pathology, Aichi Cancer Center Research Institute, 1-1 Kanokoden, Chikusa-ku, Nagoya, Aichi 464-8681, Japan
e-mail: masaoki@aichi-cc.jp

M. Aoki
Department of Cancer Genetics, Program in Function Construction Medicine, Nagoya University Graduate School of Medicine, Nagoya, Japan

Current Topics in Microbiology and Immunology (2017) 407:153–190
DOI 10.1007/82_2017_6

Published Online: 28 May 2017

1 Introduction

The PI3K (phosphatidylinositol-4,5-bisphosphate 3-kinase)/AKT (v-akt murine thymoma viral oncogene homologue)/mTOR (mechanistic target of rapamycin) pathway is frequently activated in various human cancers and has been considered a promising therapeutic target (Arteaga 2010; Maira et al. 2010; Fruman and Rommel 2014; Mayer and Arteaga 2014; Thorpe et al. 2015). Temsirolimus and everolimus, rapamycin-derivative mTOR complex 1 (mTORC1) inhibitors, have been approved for therapies of advanced renal cell carcinoma, and everolimus has also been approved for treating patients of various malignancies including advanced breast cancer (Meng and Zheng 2015). Many therapeutic agents that interfere with mTORC1 and its downstream effectors are under active development (Feldman and Shokat 2010; Bhat et al. 2015), and various pan-PI3K inhibitors, isoform-selective inhibitors, and PI3K/mTOR dual inhibitors are also currently in clinical testing (Mayer and Arteaga 2014; Yap et al. 2015). Although the efficacy of the inhibitors of this pathway in the clinic may fall short of our expectations so far (LoRusso 2016), it does not necessarily mean that the PI3K/AKT/mTOR pathway is not an essential player in carcinogenesis. Cancer cells can be intrinsically resistant to targeted therapies through additional genetic or epigenetic changes that activate the alternative pathways, or they can acquire resistance through feedback activation of such pathways in response to the treatment (Holohan et al. 2013). In this chapter, we address the oncogenic roles of the PI3K/AKT/mTOR axis, an arguably the most important pathway downstream of PI3K in carcinogenesis, by focusing our attention to studies using transformation assays in vitro and genetically engineered mouse models (GEMMs) in vivo.

2 Overview of the PI3K/AKT/MTOR Signaling Pathway

2.1 PI3Ks and PI3K-Associated Phosphatases

Today, PI3Ks are grouped into three classes, according to their structure and substrate specificity (Vanhaesebroeck et al. 1997; Fruman et al. 1998; Thorpe et al. 2015). In vivo, class IA and class IB PI3Ks phosphorylate PtdIns(4,5)P_2 into PtdIns(3,4,5)P_3, which serves as an important signaling molecule (Fig. 1).

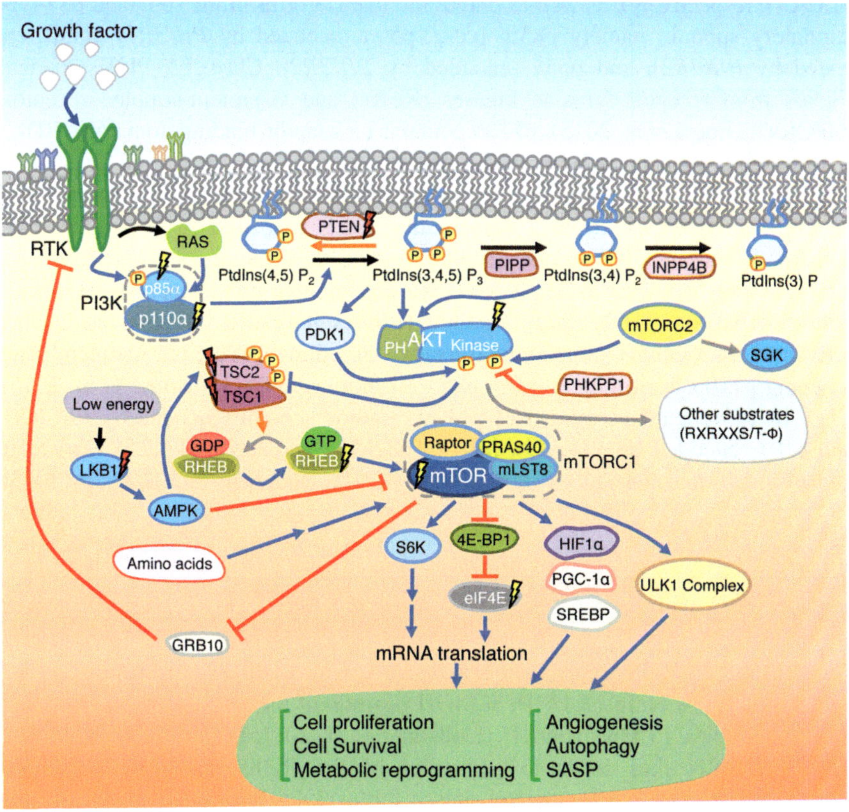

Fig. 1 Schematic view of the PI3K/AKT/mTOR signaling pathway. Upstream signals from receptor tyrosine kinases (RTKs) or RAS activate PI3K, leading to generation of PtdIns(3,4,5) P_3. PtdIns(3,4,5) P_3 recruits AKT to the cell membrane, where AKT is activated by phosphorylation. AKT phosphorylates TSC2, which relieves RHEB from inhibition by the TSC1/TSC2 complex, and GTP-loaded RHEB activates mTOR complex 1 (mTORC1). mTORC1 activation causes phosphorylation of S6K and eIF4E-binding protein 1 (4E-BP1), leading to enhanced mRNA translation of a subset of mRNAs through liberation of eIF4E. eIF4E and other mTORC1 effectors including HIF1α, PGC-1α, SREBP, and ULK1 complex regulate various cellular functions related to oncogenic phenotypes. Oncoproteins and the tumor suppressors in the PI3K/AKT/mTOR axis are marked with yellow and red lightning symbols, respectively

The class III enzyme Vps34 (PIK3C3) phosphorylate PtdIns into PtdIns(3)P on early and late endosome membranes and thereby controls membrane dynamics including autophagy (Simonsen and Tooze 2009). Although not much was known about functions of class II PI3Ks, recent studies suggest that the class II enzyme PI3K-C2a phosphorylates PtdIns into PtdIns(3)P at perinuclear recycling endocytic compartment at the base of primary cilia and regulates a PtdIns(3)P pool (Franco et al. 2014), whereas it phosphorylates PtdIns(4)P into PtdIns(3,4)P_2 at the plasma membrane and thereby controls clathrin-mediated endocytosis (Posor et al. 2013). Class IA PI3Ks form heterodimers containing one of the three isoforms of the catalytic subunit p110α, p110β, and p110δ (encoded by *PIK3CA*, *PIK3CB*, and *PIK3CD*, respectively), as well as one of the five isoforms of the so-called p85-type regulatory subunit, namely p85α, p55α, p50α (encoded by *PIK3R1*), p85β (encoded by *PIK3R2*), and p55γ (encoded by *PIK3R3*). Class 1A PI3Ks mediate signals from receptor tyrosine kinases (RTKs) and G-protein-coupled receptors (GPCRs) (Thorpe et al. 2015). p110α contains an adaptor-binding domain (ABD), a RAS-binding domain (RBD), a C2 domain, a helical domain, and a catalytic domain (Walker et al. 1999; Amzel et al. 2008), while p85α contains two SH2 domains, an inter-SH2 (iSH2) domain that binds to p110α, a N-terminal SH3 domain, and a RhoGAP domain (Mellor et al. 2011). Binding of the p85α regulatory subunit in the absence of upstream activating signals suppresses the lipid kinase activity of p110α. p85α binding to phosphorylated tyrosine residues on activated RTKs via SH2 domain recruits the heterodimers to the plasma membrane, relieving p110α from inhibition by p85α (Mellor et al. 2011; Thorpe et al. 2015). Class IB PI3K is a heterodimer of p110γ catalytic subunit (encoded by *PIK3CG*) and p101 (encoded by *PIK3R5*), p84, or p87PIKAP (encoded by *PIK3R6*) regulatory subunit and transduces signals from GPCRs (Stephens and Hawkins 2013).

2.2 *PTEN and INPP4B*

The phosphatase PTEN (phosphate and tensin homolog deleted on chromosome 10) dephosphorylates PtdIns(3,4,5)P_3 at 3-OH position of the inositol ring and thereby counteracts class I PI3Ks (Fig. 1) (Hollander et al. 2011a). The loss or decreased level of PTEN thus leads to activation of the PI3K/AKT/mTOR pathway. 5-position phosphate of the inositol ring of PtdIns(3,4,5)P_3 is dephosphorylated by the phosphatase SHIP or phosphatidylinositol 4,5-bisphosphate 5-phosphatase (PIB5PA)/proline-rich inositol polyphosphate phosphatase (PIPP), generating PtdIns(3,4)P_2 (Bunney and Katan 2010). PtdIns(3,4)P_2 and PtdIns(3,4,5)P_3 can both bind to the pleckstrin homology (PH) domains containing proteins such as AKT, PDK1. INPP4A and INPP4B (inositol polyphosphate 4-phosphatase type II) have lipid phosphatase activity and hydrolyze PtdIns(3,4)P_2 to PtdIns(3) P. INPP4A/B can both positively and negatively regulate the pathway in a cell-context-dependent manner. These negative regulators of PI3K are also involved in cancer progression (Bunney and Katan 2010).

2.3 AKT

PtdIns(3,4,5)P_3 generated by class I PI3Ks serves as a second messenger by binding to PH domain (Vivanco and Sawyers 2002) (Fig. 1). Among many PH-domain-containing proteins, of the highest importance in the PI3K signaling is the serine/threonine kinases AKT (Vasudevan and Garraway 2010; Toker and Marmiroli 2014). There are three isoforms of mammalian AKT: AKT1, AKT2, and AKT3, which are encoded by different genes. AKT1 and AKT2 are ubiquitously expressed, although AKT2 tends to be expressed highly in insulin-responsive tissues such as the liver, skeletal muscle, and adipose tissue. AKT3 expression is restricted to heart, kidney, brain, testis, lung, and skeletal muscle. Binding of PtdIns (3,4,5)P_3 or PtdIns(3,4)P_2 to the PH domain of AKT induces localization of AKT to cell membrane and induces phosphorylation and activation of AKT at serine 473 and threonine 308 by mTOR complex 2 (mTORC2) and PDK1, respectively. Activated AKT then phosphorylates its substrates. More than 200 AKT substrates have been identified, including FoxO transcription factors, GSK3β, TSC2, and PRAS40 (Manning and Cantley 2007; Vasudevan and Garraway 2010). They share a consensus the AKT phosphorylation motif RxxRxS/TΦ (where R is Arg, S is Ser, T is Thr, x can be any amino acid, and Φ is as hydrophobic amino acid) and are involved in various cellular functions (Toker and Marmiroli 2014). Negative regulation of AKT following activation can be achieved by the phosphatases PP2A and PHLPP, which dephosphorylate AKT at Thr 308 and Ser 473, respectively (Gao et al. 2005; Toker and Marmiroli 2014).

2.4 mTOR

mTOR exists complexes called mTOR complex 1 (mTORC1) and mTORC2 (Laplante and Sabatini 2012). mTORC1 and mTORC2 share mTOR, mLST8/G-protein β-subunit-like protein (GβL), and DEP domain-containing mTOR-interacting protein (DEPTOR). Whereas mTORC1 additionally contains regulatory associated protein of mTOR (RAPTOR) and 40 kDa Pro-rich Akt substrate (PRAS40), mTORC2 contains rapamycin-insensitive companion of mTOR (RICTOR), mammalian stress-activated map kinase-interacting protein 1 (mSIN1; also known as MAPKAP1), and protein observed with RICTOR (PROTOR). mTORC1 integrates four major regulatory signals, namely nutrients, growth factors, energy, and stress (Zoncu et al. 2011). AKT activates mTORC1 by phosphorylation of tuberous sclerosis complex 2 (TSC2) (Manning et al. 2002; Inoki et al. 2002; Potter et al. 2002). The phosphorylated TSC2 translocates from the membrane to the cytosol, which disrupts the TSC1/TSC2 complex and thereby relives the membrane-bound RAS homologue enriched in brain (RHEB) from repression (Cai et al. 2006). AKT can also phosphorylate PRAS40 at threonine 246, which generates a binding site for 14-3-3, leading to liberation of RAPTOR from inhibition by PRAS40 (Vander Haar et al. 2007; Sancak et al. 2007). GTP-loaded RHEB enhances mTORC1, but GTP hydrolysis to GDP stimulated by the TSC1/TSC2 complex inactivates RHEB. mTORC1 activation causes phosphorylation of ribosomal S6 protein kinase 1 (S6K1) and eukaryotic initiation factor 4E (eIF4E)-binding protein 1 (4E-BP1), leading to enhanced protein synthesis. Unphosphorylated 4E-BP1 binds to and thereby inhibits eukaryotic translation initiation factor 4E (eIF4E) (Bhat et al. 2015). eIF4E is a component of the eIF4F complex and binds to the cap structure (m7GpppN, where m is a methyl group and N is any nucleotide) found in all mRNAs transcribed in the nuclei. Phosphorylation of 4E-BP1 relieves eIF4E, which leads to eIF4F formation and translation initiation, the rate-limiting step of cap-dependent mRNA translation. The eIF4E can affect general translation, but eIF4E facilitates the translation of a subset of mRNAs that contain complex and structured 5' untranslated regions (UTRs) (Bhat et al. 2015). S6K1 phosphorylates S6, as well as other substrates including eIF4B, SKAR, eEF2K, CBP80, and PDCD4 (Ma and Blenis 2009). Although S6K1 was implicated in regulating translation of a subset of mRNAs containing 5'-TOP sequences, the role of S6K1 in their translation remains elusive. Increased protein synthesis helps cell growth (increase in cellular mass). Functional consequences of mTORC1 activation are discussed in Sect. 5 of this chapter.

3 Oncoproteins of the PI3K/AKT/MTOR Signaling Axis

3.1 PI3K

3.1.1 p110α

While association between PI3K and viral oncoproteins including Src and polyoma middle T antigen suggested an oncogenic role of PI3K (Macara et al. 1984; Sugimoto et al. 1984; Whitman et al. 1988), a direct evidence was missing. Viral oncogenes played essential roles in our understanding of the signaling pathways involved in human cancers (Vogt 2012), and the evidence for PI3K involvement in oncogenic transformation was also provided by a new oncogene found in the genome of the avian retrovirus ASV16 in 1997 (Chang 1997). ASV16 contained in its genome a cell-derived oncogene (v-p3k) encoding a truncated p110α fused to viral Gag sequences at the N-terminus (Chang 1997). In 2004, Vogelstein and colleagues reported frequent somatic mutation of the *PIK3CA* gene encoding p110α in human cancers (Samuels et al. 2004). Since then, *PIK3CA* mutations have been found in various cancers including those of the colon, breast, brain, liver, and ovary (Samuels and Waldman 2010; Thorpe et al. 2015), with mutation hot spots in the helical domain (e.g., E542K and E545K) and the kinase domain (e.g., H1047R). Oncogenic activity of the mutant class IA PI3Ks has been demonstrated both in vitro and in vivo (Vogt et al. 2010). The v-P3k oncoproteins from ASV 16 and

ASV8905, an independently isolated avian retrovirus, were shown to transform chicken embryo fibroblasts (CEF) (Chang 1997; Aoki 2000). Overexpression of mutant forms of p110α found in human cancers including E542K, E545K, and H1047R, but not wild-type p110α transformed CEF (Kang et al. 2005), as well as mammalian fibroblast or epithelial cell lines (Ikenoue 2005; Isakoff et al. 2005; Zhao et al. 2005). The p110α mutants with transforming activity have elevated associated lipid kinase activity and can constitutively activate downstream signaling when overexpressed in cells (Samuels et al. 2004; Ikenoue 2005; Isakoff et al. 2005; Kang et al. 2005; Zhao et al. 2005). Helical domain mutations such as E542K and E545K are considered to reduce inhibition of p110α by p85 as in the case of growth-factor-induced activation (Huang et al. 2007; Zhao and Vogt 2008a, b; Burke et al. 2012), or by facilitating interaction between p110α and insulin receptor substrate 1 (IRS1) (Hao et al. 2013). The kinase-domain mutations like H1047R confer a conformational change in the proposed lipid-binding region of the kinase domain, which leads to increased affinity for membranes (Mandelker et al. 2009; Hon et al. 2012).

Oncogenic activity of the H1047R and E545K mutants and an artificially generated constitutively active mutant of p110α has been addressed in vivo using genetically engineered mouse models (GEMMs) (Table 1). Conditional transgenic expression of p110α-H1047R in mammary epithelial cells using MMTV-driven Cre and ROSA26 knock-in system (R26-Pik3caH1047R; MMTV-Cre) resulted in development of mammary tumors, specifically adenosquamous carcinoma or adenomyoepithelioma (Adams et al. 2011). In ROSA26 knock-in system, expression of the transgene preceded by a *loxP*-flanked STOP cassette in the *Rosa26* knock-in allele is induced by Cre-mediated recombination. In contrast, transgenic expression of p110α-H1047R in luminal mammary epithelial cells using *WAP*-Cre resulted in development of diverse spectrum of mammary tumors, including adenosquamous carcinomas, adenomyoepitheliomas, and adenocarcinomas (Meyer et al. 2011). Intriguingly, Koren et al. and Van Keymeulen et al. recently demonstrated through lineage tracing that H1047R mutation causes loss of lineage restriction. Namely, induction of p110α-H1047R expression in luminal epithelial cells led to basal epithelial progeny, and vice versa, generating multipotency and tumor heterogeneity (Koren et al. 2015; Van Keymeulen et al. 2015). Engelman et al. reported that transgenic expression of p110α-H1047R under control of tet-inducible promoter and *Scgb1a1*-promoter resulted in development of mixed adenocarcinomas with bronchioloalveolar features (Engelman et al. 2008). However, Trejo et al. reported that p110α-H1047R expression in the lung epithelium failed to induce tumor formation (Trejo et al. 2013). The reason for the apparent discrepancy between reports from two groups is currently unclear. The same group showed that p110α-H1047R cooperated with BRAF-V600E in lung tumorigenesis (Trejo et al. 2013), and also enhanced tumorigenesis induced by KRAS-G12D (Green et al. 2015). Transgenic expression of p110*, a constitutively active chimeric construct in which the iSH2 domain of p85α is connected to the N-terminus of p110α via a flexible linker sequence, in intestinal epithelial cells using ROSA26 knock-in system with *FABP*-Cre resulted in development of hyperplasia and invasive

Table 1 Selected genetically engineered mouse models of cancer induced by the oncoproteins in the PI3K/AKT/mTOR axis

Oncoprotein	Expression mode	Promoter and/or driver	AKT/mTOR signaling status	Reference	Phenotypes
Mammary gland					
p110α-H1047R	Transgenic (Doxycycline inducible)	MMTV-tet on	pAKT, pS6	Liu et al. (2011)	Heterogeneous mammary tumors
	Rosa26-knock-in LSL**	MMTV-Cre		Adams et al. (2011)	Mammary tumors. p53-loss decreases survival
	Knock-in LSL	MMTV-Cre		Tikoo et al. (2012), Yuan et al. (2013)	Mammary gland tumors (ER-positive) with a long latency
	Transgenic LSL	CAG; MMTV-Cre		Meyer et al. (2011)	Heterogeneous multilineage mammary tumors
		Krt8-CreERT2		Koren et al. (2015)	
	Knock-in LSL			Van Keymeulen et al. (2015)	
Akt1-T308D/S473D	Transgenic	MMTV	pAKT	Hutchinson et al. (2001), Hutchinson et al. (2004b)	Accelerates mammary tumor formation by polyoma middle T antigen or MMTV-ErbB2
Myr-AKT1	Transgenic	MMTV	pAKT	Young et al. (2008)	Accelerates mammary tumor formation in MMTV-c-ErbB2 transgenic mice
AKT2-T309D/S474D	Transgenic	MMTV	pAKT	Dillon et al. (2009)	Accelerates mammary tumor metastasis in ErbB2 or polyoma middle T transgenic mice
AKT1- E17K	Transgenic (Tetracycline-controllable)	MMTV-tet off	pAKT	Mancini et al. (2016)	Mammary gland hyperplasia. Suppresses ErbB2/HER2-mediated tumorigenesis
Lung					
p110α-H1047R	Transgenic (Doxycycline inducible)	Scgb1a -tet on	pAKT, pS6, p4E-BP1	Engelman et al. (2008)	Lung tumors upon Doxycycline induction
	Knock-in LSL	Adeno-Cre		Green et al. (2015)	p110α-H1047R enhances Kras-induced lung adenocarcinoma formation
				Trejo et al. (2013)	p110α-H1047R enhances BRAF V600E-induced lung adenocarcinoma formation

(continued)

Table 1 (continued)

Oncoprotein	Expression mode	Promoter and/or driver	AKT/mTOR signaling status	Reference	Phenotypes
AKT1-E17K	Rosa26-knock-in LSL	Adeno-Cre	pAKT	Malanga et al. (2016)	Bronchial and/or bronchiolar hyperplastic lesions
Ovary					
p110α-E545K	Knock-in LSL	Adeno-Cre	pAKT	Wu et al. (2013)	No tumor formed with p110α-E545 K alone; requires loss of Apc and/or Pten for tumor formation
			N.D.	Zhai et al. (2016)	
p110α-H1047R	Knock-in LSL		pAKT, pS6	Kinross et al. (2012)	Serous papillary hyperplasia, but no tumors. Tumor formation with Pten loss
Myr-p110α	Transgenic	MISIIR	pAKT	Liang et al. (2009)	Hyperplasia of ovarian surface epithelium, but not tumors
Intestines					
p110*	Rosa26-knock-in LSL	Fabp-Cre	pS6, p4EBP1, pAKT	Deming et al. (2014)	Increases intestinal tumor multiplicity and tumor size of $Apc^{Min/+}$ mice
				Leystra et al. (2012)	Invasive adenocarcinoma
				Deming et al. (2013)	Rapamycin treatment suppresses tumor formation
p110α-H1047R	Knock-in LSL	Gpa33-Cre	pS6	Hare et al. (2014)	Invasive adenocarcinoma combined with Apc *loss*
Endometrium					
p110α-E545 K	Knock-in LSL	Wnt7a or Ksp1.3-Cre	pAKT	Joshi et al. (2015)	No tumors, but carcinoma with Pten loss in a synergistic manner
Prostate					
Myr-AKT1	Transgenic	Probasin	pAKT, pS6	Majumder et al. (2003), Majumder et al. (2004)	Prostatic intraepithelial neoplasia (PIN). Everolimus treatment suppresses PIN formation
RHEB	Transgenic	Probasin	pS6, p4EBP1	Nardella et al. (2008)	Hyperplasia and a low-grade neoplastic phenotype in prostate

(continued)

Table 1 (continued)

Oncoprotein	Expression mode	Promoter and/or driver	AKT/mTOR signaling status	Reference	Phenotypes
Kidney					
RHEB	Transgenic	CAG	pS6K, p4EBP1	Jiang et al. (2013)	Progressive kidney fibrosis
Epithelial basal cells or skin epithelial cells					
Myr-AKT1	Transgenic	Krt5	pAKT pS6K	Segrelles et al. (2007)	Squamous cell carcinoma or head and neck squamous cell carcinoma in some mice
RHEB	Transgenic	K14	pS6K, pS6	Lu et al. (2010)	Skin neoplasia
Melanocytes					
p110α-H1047R	Knock-in LSL	Tyr-Cre/ERT2	pAKT	Deuker et al. (2015), Durban et al. (2013)	No tumors, but enhances BRAF-V600E-induced melanoma
Thyroid					
p110α-H1047R	Knock-in LSL	Thyroglobulin-CreERT2	N.D.	Charles et al. (2014)	No tumors, but enhances BRAF-V600E-induced thyroid carcinoma
Pancreas					
p110α-H1047R	Knock-in LSL	Ptf1a-Cre	pAKT, pGSK3β	Eser et al. (2013)	ADM and premalignant PanIN
		Pdx1-Cre	N.D	Collisson et al. (2012)	No PanIN formation
Myr-AKT1	Transgenic LSL	Pdx1-Cre Elastase-Cre	N.D.	Elghazi et al. (2009)	Induces malignant transformation in pancreatic progenitor cells
Lymphocytes					
p110*	Rosa26-knock-in	Cγ1-Cre	N.D.	Sander et al. (2012)	No tumors, but lymphomagenesis with MYC
Myr-AKT1	Transgenic	Lck	N.D.	Malstrom et al. (2001)	Thymic lymphomas
AKT1-E4 K					

(continued)

Table 1 (continued)

Oncoprotein	Expression mode	Promoter and/or driver	AKT/mTOR signaling status	Reference	Phenotypes
Myr-AKT1	Introduced into Eμ-Myc HSCs by retrovirus		pAKT, pS6, p4EBP1	Wendel et al. (2004)	Earlier onset of Eμ-Myc B-cell lymphoma promotes chemoresistance that can be reversed by rapamycin
Myr-AKT2	Transgenic	Lck	pAKT, pS6, p4EBP1	Mende et al. (2001), Hsieh et al. (2010)	Thymic lymphomas
Brain					
p110α-E545 K	Knock-in LSL	Blbp-Cre	pS6, p4EBP1	Robinson et al. (2012a)	No tumors, but accelerates medulloblastoma in *Ctnnb1*$^{+/lox(Ex3)}$;*Tp53*$^{+/flox}$ mice
Systemic					
eIF4E	Transgenic	β-actin promoter	N.D.	Ruggero et al. (2004)	Various tumors including B-cell lymphomas and lung adenocarcinoma. Accelerated MYC-induced B-cell lymphomagenesis

p110 is a constitutively active chimeric construct in which the iSH2 domain of p85α is connected to the N-terminus of p110α via a flexible linker sequence
**loxP-STOP-loxP

mucinous adenocarcinomas in the proximal colon (Leystra et al. 2012). However, their intestinal tumors did not show nuclear localization of β-catenin, suggesting no aberration of Wnt signaling, and the mice did not develop polypoid precursor lesions. The same group crossed the mice with *Apc*$^{MIN/+}$ mice, a model of familial adenomatous polyposis, and showed increased number and size of the intestinal polyps in the combined mutant mice as compared to *Apc*$^{MIN/+}$ mice, as well as development of invasive adenocarcinomas (Deming et al. 2014). Hare et al. also demonstrated the development of invasive adenocarcinomas in combined mutant mice in which conditional knockout of *Apc* (*Apc*$^{580D/580D}$) and conditional expression of p110α-H1047R from the knock-in *Pik3ca* locus were induced by *Gpa33*-Cre, indicating the synergistic effect of p110α mutation with Wnt activation in intestinal tumor progression (Hare et al. 2014). Ovarian cancer models suggested differential oncogenic activity between H1047R and E545K mutants. Conditional expression of p110α-H1047R from the targeted knock-in allele or transgenic expression of Myr-p110α induced premalignant hyperplasia of the ovarian surface epithelium, but no tumors (Liang et al. 2009; Kinross et al. 2012). In contrast, expression of E545K from the knock-in allele did not induce hyperplastic growth of ovarian surface epithelial cells (Wu et al. 2013). The oncogenic role of PIK3CA was also studied in GEMMs of thyroid cancer and melanoma, tumor types with high frequency of BRAF mutation in clinical samples. Although expression of PIK3CA in thyroid cells or melanocytes by itself did not result in tumor formation, it cooperated with BRAF-V600E in enhance its tumorigenic potentials (Durban et al. 2013; Charles et al. 2014). Conditional expression of p110α-H1047R from the targeted knock-in *Pik3ca* allele using *Pdx1*-CreERT2 did not result in tumor formation in the pancreas (Collisson et al. 2012). In clear contrast, transgenic expression of p110α-H1047R from the *Rosa26* locus in pancreatic acini, ducts, and islets induced by *Ptf1a*-Cre driver resulted in massive induction of acinar-to-ductal metaplasia, as well as pancreatic intraepithelial neoplasia (PanIN), a precursor of pancreatic ductal adenocarcinoma (PDAC) (Eser et al. 2013). All the aged mice developed PDAC. The reason for the discrepancy between the two studies is not clear, but it can be due to different target cells of the Cre-driving promoters, different expression levels of mutant p110α, or a slight difference in the genetic background of the mice (e.g., number of backcross). Conditional expression of E545K mutant from the targeted knock-in allele by *Blbp*-Cre driver, which induces efficient recombination in progenitor cell populations across the hindbrain, did not result in tumor formation, but accelerated medulloblastoma formation induced by β-catenin stabilization (Robinson et al. 2012b). The roles of class IA PI3Ks in tumorigenesis induced by RTK or RAS oncoproteins have also been determined using GEMMs. Gupta et al. generated mice with knock-in alleles for *Pik3ca* gene in which the region encoding RAS-binding domain was mutated to disrupt the interaction with RAS, and showed that binding or KRAS to p110α is required for KRAS-driven carcinogenesis in the lungs of *KRAS*-LA2 transgenic mice (Gupta et al. 2007). The same group also demonstrated later that the disruption of the interaction between p110α and KRAS is required not only for development, but also for maintenance of the established KRAS-driven lung tumors (Castellano et al. 2013).

Consistently, Engelman et al. showed that ablation of p85 inhibited tumorigenesis in the KRAS-G12D lung cancer model (Engelman et al. 2008). Baer et al. showed that the p110α-selective inhibitor A66 blocked the formation of tissue-injury-induced acinar-to-ductal metaplasia (ADM) and PanIN in *KRAS* LSL-G12D; *Pdx1*-Cre mice (Baer et al. 2014). Consistently, tissue-specific ablation of p110α prevented the occurrence of lesions induced by mutant KRAS, whereas ablation of p110β failed to prevent the formation of preneoplastic lesions (Baer et al. 2014). Wu et al. employed *KRAS* LSL-G12D; *Ptf1a*-Cre mice and showed that ablation of p110α, but not p110β, prevents pancreatic tumor formation (Wu et al. 2014a). However, they show that the oncogenic function of p110α in this KRAS-induced tumor formation is dependent on RAC1 and independent of AKT. The role of p110α in RAS-induced leukemogenesis was also addressed. Gritsman et al. showed that conditional deletion of p110α in hematopoietic stem cells markedly prolonged the latency of myeloproliferative neoplasm (MPN) induced by mutant KRAS (Gritsman et al. 2014). Tissue-specific ablation of p110α was also shown to retard mammary cancers induced by polyoma middle T antigen and ERBB2 (Utermark et al. 2012). Interestingly, p110β ablation accelerated mammary tumor formation in these models (Utermark et al. 2012).

3.1.2 Other Isoforms of PI3Ks

Somatic mutations in human cancers have also been found in genes encoding regulatory subunits of class I PI3Ks, namely *PIK3R1* encoding p85α, p55α, and p50α; *PIK3R2* encoding p85β; and *PIK3R5* encoding p101 (Thorpe et al. 2015). *PIK3R1* mutations have been identified in several cancers including endometrial carcinomas and pancreatic cancers, and most of them are clustered in the inter-SH2 (iSH2)-domain-mediating interaction with C2 domain of p110 (Thorpe et al. 2015). The mutations relieve p110α from inhibitory activity of p85, causing increased lipid kinase activity which leads to oncogenic transformation (Jaiswal et al. 2009; Wu et al. 2009; Sun et al. 2010). Mutations in *PIK3R2*, *PIK3R4*, and *PIK3R5* genes encoding p85β, p150, and p101, respectively, have also been identified in endometrial and colorectal cancers (Jaiswal et al. 2009; Cheung et al. 2011). Transforming activities of the catalytic subunits of the class IA PI3K isoforms other than p110α, namely p110β and p110δ, as wells as the catalytic subunit of the class1B PI3K, namely p110γ, were also evaluated in CEF (Zhao and Vogt 2008a). Whereas overexpression of wild-type p110α failed to transform CEF, overexpression of wild-type p110β, p110δ, or p110γ efficiently transformed CEF, probably because of their higher basal lipid kinase activity (Kang et al. 2006). However, they differ in activation of downstream pathways, interaction with RAS, and sensitivity to MEK inhibition (Denley et al. 2008). Although mutations of *PIK3CB*, *PIK3CD*, and *PIK3CG* are rare in human cancers, their overexpression by copy number gain or amplification is found frequently in some cancer types (Thorpe et al. 2015). p110δ is a major subclass in B lymphocytes and plays an essential role in B-cell receptor signaling. Heterozygous *PIK3CD* mutations have been identified in patients of activated phosphoinositide-3 kinase δ syndrome (APDS), a primary immunodeficiency disease (Angulo et al. 2013). E1021K mutation in p110δ conferred increased lipid kinase activity, like H1047R mutation in p110α. One study showed that 75% of APDS patients suffer nonneoplastic lymphoproliferation, and 13% of them develop cutaneous T-cell lymphoma (Coulter et al. 2017). p110δ is also essential in the pathogenesis of B-cell malignancies, the p110δ-specific inhibitor idelalisib has been approved by FAD for treating patients with B-cell malignancies including relapsed follicular B-cell non-Hodgkin lymphoma and small lymphocytic lymphoma, and chronic lymphocytic leukemia (CLL) (Furman et al. 2014; Gopal et al. 2014; Yang et al. 2015).

3.2 AKT

More than a decade after identification of AKT as an important effector of PI3K, a somatic mutation of AKT1 (E17K) was identified in breast cancer, as well as in lung, endometrial, colorectal, and prostate cancers (Carpten et al. 2007). An analogous mutation has been found in AKT2 and AKT3 in various cancers including breast cancer and melanoma (Parsons et al. 2005; Sasaki et al. 2008; Banerji et al. 2012; Stephens et al. 2012). One of the three paths that led to the original identification and molecular cloning of AKT (or PKB) was through finding the retroviral oncogene *Akt* (Bellacosa et al. 1991). Transforming activity of constitutively active forms of AKT, namely v-AKT, AKT with a myristoylation signal on the N-terminus (Myr-AKT), and E40K with increased affinity for PI3,4,5P3 (AKT-E40K), was demonstrated in chicken embryo fibroblasts (Aoki et al. 1998), as well as in Rat1 cells (Mirza et al. 2000) and NIH3T3 cells (Sun et al. 2001). AKT-E17K, a mutant found in human cancers, also transformed Rat1 cells in culture (Carpten et al. 2007). Oncogenic roles of AKT1 in vivo have been addressed by GEMMs (Table 1). Transgenic mice that over express Myr-AKT1 or AKT1-E40K under control of *Lck* promoter developed lymphomas (Malstrom et al. 2001). Although transgenic expression of Myr-AKT1 or AKT1-T308D/S473D, a phospho-mimetic constitutively active mutant, in mammary epithelial cells driven by MMTV-promoter did not result in tumor formation, transgenic expression of Myr-AKT1 or AKT1-T308D/S473D accelerated tumorigenesis induced by ErbB2/HER2 (Hutchinson et al. 2004a; Young et al. 2008). However, in clear contrast with this mammary tumor-promoting effect of the two AKT mutants, transgenic mice expressing AKT1-E17K harboring a mutation found in human cancers developed mammary hyperplasia with increased expression of estrogen receptor, but suppressed ErbB2/HER2-mediated tumorigenesis (Mancini et al. 2016). Mancini et al. proposes that AKT1-mediated negative feedback on the receptor tyrosine kinase ErbB2 can be overcome by strong kinase activity of Myr-AKT1 or AKT1-T308D/S473D, but not by relatively weak kinase activity of AKT1-E17K (Mancini et al. 2016). Their finding is consistent with the finding that AKT1-E17K mutation in breast cancer is mutually exclusive with HER2-positive status (Stemke-Hale et al. 2008). Further studies will clarify the role of AKT1-E17K mutation in mammary carcinogenesis. Notably, co-expression of AKT1-T308D/S473D suppressed the metastasis of mammary tumors to the lungs induced by transgenic expression of ErbB2 or polyoma middle T antigen (Hutchinson et al. 2004a), whereas co-expression of AKT2-T309D/S474D, the AKT2 version of AKT1-T308D/S473D, did not affect tumor formation but markedly enhanced the metastasis in the same models. Distinct roles of AKT1 and AKT2 in mammary tumor induction and metastasis have been supported by extensive studies from multiple laboratories (Toker 2012). Transgenic expression of Myr-AKT1 in prostate epithelial cells driven by the probasin promoter resulted in formation of prostate intraepithelial neoplasia (PIN) (Majumder et al. 2003). Co-expression of Myr-AKT1 and Myc led to accelerated progression to microinvasive adenocarcinoma (Clegg et al. 2011). Overexpression of wild-type AKT1 or Myr-AKT1 in the basal layer of stratified epithelia using the bovine keratin K5 (Krt5) promoter caused epidermal hyperplasia and a variety of tumors in epithelial organs, including prostate, mammary, cervix, and skin tumors (Segrelles et al. 2007). The same group reported that *K14*-Cre-mediated conditional knockout of p53 in the Krt5-Myr-AKT1 mice resulted in development of oral cavity tumors that phenocopy human head and neck squamous cell carcinoma (HNSCC) (Moral et al. 2009). Adeno-Cre-mediated expression of AKT-E17K in bronchioepithelial cells resulted in hyperplasia of bronchi and bronchiole, which can progress to overt adenocarcinoma at low frequency (Malanga et al. 2016). *Pdx1*-Cre-driven expression of a constitutively active AKT1 (no detailed information provided) in pancreatic progenitor cells resulted in development of ADM and intraductal papillary mucinous neoplasia (IPMN) in younger mice ($\sim$3 months old), as well as acinar cell carcinoma and pancreatic ductal adenocarcinoma in older mice ($\sim$8 months old) (Elghazi et al. 2009). Myr-AKT2 and Myr-AKT3 transformed CEF in culture, and transgenic expression of Myr-AKT2 under control of the *Lck* promoter resulted in lymphomagenesis, suggesting that constitutively active AKT1, AKT2, and AKT3 have similar oncogenic activity (Mende et al. 2001). However, a recent study suggests differential roles of AKT isoforms in glioma progression. Turner et al. employed RCAS/tv-a glioma model, in which avian retroviral vectors are injected intracranially into transgenic mice expressing the avian retrovirus receptor tv-a (Turner et al. 2015). They tested the effect of co-expression of the constitutively active forms of the three AKT isoforms on platelet-derived growth factor B (PDGFB)-driven low-grade glioma and identified AKT3 as the dominant AKT isoform that robustly stimulates glioma progression (Turner et al. 2015). AKT activation has been shown to enhance chemical carcinogenesis in vivo. Expression of wild-type AKT1 or Myr-AKT1 in the basal layer of stratified epithelia enhanced susceptibility to carcinogenesis by 7,12-dimethylbenz(a)anthracene (DMBA) and 12-O-tetradecanoyl-phorbol-13-acetate (TPA) (Segrelles et al. 2007). Increased susceptibility to DMBA-induced carcinogenesis was also observed in mammary tumor mouse models that overexpress Myr-AKT1 under control of the MMTV-promoter (Blanco-Aparicio et al. 2007; Wu et al. 2014b). AKT-E17K overexpression in mouse lung epithelial cells also cooperated with urethane to induce fully malignant lung cancer (Malanga et al. 2016). These findings indicate that constitutive activation of AKT is sufficient for oncogenic transformation. There are also evidences that suggest AKT is required for transformation induced by PI3K-activating signals. Primary AKT1-deleted mouse embryo fibroblasts (MEFs) showed a dramatically reduced susceptibility to oncogenic transformation by combination of a mutant HRAS with MYC or a dominant-negative p53 (Skeen et al. 2006). MEFs with double knockout for AKT1 and AKT2 were almost completely resistant to oncogenic transformation (Skeen et al. 2006). They further showed that AKT1 deletion was sufficient to impair mammary tumorigenesis induced by MMTV-v-HRAS in vivo (Skeen et al. 2006). AKT1 deficiency suppressed tumor development in *Pten*$^{+/-}$ mice (Chen et al. 2006), whereas AKT2 deletion had little effect on the incidence of tumors in prostate, uterus, intestines, and adrenals (Xu et al. 2012). However, AKT2 deficiency significantly decreased the incidence of thyroid tumors in *Pten*$^{+/-}$ mice (Xu et al. 2012). Differential requirement for AKT isoforms has also been demonstrated in lung tumorigenesis induced by mutant KRAS. Hollander et al. showed that ablation of AKT1, but not AKT2 or AKT3 prevents lung tumorigenesis in *KRAS* LA2 mice (Hollander et al. 2011b).

3.3 mTOR and RHEB

Roles of mTORC1 in mediating the oncogenic signal from the PI3K/AKT pathway have been demonstrated using the mTORC1 inhibitor rapamycin and its derivatives (rapalogs), or more recently with mTOR kinase inhibitors (Zaytseva et al. 2012). The mTORC1 pathway was shown to be hyperactivated in CEF transformed by Myr-p110α or Myr-AKT, and rapamycin treatment efficiently inhibited the oncogenic transformation induced by these oncoproteins, but not that by other oncoproteins such as v-Src or v-Jun (Aoki et al. 2001). Soon after, treatment of *Pten*$^{+/-}$ mice with temsirolimus (CCI-779), a rapamycin derivative currently used in the clinic, was shown to prevent growth and proliferation of uterine and adrenal tumors (Podsypanina et al. 2001), and to suppress the growth of *Pten*-null MEF in vitro, as well as that of *Pten*-null ES cells in vivo (Neshat et al. 2001). Suppressive effect of rapalogs and mTOR kinase inhibitors on tumor formation and progression has also been demonstrated in other tumor models with *Pten* mutations. The growth of ovarian endometrioid adenocarcinomas induced by conditional inactivation of PTEN and APC in ovarian surface epithelium was significantly reduced by rapamycin treatment (Wu et al. 2011). The growth of invasive mucinous adenocarcinomas induced by conditional expression of p110* was also inhibited by rapamycin (Deming et al. 2013). In the above-mentioned model of PIN induced by Myr-AKT1, treatment with everolimus led to the complete reversal of the neoplastic phenotype (Majumder et al. 2003). Interestingly, prostate tumors induced by co-expression of Myr-AKT1 and Myc were resistant to everolimus treatment (Clegg et al. 2011). Genetic evidence for the requirement of mTOR signaling activation in oncogenic transformation induced by the PI3K/AKT pathway was provided for prostate tumors. Namely, Nardella et al. showed that conditional knockout of mTOR in adult mouse prostate markedly suppressed PTEN-loss-induced tumor initiation and progression in the prostate (Nardella et al. 2009). The suppressive effect of mTOR deletion was more pronounced than that of everolimus, which was explained by the finding by Guertin et al. which showed requirement of mTORC2 in the PTEN-loss-induced tumorigenesis in the prostate (Guertin et al. 2009). In addition to the requirement of mTORC1 in oncogenic transformation as described above, several lines of evidence suggest that mTORC1 activation is sufficient for transformation. Such evidences include transforming activity of RHEB, an activator of mTORC1, eIF4E, a downstream effector of mTORC1, and more recently mTOR itself. Jiang and Vogt showed that overexpression of two constitutively active mutants of RHEB, namely RHEB/Q64L and RHEB/N153T, induced oncogenic transformation of CEF in culture (Jiang and Vogt 2008). Transformation induced by RHEB/Q64L was markedly suppressed by rapamycin treatment, indicating RHEB-induced transformation was mediated by mTORC1 activation. Overexpression of wild-type RHEB barely transformed CEF (Jiang and Vogt 2008). Lu et al. reported frequent overexpression of RHEB in human cancers and showed that transgenic expression of RHEB in basal epidermal keratinocytes driven by K14 promoter resulted in epidermal neoplasia formation. Unlike *PI3K3CA* or *AKT1*, *mTOR* was not well accepted as a bona fide oncogene until recently. Edinger and Thompson demonstrated that expression of an mTOR mutant with a deletion in its repressor domain conferred the ability to grow in soft agar (anchorage-independent growth) in p53-mutant mouse embryonic fibroblasts (Edinger and Thompson 2004). Ohne et al. then identified hyperactive mTOR mutants by a genetic screen, but those mutants did not transform NIH3T3 cells (Ohne et al. 2009). Thereafter, Sato et al. found mTOR mutations in human cancers using public database and showed that two of them, namely S2215Y and R2505P, conferred constitutive activation of the mTOR signaling (Sato et al. 2010). However, they showed that mTOR/S2215Y did not induce oncogenic transformation of Rat1 cells. More recently, Murugan et al. generated constitutively active mutants of mTOR and demonstrated that overexpression of P2273S and E2288 K, the two most catalytically active mutants, in NIH3T3 cells induced oncogenic transformation and enhanced their invasion activity (Murugan et al. 2013). Subcutaneous inoculation of the transfected cells induced tumor formation in nude mice (Murugan et al. 2013). Yamaguchi et al. recently identified L2209 V mutation in large cell neuroendocrine carcinoma and showed a marked transforming activity of mTOR-L2209V in a focus formation assay using mouse 3T3 fibroblasts (Yamaguchi et al. 2015). They identified more mutations in various cancers using public databases and showed that such mTOR mutants (T1977K, T1977R, S2215F, and S2215Y) efficiently transformed 3T3 cells in vitro. The transforming mutants of mTOR conferred constitutive activation of the mTOR signaling, and the oncogenic transformation by these mutants was sensitive to rapamycin (Yamaguchi et al. 2015). Discrepancy on the transforming activity of mTOR-S2215Y between Sato et al. and Yamaguchi et al. was not discussed, but it may be due to a differential sensitivity for transformation between Rat1 cells and mouse 3T3 cells, or to the expression level of transduced mTOR. It should be noted that the 3T3 cells transformed by these mTOR mutants did not form tumors when transplanted into immunocompromised mice in vivo (Yamaguchi et al. 2015). In addition to the somatic mutations in the mTOR gene found by Sato et al., Gerlinger et al. reported L2431P mutation in a subset of a primary renal cell carcinoma (Gerlinger et al. 2012), and many more somatic mTOR mutations have been identified by cancer genome sequencing projects recently; Grabiner et al. cataloged those mutations and showed that 33 of them conferred hyperactivation of the mTORC1 pathway (Grabiner et al. 2014). Although their transforming activities were not assessed, they were shown to remain sensitive to inhibition by rapamycin (Grabiner et al. 2014). No reports have demonstrated the oncogenic activity of these mTOR mutants in genetically engineered mouse models.

3.4 eIF4E

Among downstream effectors of mTORC1, eIF4E has been shown to behave as an oncogene. Lazaris-Karatzas et al. demonstrated oncogenic transformation of NIH3T3 cells by overexpression of eIF4E (Lazaris-Karatzas et al. 1990). Overexpression of eIF4E was also shown to induce clonal expansion and anchorage-independent growth in immortalized 184-A1 human mammary epithelial cells (Avdulov et al. 2004). Ruggero et al. showed that transgenic expression of eIF4E under control of β-actin promoter resulted formation of various tumors, including B-cell lymphomas, angiosarcomas, lung adenocarcinomas, beginning at 16 months of age (Ruggero et al. 2004). They also showed that crossing the eIF4E transgenic mice with Eμ-Myc mice resulted in accelerated lymphomagenesis. Similar acceleration of MYC-induced lymphomagenesis by eIF4E was also demonstrated by another group, using adoptive transfer of retrovirally transduced hematopoietic progenitor cells (Wendel et al. 2004; Mavrakis et al. 2008). It should be noted that the oncogenic transformation by eIF4E requires phosphorylation at Ser209 by MNK1. Overexpression of eIF4E/S209A failed to transform NIH3T3 cells and to accelerate lymphoma genesis in the Eμ-Myc model (Topisirovic et al. 2004). Subsequent studies using eIF4E/S209A knock-in mice showed that S209A phosphorylation is essential for PTEN-deletion-induced prostate tumor formation (Furic et al. 2010). Hsieh et al. genetically dissected the pathways downstream of mTORC1 using tet-inducible transgenic mice that express 4E-BP1 mutant in which all five rapamycin-sensitive 4E-BP1 phosphorylation sites (T37, T46, S65, T70, and S82) were mutated to alanine. They demonstrated the importance of eIF4E hyperactivation, rather than S6 phosphorylation, in lymphomagenic activity of Myr-AKT2 driven by Lck promoter (Hsieh et al. 2010). Truitt et al. nicely addressed the requirement of eIF4E in oncogenic transformation. Namely, they demonstrated that eIF4E heterozygous mice developed normally, but *Eif4e*$^{+/-}$ MEFs that suffered 50% reduction in eIF4E protein level were dramatically resistant to oncogenic transformation induced by HRASV12 and Myc (Truitt et al. 2015).

They also showed that crossing of *KRAS* LA2 model mice for lung cancer with *Eif4e*$^{+/-}$ mice resulted in marked reduction in tumor number and burden, compared with *Eif4e*-wild-type *KRAS* LA2 mice (Truitt et al. 2015).

4 Negative Regulators of the PI3K/AKT Axis in Oncogenic Transformation

4.1 PTEN

PTEN is the most frequently mutated gene among cancer-related genes encoding signaling molecules of the PI3K/AKT/mTOR pathway and was originally identified as a candidate tumor suppressor gene in a region of human chromosome 10q23 (Song et al. 2012). Germ-line mutations or deletions of PTEN are associated with autosomal dominant tumor predisposition syndromes (collectively called the PTEN hamartoma tumor syndrome), including Cowden syndrome, Bannayan-Riley-Ruvalcaba syndrome, Lhermitte-Duclos disease, PTEN-related Proteus syndrome, and Proteus-like syndrome (Leslie and Longy 2016). Roles of PTEN loss/mutation in oncogenic transformation have been extensively studied using GEMMs. Mice heterozygous for null-mutant allele of *Pten* develop cancers in multiple organs including prostate, mammary gland, uterus, and colon (Cristofano et al. 1998; Podsypanina et al. 1999; Stambolic et al. 2000). Subsequently, a variety of conditional knockout mice for tissue-specific PTEN deletion have been generated for modeling PTEN-loss-driven neoplasias (Nardella et al. 2010). A recent study demonstrated that PTEN functions as lipid phosphatase as a homodimer and that cancer-associated PTEN mutants heterodimerize with wild-type PTEN and acts in a dominant-negative manner to suppress the phosphatase activity (Papa et al. 2014). Consistently, while tumors that develop in mice heterozygous for a null-mutant allele of *Pten* tend to lose the wild-type allele, the wild-type *Pten* allele was retained in most tumors in mice heterozygous for point-mutation alleles of *Pten* (i.e., C124S or G129E). Differential roles of p110α and p110β in PTEN-loss-induced tumors have been uncovered. Namely, tumor formation in anterior prostate induced by conditional knockout of *Pten* using *PB*-Cre driver was suppressed by ablating p110β but not p110α (Jia et al. 2008). In contrast, ablation of either p110β or p110α did not block formation of *Pten*-deficient tumors in ventral prostate, but co-ablation of p110β or p110α efficiently suppressed tumorigenesis in the ventral prostate of these mice (Jia et al. 2013). Leukemogenesis induced by conditional knockout of *Pten* in hematopoietic stem cells using *Mx1*-Cre driver was dependent on p110β, but independent of p110α (Yuzugullu et al. 2015). Formation of dermal hamartomas induced by deleting Pten using *K14*-Cre driver were slightly inhibited by ablating p110α alone and strongly suppressed by ablating both p110α and p110β. A recent study using an ovarian cancer model demonstrated that the apparent p110β dependence of *Pten*-null tumors could be altered by the genetic context

(Schmit et al. 2014). Specifically, development of ovarian endometrioid adenocarcinoma induced by PTEN loss and KRAS activation was blocked by ablation of p110α. Intriguingly, Nakanishi et al. reported recently that treatment of PTEN-mutant cell lines with the pan-PI3K inhibitor GDC-0941 caused emergence of resistant cells that harbor D1067Y mutation in p110β. They then found recurrent mutations of p110β at D1067 in diverse cancers using TCGA database. Prior to this report, only one report showed p110β mutation in a single case of breast cancer. Certain PTEN-deficient tumors depend on p110β (Jia et al. 2008; Ni et al. 2012; Wee et al. 2008).

4.2 INPP4B

INPP4B, a phosphatase that terminates PtdIns(3,4)P_2 signals by hydrolyzing 4'-phosphate of the inositol ring of PtdIns(3,4)P_2, was identified as a putative tumor suppressor in a functional shRNA-library screen in human mammary epithelial cells. Knockdown of *INPP4B* in human mammary epithelial cells resulted in anchorage-independent growth and increased migratory/invasive activity (Gewinner et al. 2009). Studies on breast cancer clinical samples showed that the reduced level of INPP4B was associated with higher tumor grade and poor survival. LOH of the region containing *INPP4B* locus was found in basal-like breast tumors, as well as in ovarian cancers and melanomas (Gewinner et al. 2009; Fedele et al. 2010). Although these data were consistent with a tumor suppressor function of INPP4B, recent studies suggest oncogenic function of INPP4B in leukemia, as well as in breast and colorectal cancers (Gasser et al. 2014; Dzneladze et al. 2015; Guo et al. 2015; Rijal et al. 2015). Importantly, Gasser et al. showed that PI3P, a lipid product of INPP4B, binds to PX domain of SGK3 and thereby activates the SGK3 signaling, which is required for proliferation, invasion/migration, and tumorigenesis of breast cancer cells with *PIK3CA* mutation (Gasser et al. 2014). These findings suggest that INPP4B can be oncogenic or tumor-suppressive, depending on the cellular context.

4.3 PIPP

PIPP dephosphorylates PtdIns(3,4,5)P_3 and generates PtdIns(3,4)P_2. Denley et al. showed that the oncogenic activity of class I PI3K is linked to the ability to produce PtdIns(3,4,5)P_3 and that reduction of PtdIns(3,4,5)P_3 levels by PIPP interferes with oncogenic transformation of PI3K (Denley et al. 2009). Consistently with this finding, Ooms et al. showed that crossing *Pipp* knockout mice with MMTV-PyMT mice resulted in increased mammary hyperplasia, accompanied by enhanced AKT activation, suggesting essential roles of PtdIns(3,4,5)P_3 in oncogenicity of PI3K (Ooms et al. 2015). They also found that *Pipp* knockout reduced the metastatic activity of mammary tumors of MMTV-PyMT mice. This apparent paradox is reminiscent of the opposing roles of AKT1 and AKT2 in breast cancer metastasis, and the authors reason that PIPP loss may result in greater activation of AKT1 over AKT2 (Ooms et al. 2015).

4.4 PHLPP1

PHLPP1 (protein phosphatase 2 (PP2A) complex and PH-domain and leucine-rich repeat protein phosphatase 1) and PHLPP2 were shown to inactivate Akt by dephosphorylating S473. PHLPP1 was subsequently demonstrated to serve as a prostate cancer suppressor in GEMM (Chen et al. 2011). Specifically, Chen et al. showed that *Phlpp1* knockout caused formation of high-grade prostatic intraepithelial neoplasia (HGPIN) and that crossing of *Pten*$^{+/-}$ mice with *Phlpp1*$^{-/-}$ mice resulted in progression to carcinoma. Comprehensive genomic analysis revealed that *PHLPP1* gene is frequently deleted in primary and advanced prostate cancer.

4.5 TSC1 and TSC2

Germ-line mutations of *TSC1* or *TSC2* genes are involved in TSC, an autosomal dominant tumor syndrome from which their names are derived; the mutations affect multiple soft tissues of mesenchymal origin, resulting in formation of hamartomas and other benign tumors in various organs including the brain, kidney, skin, heart, and lungs (Huang and Manning 2008). Consistently, heterozygous knockout mice for *Tsc1* or *Tsc2* developed tumors in multiple organs, accompanied by the loss of the wild-type alleles due to LOH (Onda et al. 1999; Kobayashi et al. 1999, 2001). More recently, Yang et al. showed that rapamycin treatment efficiently blocked kidney tumor formation in *Tsc2*$^{+/-}$ mice, indicating the role of mTORC1 (Yang et al. 2014).

4.6 LKB1

mTORC1 integrates multiple signals involved in cell growth regulation, namely the growth factor signals from PI3K/AKT pathway, nutritional (amino acids) signals, and energy signals (Ma and Blenis 2009; Dazert and Hall 2011). AMP-activated protein kinase (AMPK) is an energy sensor and negatively regulates mTORC1 by phosphorylating TSC2 and/or RAPTOR (Inoki et al. 2003; Gwinn et al. 2008). The tumor suppressor LKB1 (liver kinase B1) encoded by *STK11* phosphorylates and thereby activates AMPK, and the loss of LKB1 function leads to mTORC1 activation (Shaw et al. 2004). Somatic inactivation of LKB1 has also been found in sporadic cancers including lung cancer (Korsse et al. 2013).

5 Mechanisms of Oncogenic Transformation by the PI3K/AKT/mTORC1 Signaling

The molecular mechanisms by which activation of the PI3K/AKT/mTORC1 signaling induces oncogenic transformation are not fully understood. However, our knowledge on the outputs from mTORC1 in cellular functions is accumulating, and many of them are related to hallmarks of cancer (Hanahan and Weinberg 2011).

5.1 Cell Proliferation and Survival

Accelerated cell proliferation or increased cell survival by mTORC1 activation may be explained by enhanced translation of a subset of mRNAs by eIF4E activation. Overexpression of eIF4E can transform cells as described above, and protect cells against apoptosis (Li et al. 2004; Wendel et al. 2004; Mamane et al. 2007; Wendel et al. 2007; Hsieh et al. 2010). Consistently, eIF4E enhances translation of mRNAs coding for various regulators of cell cycle or apoptosis, including cyclin D1, cyclin D3, CDK2, MYC, PIM1, Bcl-2, Bcl-xL, MCL1, VEGF, ornithine decarboxylase, osteopontin, and survivin (Graff and Zimmer 2003; Mamane et al. 2004; Larsson et al. 2007; Mamane et al. 2007). They belong to the so-called eIF4E-sensitive mRNAs, most of which have long and highly structured 5'UTRs and are more dependent on the unwinding activity of eIF4A, an RNA helicase subunit of eIF4F complex (Siddiqui and Sonenberg 2015). eIF4E is limiting for eIF4F complex formation, and its increase leads to enhanced unwinding by eIF4A. eIF4E has also been shown to stimulate the helicase activity of eIF4A by another mechanism. Specifically, eIF4E binding to eIF4G alleviates the inhibitory effect of eIF4G on eIF4A helicase activity (Siddiqui and Sonenberg 2015). More recently, important progress has been made regarding the role of eIF4E-mediated translational control in cancer cell survival. In the study mentioned above, in which 50% reduction of the eIF4E level resulted in resistance of MEF to transformation, Truitt et al. performed unbiased translational profiling approach and found that translation of a subset of transcripts is sensitive to the eIF4E dose during oncogenic transformation (Truitt et al. 2015). The affected gene classes included those involved in cell signaling, apoptosis, ribosome biogenesis, the proteasome, nucleotide biosynthesis, oxidative phosphorylation, and the oxidative stress response (Truitt et al. 2015). They further demonstrated that ferritin heavy chain (Fth1) and the catalytic subunit of glutamate-cysteine ligase (Gclc), key eIF4E-dependent ROS targets, are required for cell transformation and that the reduced level of eIF4E increased oncogene-induced intracellular ROS accumulation and sensitized tumor cells to ROS-induced apoptosis (Truitt et al. 2015).

5.2 *Metabolic Reprogramming*

Recent studies revealed that oncogenic transformation is accompanied by drastic shift in cell metabolism, which enables efficient production of biomass (cellular building blocks such as nucleotides, amino acids, and lipids) required for rapid growth and proliferation of neoplastic cells (Vander Heiden et al. 2009). This so-called metabolic reprogramming is now considered one of the hallmarks of cancer (Hanahan and Weinberg 2011; Ward and Thompson 2012) and is coupled with aerobic glycolysis (the Warburg effect). Specifically, neoplastic cells shift from oxidative phosphorylation to less effective glycolysis for ATP production and thereby drive the pentose-phosphate pathway, which leads to efficient production of nucleotides and amino acids, as well as NADPH, helping synthesize lipid, together with acetyl-CoA generated by enhanced glutamine metabolism. The link of mTORC1 activation to the metabolic reprogramming has been suggested by various studies (Lien et al. 2016). mTORC1 can increase expression of glucose transporter and glycolysis genes by inducing HIF1α and/or MYC, important transcription factors implicated in metabolic reprogramming of cancer cells (Hudson et al. 2002; Düvel et al. 2010; Dodd et al. 2015; Stine et al. 2015). mTORC1 enhances protein synthesis through eIF4E as discussed above. mTORC1 is also involved in glutamine metabolism. Glutamine plays essential roles in highly proliferating cells including cancer cells; in a process called glutamine anaplerosis, glutamine is converted to glutamate by glutaminase, which is then converted to α-ketoglutarate by glutamate dehydrogenase. α-ketoglutarate then enters TCA cycle, the critical anaplerotic step for production of biomass (DeBerardinis et al. 2008). Cancer cells are often addicted to glutamine, and Csibi et al. showed that mTORC1 stimulates glutamine anaplerosis by activating GDH through transcriptional repression of SIRT4, an inhibitor of GDH, by inducing degradation of CREB2 (Csibi et al. 2013). mTORC1 also affects energy metabolism by activating important transcriptional regulators including YY1-PGC-1α complex, SRE-BP1/2 and HIF-1α (Cunningham et al. 2007; Porstmann et al. 2008; Düvel et al. 2010). Morita et al. demonstrated that mTORC1 enhances translation of a subset of nucleus-encoded mitochondria-related mRNAs through eIF4E activation and thereby regulates mitochondrial activity and biogenesis (Morita et al. 2013). Such mRNAs included those encoding components of complex V such as ATP5D, 5G1, 5L, and 5O, as well as TFAM, which promotes mitochondrial DNA replication and transcription. mTOR inhibition can lead to decreased mitochondrial biogenesis and respiration in a manner dependent on the 4E-BP1/eIF4E axis (Morita et al. 2013). mTORC1 thus coordinates protein synthesis, mitochondria biogenesis, and proliferation (Morita et al. 2015). mTORC1 stimulates lipid synthesis by activating SREBP1/2, a critical regulator of fatty acid and cholesterol biosynthetic gene expression. Peterson et al. demonstrated that mTORC1 regulates SREBP by controlling the nuclear entry of lipin 1, a phosphatidic acid phosphatase (Peterson et al. 2011). mTORC1 also stimulates nucleotide synthesis by activating carbamoyl-phosphate synthase 2, aspartate transcarbamylase, dihydroorotase (CAD), which catalyzes the first three steps in de novo pyrimidine synthesis (Ben-Sahra et al. 2013; Robitaille et al. 2013). More recently, Ben-Sahra et al. showed that mTORC1 also induces purine synthesis through ATF4-mediated transcriptional upregulation of the mitochondrial tetrahydrofolate (mTHF) cycle enzyme methylenetetrahydrofolate dehydrogenase 2 (MTHFD2) (Ben-Sahra et al. 2016).

5.3 Suppressing Autophagy and Senescence

In addition to stimulating anabolic processes as described above, mTORC1 can also inhibit autophagy, a major catabolic process. mTORC1 phosphorylates and thereby inhibits two autophagy-promoting factors, namely unc-51-like kinase 1 (ULK1) and autophagy-related gene 13 (ATG13) (Jung et al. 2009; Ganley et al. 2009; Utani 2010; Kim et al. 2011). mTORC1 has also been shown to inhibit autophagy by blocking lysosome biogenesis through inhibition of nuclear translocation of transcription factor EB (TFEB). Although the role of autophagy in cancer cells remains elusive and is probably context specific, it is considered anti-neoplastic at early steps of carcinogenesis (Laplante and Sabatini 2012). Suppression of autophagy may thus contribute to oncogenic transformation of normal cells induced by PI3K/AKT/mTORC1 activation. Like autophagy, cellular senescence is a double-edged sword that can promote or suppress oncogenic transformation in vivo. Excessive proliferative signals can trigger senescence, and such cell-autonomous senescence can be regarded as a tumor-suppressive mechanism. On the other hand, senescent cells are known to produce large amounts of secretory proteins including pro-inflammatory cytokines and chemokines. This so-called senescence-associated secretory phonotype (SASP) can help generate both tumor-suppressive and tumorigenic tumor microenvironment (Pérez-Mancera et al. 2014). The link between mTOR and SASP was first provided by Narita et al., who identified a cellular compartment at the trans side of the Golgi apparatus where lysosomes and mTOR accumulated during RAS-induced senescence (Narita et al. 2011). They termed this compartment the TOR-autophagy spatial coupling compartment (TASCC) and showed that mTOR recruitment to TASCC, mediated by autolysosome-derived amino acids, was required for SASP. Iglesias-Bartolome demonstrated that mTORC1 inhibition by rapamycin increased the clonogenic capacity of primary human oral keratinocytes and resident self-renewing cells by preventing stem cell senescence through increased expression of mitochondrial superoxide dismutase (Iglesias-Bartolome et al. 2012). mTORC1 inhibition also blocked the loss of proliferative basal epithelial stem cells upon ionizing radiation in vivo, and rapamycin reduced the secretion of most senescence-associated cytokines. More recently, two groups provided molecular mechanisms by which mTORC1 inhibition suppresses SASP. Namely, Laberge et al. and Herranz et al. independently showed that rapamycin suppressed the secretion of inflammatory cytokines by senescent cells, and demonstrated that IL1α was the cytokine whose translation was the most affected by mTOR inhibition. While Laberge et al. claimed that IL-1α was responsible for SASP, Herranz et al. further demonstrated that mTOR controls SASP by regulating the translation of mitogen-activated protein kinase-activated protein kinase 2 (MAPKAP2) through 4E-BP1. MAPKAP2 phosphorylates the ZFP36L1, an RNA-binding protein, and thereby inhibits ZFP36L1-mediated degradation of the mRNAs encoding many SASP components.

5.4 *Invasion/Metastasis, EMT*

Several studies using cancer cell lines have suggested the involvement of activated PI3K/AKT/mTORC1 signaling in cancer cell invasion and metastasis. For example, TGF-β-induced EMT in NMuMG mammary epithelial cells accompanied PI3K/AKT/mTORC1 activation, and mTORC1 inhibition suppressed the invasion activity of TGF-β-treated cells without affecting EMT itself (Lamouille and Derynck 2007). Gulhati et al. reported that mTOR signaling inhibition by rapamycin or knockdown of RAPTOR and RICTOR suppressed migration and invasion activity of colorectal cancer cells (Gulhati et al. 2011). They also showed that knockdown of mTORC1 and mTORC2 induced a mesenchymal–epithelial transition (MET) (Gulhati et al. 2011). Cai et al. demonstrated that *4E-BP1* knockdown causes colorectal and breast cancer cells to undergo EMT, accompanied by upregulation of SNAIL, and enhanced migration and invasion activities as well as metastasis, while a dominant active mutant 4E-BP1 effectively downregulated SNAIL and suppressed cell migration and invasion (Cai et al. 2014). Intriguingly, Hsieh et al. performed a sophisticated ribosome profiling using prostate cancer cells and mTOR inhibitors and identified a class of translationally controlled pro-invasion messenger RNAs that direct prostate cancer invasion and metastasis downstream of mTORC1 (Hsieh et al. 2012). They found that the largest fraction of mTOR-responsive mRNAs clustered into a node consisting of key components of the translational apparatus, whereas the second largest node comprised bona fide or putative regulators of cell invasion and metastasis, such as Y-box binding protein 1 (YB1), vimentin, metastasis associated 1 (MTA1), and CD44 (Hsieh et al. 2012). Defining the roles of mTORC1 signaling in cancer metastasis using GEMMs can be challenging, because it is generally difficult to distinguish the effect on growth of primary tumors and metastasis itself. Further technical advances are awaited to address this important question.

6 Conclusions

Almost 30 years have passed since Cantley and colleagues discovered PI3K activity associated with polyoma middle T antigen (Whitman et al. 1988). The PI3K/AKT/mTOR axis is now considered one of the most frequently activated signaling pathways in human cancers, and major efforts are made in developing preventive or therapeutic agents targeting the pathway (Arteaga 2010; Maira et al. 2010; Fruman and Rommel 2014; Mayer and Arteaga 2014; Thorpe et al. 2015; Yap et al. 2015). However, we still do not clearly understand how activation of the PI3K/AKT/mTOR axis induces oncogenic transformation. Elucidating the molecular mechanisms by which downstream effectors of mTORC1 contribute to each trait of transformed cells may facilitate development of novel strategies for intervening the pathway with less probability of causing drug resistance, a major hurdle for targeted cancer therapies.

Acknowledgments The authors thank Tatsuhiro Sato for critical reading of the manuscript. M. Aoki would like to thank Peter K. Vogt for his excellent mentorship during my years in his laboratory and for continuous encouragement.

References

Adams JR, Xu K, Liu JC et al (2011) Cooperation between Pik3ca and p53 mutations in mouse mammary tumor formation. Cancer Res 71:2706–2717. doi:10.1158/0008-5472.CAN-10-0738

Amzel LM, Huang CH, Mandelker D et al (2008) Structural comparisons of class I phosphoinositide 3-kinases. Nat Rev Cancer 8:665–669. doi:10.1038/nrc2443

Angulo I, Vadas O, Garçon F et al (2013) Phosphoinositide 3-kinase δ gene mutation predisposes to respiratory infection and airway damage. Science 342:866–871. doi:10.1126/science.1243292

Aoki M (2000) The Catalytic Subunit of Phosphoinositide 3-Kinase: Requirements for Oncogenicity. J Biol Chem 275:6267–6275. doi:10.1074/jbc.275.9.6267

Aoki M, Batista O, Bellacosa A et al (1998) The Akt kinase: Molecular determinants of oncogenicity. Proc Natl Acad Sci 95:14950–14955. doi:10.1073/pnas.95.25.14950

Aoki M, Blazek E, Vogt PK (2001) A role of the kinase mTOR in cellular transformation induced by the oncoproteins P3k and Akt. Proc Natl Acad Sci U S A 98:136–141. doi:10.1073/pnas.011528498

Arteaga CL (2010) Clinical development of phosphatidylinositol-3 kinase pathway inhibitors. Curr Top Microbiol Immunol 347:189–208. doi:10.1007/82_2010_54

Avdulov S, Li S, Michalek V et al (2004) Activation of translation complex eIF4F is essential for the genesis and maintenance of the malignant phenotype in human mammary epithelial cells. Cancer Cell 5:553–563. doi:10.1016/j.ccr.2004.05.024

Baer R, Cintas C, Dufresne M et al (2014) Pancreatic cell plasticity and cancer initiation induced by oncogenic Kras is completely dependent on wild-type PI 3-kinase p110α Genes Dev 28:2621–2635. doi:10.1101/gad.249409.114

Banerji S, Cibulskis K, Rangel-Escareno C et al (2012) Sequence analysis of mutations and translocations across breast cancer subtypes. Nature 486:405–409. doi:10.1038/nature11154

Bellacosa A, Testa JR, Staal SP, Tsichlis PN (1991) A retroviral oncogene, akt, encoding a serine-threonine kinase containing an SH2-like region. Science 254:274–277. doi:10.1126/science.1833819

Ben-Sahra I, Howell JJ, Asara JM, Manning BD (2013) Stimulation of de novo pyrimidine synthesis by growth signaling through mTOR and S6K1. Science 339:1323–1328. doi:10.1126/science.1228792

Ben-Sahra I, Hoxhaj G, Ricoult SJH et al (2016) mTORC1 induces purine synthesis through control of the mitochondrial tetrahydrofolate cycle. Science 351:728–733. doi:10.1126/science.aad0489

Bhat M, Robichaud N, Hulea L et al (2015) Targeting the translation machinery in cancer. Nat Rev Drug Discov 14:261–278. doi:10.1038/nrd4505

Blanco-Aparicio C, Pérez-Gallego L, Pequeño B et al (2007) Mice expressing myrAKT1 in the mammary gland develop carcinogen-induced ER-positive mammary tumors that mimic human breast cancer. Carcinogenesis 28:584–594. doi:10.1093/carcin/bgl190

Bunney TD, Katan M (2010) Phosphoinositide signalling in cancer: beyond PI3K and PTEN. Nat Rev Cancer 10:342–352. doi:10.1038/nrc2842

Burke JE, Perisic O, Masson GR et al (2012) Oncogenic mutations mimic and enhance dynamic events in the natural activation of phosphoinositide 3-kinase p110α (PIK3CA). Proc Natl Acad Sci U S A 109:15259–15264. doi:10.1073/pnas.1205508109

Cai SL, Tee AR, Short JD et al (2006) Activity of TSC2 is inhibited by AKT-mediated phosphorylation and membrane partitioning. J Cell Biol 173:279–289. doi:10.1083/jcb.200507119

Cai W, Ye Q, She Q-B (2014) Loss of 4E-BP1 function induces EMT and promotes cancer cell migration and invasion via cap-dependent translational activation of snail. Oncotarget 5:6015–6027. doi:10.18632/oncotarget.2109

Carpten JD, Faber AL, Horn C et al (2007) A transforming mutation in the pleckstrin homology domain of AKT1 in cancer. Nature 448:439–444. doi:10.1038/nature05933

Castellano E, Sheridan C, Thin MZ et al (2013) Requirement for interaction of PI3-kinase p110α with RAS in lung tumor maintenance. Cancer Cell 24:617–630. doi:10.1016/j.ccr.2013.09.012

Chang HW (1997) Transformation of Chicken cells by the gene encoding the catalytic subunit of PI 3-Kinase. Science 276:1848–1850. doi:10.1126/science.276.5320.1848

Charles R-P, Silva J, Iezza G et al (2014) Activating BRAF and PIK3CA Mutations Cooperate to Promote Anaplastic Thyroid Carcinogenesis. Mol Cancer Res 12:979–986. doi:10.1158/1541-7786.MCR-14-0158-T

Chen M, Pratt CP, Zeeman ME et al (2011) Identification of PHLPP1 as a tumor suppressor reveals the role of feedback activation in PTEN-mutant prostate cancer progression. Cancer Cell 20:173–186. doi:10.1016/j.ccr.2011.07.013

Chen ML, Xu PZ, Peng XD et al (2006) The deficiency of Akt1 is sufficient to suppress tumor development in Pten± mice. Genes Dev 20:1569–1574. doi:10.1101/gad.1395006

Cheung LWT, Hennessy BT, Li J et al (2011) High frequency of PIK3R1 and PIK3R2 mutations in endometrial cancer elucidates a novel mechanism for regulation of PTEN protein stability. Cancer Discov 1:170–185. doi:10.1158/2159-8290.CD-11-0039

Clegg NJ, Couto SS, Wongvipat J et al (2011) MYC cooperates with AKT in prostate tumorigenesis and alters sensitivity to mTOR inhibitors. PLoS ONE. doi:10.1371/journal.pone.0017449

Collisson EA, Trejo CL, Silva JM et al (2012) A Central role for RAF → MEK → ERK signaling in the genesis of pancreatic ductal adenocarcinoma. Cancer Discov 2:685–693. doi:10.1158/2159-8290.CD-11-0347

Coulter TI, Chandra A, Bacon CM et al (2017) Clinical spectrum and features of activated phosphoinositide 3-kinase δ syndrome: a large patient cohort study. J Allergy Clin Immunol 139(597–606):e4. doi:10.1016/j.jaci.2016.06.021

Cristofano AD, Pesce B, Cordon-Cardo C, Pandolfi PP (1998) Pten is essential for embryonic development and tumour suppression. Nat Genet 19:348–355. doi:10.1038/1235

Csibi A, Fendt S-M, Li C et al (2013) The mTORC1 pathway stimulates glutamine metabolism and cell proliferation by repressing SIRT4. Cell 153:840–854. doi:10.1016/j.cell.2013.04.023

Cunningham JT, Rodgers JT, Arlow DH et al (2007) mTOR controls mitochondrial oxidative function through a YY1-PGC-1α transcriptional complex. Nature 450:736–740. doi:10.1038/nature06322

Dazert E, Hall MN (2011) mTOR signaling in disease. Curr Opin Cell Biol 23:744–755. doi:10.1016/j.ceb.2011.09.003

DeBerardinis RJ, Lum JJ, Hatzivassiliou G, Thompson CB (2008) The biology of cancer: metabolic reprogramming fuels cell growth and proliferation. Cell Metab 7:11–20. doi:10.1016/j.cmet.2007.10.002

Deming DA, Leystra AA, Farhoud M et al (2013) mTOR inhibition elicits a dramatic response in PI3K-dependent colon cancers. PLoS ONE. doi:10.1371/journal.pone.0060709
Deming DA, Leystra AA, Nettekoven L et al (2014) PIK3CA and APC mutations are synergistic in the development of intestinal cancers. Oncogene 33:2245–2254. doi:10.1038/onc.2013.167
Denley A, Kang S, Karst U, Vogt PK (2008) Oncogenic signaling of class I PI3K isoforms. Oncogene 27:2561–2574. doi:10.1038/sj.onc.1210918
Denley A, Gymnopoulos M, Kang S et al (2009) Requirement of phosphatidylinositol(3,4,5) trisphosphate in phosphatidylinositol 3-kinase-induced oncogenic transformation. Mol Cancer Res 7:1132–1138. doi:10.1158/1541-7786.MCR-09-0068
Deuker MM, Durban VM, Phillips WA, McMahon M (2015) PI3′-kinase inhibition forestalls the onset of MEK1/2 inhibitor resistance in BRAF-mutated melanoma. Cancer Discov 5:143–153. doi:10.1158/2159-8290.CD-14-0856
Dillon RL, Marcotte R, Hennessy BT et al (2009) Akt1 and Akt2 play distinct roles in the initiation and metastatic phases of mammary tumor progression. Cancer Res 69:5057–5064. doi:10.1158/0008-5472.CAN-08-4287
Dodd KM, Yang J, Shen MH et al (2015) mTORC1 drives HIF-1α and VEGF-A signalling via multiple mechanisms involving 4E-BP1, S6K1 and STAT3. Oncogene 34:2239–2250. doi:10.1038/onc.2014.164
Durban VM, Deuker MM, Bosenberg MW et al (2013) Differential AKT dependency displayed by mouse models of BRAF V600E-Initiated melanoma. J Clin Invest 123:5104–5118. doi:10.1172/JCI69619
Düvel K, Yecies JL, Menon S et al (2010) Activation of a metabolic gene regulatory network downstream of mTOR complex 1. Mol Cell 39:171–183. doi:10.1016/j.molcel.2010.06.022
Dzneladze I, He R, Woolley JF et al (2015) INPP4B overexpression is associated with poor clinical outcome and therapy resistance in acute myeloid leukemia. Leukemia 29:1485–1495. doi:10.1038/leu.2015.51
Edinger AL, Thompson CB (2004) An activated mTOR mutant supports growth factor-independent, nutrient-dependent cell survival. Oncogene 23:5654–5663. doi:10.1038/sj.onc.1207738
Elghazi L, Weiss AJ, Barker DJ et al (2009) Regulation of pancreas plasticity and malignant transformation by Akt signaling. Gastroenterology 136(1091–1103):e8. doi:10.1053/j.gastro.2008.11.043
Engelman JA, Chen L, Tan X et al (2008) Effective use of PI3K and MEK inhibitors to treat mutant Kras G12D and PIK3CA H1047R murine lung cancers. Nat Med 14:1351–1356. doi:10.1038/nm.1890
Eser S, Reiff N, Messer M et al (2013) Selective requirement of PI3K/PDK1 signaling for kras oncogene-driven pancreatic cell plasticity and cancer. Cancer Cell 23:406–420. doi:10.1016/j.ccr.2013.01.023
Fedele CG, Ooms LM, Ho M et al (2010) Inositol polyphosphate 4-phosphatase II regulates PI3K/Akt signaling and is lost in human basal-like breast cancers. Proc Natl Acad Sci U S A 107:22231–22236. doi:10.1073/pnas.1015245107
Feldman ME, Shokat KM (2010) New inhibitors of the PI3K-Akt-mTOR pathway: insights into mTOR signaling from a new generation of tor kinase domain inhibitors (TORKinibs). Curr Top Microbiol Immunol 347:241–262. doi:10.1007/82-2010-64
Franco I, Gulluni F, Campa CC et al (2014) PI3K class II α controls spatially restricted endosomal PtdIns3P and Rab11 activation to promote primary cilium function. Dev Cell 28:647–658. doi:10.1016/j.devcel.2014.01.022
Fruman DA, Meyers RE, Cantley LC (1998) Phosphoinositide kinases. Annu Rev Biochem 67:481–507. doi:10.1146/annurev.biochem.67.1.481
Fruman DA, Rommel C (2014) PI3K and cancer: lessons, challenges and opportunities. Nat Rev Drug Discov 13:140–156. doi:10.1038/nrd4204
Furic L, Rong L, Larsson O et al (2010) eIF4E phosphorylation promotes tumorigenesis and is associated with prostate cancer progression. Proc Natl Acad Sci U S A 107:14134–14139. doi:10.1073/pnas.1005320107

Furman RR, Sharman JP, Coutre SE et al (2014) Idelalisib and Rituximab in relapsed chronic lymphocytic leukemia. N Engl J Med 370:997–1007. doi:10.1056/NEJMoa1315226

Ganley IG, Lam DH, Wang J et al (2009) ULK1.ATG13.FIP200 complex mediates mTOR signaling and is essential for autophagy. J Biol Chem 284:12297–12305. doi:10.1074/jbc.M900573200

Gao T, Furnari F, Newton AC (2005) PHLPP: A phosphatase that directly dephosphorylates Akt, promotes apoptosis, and suppresses tumor growth. Mol Cell 18:13–24. doi:10.1016/j.molcel.2005.03.008

Gasser JA, Inuzuka H, Lau AW et al (2014) SGK3 mediates INPP4B-Dependent PI3K signaling in breast cancer. Mol Cell 56:595–607. doi:10.1016/j.molcel.2014.09.023

Gerlinger M, Rowan AJ, Horswell S et al (2012) Intratumor heterogeneity and branched evolution revealed by multiregion sequencing. N Engl J Med 366:883–892. doi:10.1056/NEJMoa1113205

Gewinner C, Wang ZC, Richardson A et al (2009) Evidence that inositol polyphosphate 4-phosphatase type II is a tumor suppressor that inhibits PI3K signaling. Cancer Cell 16:115–125. doi:10.1016/j.ccr.2009.06.006

Gopal AK, Kahl BS, de Vos S et al (2014) PI3Kδ inhibition by idelalisib in patients with relapsed indolent lymphoma. N Engl J Med 370:1008–1018. doi:10.1056/NEJMoa1314583

Grabiner BC, Nardi V, Birsoy K et al (2014) A diverse array of cancer-associated MTOR mutations are hyperactivating and can predict rapamycin sensitivity. Cancer Discov 4:554–563. doi:10.1158/2159-8290.CD-13-0929

Graff JR, Zimmer SG (2003) Translational control and metastatic progression: enhanced activity of the mRNA cap-binding protein eIF-4E selectively enhances translation of metastasis-related mRNAs. Clin Exp Metastasis 20:265–273

Green S, Trejo CL, McMahon M (2015) PIK3CAH1047R accelerates and enhances KRASG12D-driven lung tumorigenesis. Cancer Res 75:5378–5391. doi:10.1158/0008-5472.CAN-15-1249

Gritsman K, Yuzugullu H, Von T et al (2014) Hematopoiesis and RAS-driven myeloid leukemia differentially require PI3K isoform p110α. J Clin Invest 124:1794–1809. doi:10.1172/JCI69927

Guertin DA, Stevens DM, Saitoh M et al (2009) mTOR complex 2 is required for the development of prostate cancer induced by Pten loss in mice. Cancer Cell 15:148–159. doi:10.1016/j.ccr.2008.12.017

Gulhati P, Bowen KA, Liu J et al (2011) mTORC1 and mTORC2 regulate EMT, motility, and metastasis of colorectal cancer via RhoA and Rac1 signaling pathways. Cancer Res 71:3246–3256. doi:10.1158/0008-5472.CAN-10-4058

Guo ST, Chi MN, Yang RH, et al (2015) INPP4B is an oncogenic regulator in human colon cancer. Oncogene 1–13. doi:10.1038/onc.2015.361

Gupta S, Ramjaun AR, Haiko P et al (2007) Binding of ras to phosphoinositide 3-kinase p110α is required for ras-driven tumorigenesis in mice. Cell 129:957–968. doi:10.1016/j.cell.2007.03.051

Gwinn DM, Shackelford DB, Egan DF et al (2008) AMPK phosphorylation of raptor mediates a metabolic checkpoint. Mol Cell 30:214–226. doi:10.1016/j.molcel.2008.03.003

Hanahan D, Weinberg RA (2011) Hallmarks of cancer: the next generation. Cell 144:646–674. doi:10.1016/j.cell.2011.02.013

Hao Y, Wang C, Cao B et al (2013) Gain of interaction with IRS1 by p110α-helical domain mutants is crucial for their oncogenic functions. Cancer Cell 23:583–593. doi:10.1016/j.ccr.2013.03.021

Hare LM, Phesse TJ, Waring PM et al (2014) Physiological expression of the PI3K-activating mutation Pik3ca(H1047R) combines with Apc loss to promote development of invasive intestinal adenocarcinomas in mice. Biochem J 458:251–258. doi:10.1042/BJ20131412

Hollander MC, Blumenthal GM, Dennis PA (2011a) PTEN loss in the continuum of common cancers, rare syndromes and mouse models. Nat Rev Cancer 11:289–301. doi:10.1038/nrc3085

Hollander MC, Maier CR, Hobbs EA et al (2011b) Akt1 deletion prevents lung tumorigenesis by mutant K-ras. Oncogene 30:1812–1821. doi:10.1038/onc.2010.556

Holohan C, Van Schaeybroeck S, Longley DB, Johnston PG (2013) Cancer drug resistance: an evolving paradigm. Nat Rev Cancer 13:714–726. doi:10.1038/nrc3599

Hon WC, Berndt A, Williams RL (2012) Regulation of lipid binding underlies the activation mechanism of class IA PI3-kinases. Oncogene 31:3655–3666. doi:10.1038/onc.2011.532

Hsieh AC, Costa M, Zollo O et al (2010) Genetic dissection of the oncogenic mTOR pathway reveals druggable addiction to translational control via 4EBP-eIF4E. Cancer Cell 17:249–261. doi:10.1016/j.ccr.2010.01.021

Hsieh AC, Liu Y, Edlind MP et al (2012) The translational landscape of mTOR signalling steers cancer initiation and metastasis. Nature 485:55–61. doi:10.1038/nature10912

Huang C-H, Mandelker D, Schmidt-Kittler O et al (2007) The structure of a human p110α/p85α complex elucidates the effects of oncogenic PI3Kα mutations. Science 318:1744–1748. doi:10.1126/science.1150799

Huang J, Manning BD (2008) The TSC1-TSC2 complex: a molecular switchboard controlling cell growth. Biochem J 412:179–190. doi:10.1042/BJ20080281

Hudson CC, Liu M, Chiang GG et al (2002) Regulation of hypoxia-inducible factor 1α expression and function by the mammalian target of rapamycin. Mol Cell Biol 22:7004–7014. doi:10.1128/MCB.22.20.7004

Hutchinson J, Jin J, Cardiff RD et al (2001) Activation of Akt (protein kinase B) in mammary epithelium provides a critical cell survival signal required for tumor progression. Mol Cell Biol 21:2203–2212. doi:10.1128/MCB.21.6.2203-2212.2001

Hutchinson JN, Jin J, Cardiff RD et al (2004a) Activation of Akt-1 (PKB-α) Can accelerate ErbB-2-mediated mammary tumorigenesis but suppresses tumor invasion activation of Akt-1 (PKB-␣) can accelerate ErbB-2-mediated mammary tumorigenesis but suppresses tumor invasion. 1:3171–3178

Hutchinson JN, Jin J, Cardiff RD et al (2004b) Activation of Akt-1 (PKB-α) can accelerate ErbB-2-mediated mammary tumorigenesis but suppresses tumor invasion. Cancer Res 64:3171–3178. doi:10.1158/0008-5472.can-03-3465

Iglesias-Bartolome R, Patel V, Cotrim A et al (2012) MTOR inhibition prevents epithelial stem cell senescence and protects from radiation-induced mucositis. Cell Stem Cell 11:401–414. doi:10.1016/j.stem.2012.06.007

Ikenoue T (2005) Functional analysis of PIK3CA gene mutations in human colorectal cancer. Cancer Res 65:4562–4567. doi:10.1158/0008-5472.CAN-04-4114

Inoki K, Li Y, Zhu T et al (2002) TSC2 is phosphorylated and inhibited by Akt and suppresses mTOR signalling. Nat Cell Biol 4:648–657. doi:10.1038/ncb839

Inoki K, Zhu T, Guan K-L (2003) TSC2 mediates cellular energy response to control cell growth and survival. Cell 115:577–590 (S0092867403009292 [pii])

Isakoff SJ, Engelman JA, Irie HY et al (2005) Breast cancer-associated PIK3CA mutations are oncogenic in mammary epithelial cells. Cancer Res 65:10992–11000. doi:10.1158/0008-5472.CAN-05-2612

Jaiswal BS, Janakiraman V, Kljavin NM et al (2009) Somatic mutations in p85α promote tumorigenesis through class IA PI3K activation. Cancer Cell 16:463–474. doi:10.1016/j.ccr.2009.10.016

Jia S, Gao X, Lee SH et al (2013) Opposing effects of androgen deprivation and targeted therapy on prostate cancer prevention. Cancer Discov 3:44–51. doi:10.1158/2159-8290.CD-12-0262

Jia S, Liu Z, Zhang S et al (2008) Essential roles of PI(3)K-p110β in cell growth, metabolism and tumorigenesis. Nature 454:776–779. doi:10.1038/nature07091

Jiang H, Vogt PK (2008) Constitutively active Rheb induces oncogenic transformation. Oncogene 27:5729–5740. doi:10.1038/onc.2008.180

Jiang L, Xu L, Mao J et al (2013) Rheb/mTORC1 signaling promotes kidney fibroblast activation and fibrosis. J Am Soc Nephrol 24:1114–1126. doi:10.1681/ASN.2012050476

Joshi A, Miller C, Baker SJ, Ellenson LH (2015) Activated mutant p110α causes endometrial carcinoma in the setting of biallelic Pten deletion. Am J Pathol 185:1104–1113. doi:10.1016/j.ajpath.2014.12.019

Jung CH, Jun CB, Ro S-H et al (2009) ULK-Atg13-FIP200 complexes mediate mTOR signaling to the autophagy machinery. Mol Biol Cell 20:1992–2003. doi:10.1091/mbc.E08-12-1249

Kang S, Bader AG, Vogt PK (2005) Phosphatidylinositol 3-kinase mutations identified in human cancer are oncogenic. Proc Natl Acad Sci U S A 102:802–807. doi:10.1073/pnas.0408864102

Kang S, Denley A, Vanhaesebroeck B, Vogt PK (2006) Oncogenic transformation induced by the p110β, -γ, and -δ isoforms of class I phosphoinositide 3-kinase. Proc Natl Acad Sci U S A 103:1289–1294. doi:10.1073/pnas.0510772103

Kim J, Kundu M, Viollet B, Guan K-L (2011) AMPK and mTOR regulate autophagy through direct phosphorylation of Ulk1. Nat Cell Biol 13:132–141. doi:10.1038/ncb2152

Kinross KM, Montgomery KG, Kleinschmidt M et al (2012) An activating Pik3ca mutation coupled with Pten loss is sufficient to initiate ovarian tumorigenesis in mice. J Clin Invest 122:553–557. doi:10.1172/JCI59309

Kobayashi T, Minowa O, Kuno J et al (1999) Renal carcinogenesis, hepatic hemangiomatosis, and embryonic lethality caused by a germ-line Tsc2 mutation in mice. Cancer Res 59:1206–1211

Kobayashi T, Minowa O, Sugitani Y et al (2001) A germ-line Tsc1 mutation causes tumor development and embryonic lethality that are similar, but not identical to, those caused by Tsc2 mutation in mice. Proc Natl Acad Sci U S A 98:8762–8767. doi:10.1073/pnas.151033798

Koren S, Reavie L, Couto JP et al (2015) PIK3CA(H1047R) induces multipotency and multi-lineage mammary tumours. Nature 525:114–118. doi:10.1038/nature14669

Korsse SE, Peppelenbosch MP, Van Veelen W (2013) Targeting LKB1 signaling in cancer. Biochim Biophys Acta - Rev Cancer 1835:194–210. doi:10.1016/j.bbcan.2012.12.006

Lamouille S, Derynck R (2007) Cell size and invasion in TGF-β-induced epithelial to mesenchymal transition is regulated by activation of the mTOR pathway. J Cell Biol 178:437–451. doi:10.1083/jcb.200611146

Laplante M, Sabatini DM (2012) MTOR signaling in growth control and disease. Cell 149:274–293. doi:10.1016/j.cell.2012.03.017

Larsson O, Li S, Issaenko OA et al (2007) Eukaryotic translation initiation factor 4E-induced progression of primary human mammary epithelial cells along the cancer pathway is associated with targeted translational deregulation of oncogenic drivers and inhibitors. Cancer Res 67:6814–6824. doi:10.1158/0008-5472.CAN-07-0752

Lazaris-Karatzas A, Montine K, Sonenberg N (1990) Malignant transformation by a eukaryotic initiation factor subunit that binds to mRNA 5'cap. Nature 345:544–547. doi:10.1038/345544a0

Leslie NR, Longy M (2016) Inherited PTEN mutations and the prediction of phenotype. Semin Cell Dev Biol 52:30–38. doi:10.1016/j.semcdb.2016.01.030

Leystra AA, Deming DA, Zahm CD et al (2012) Mice expressing activated PI3K rapidly develop advanced colon cancer. Cancer Res 72:2931–2936. doi:10.1158/0008-5472.CAN-11-4097

Li S, Perlman DM, Peterson MS et al (2004) Translation initiation factor 4E blocks endoplasmic reticulum-mediated apoptosis. J Biol Chem 279:21312–21317. doi:10.1074/jbc.M312467200

Liang S, Yang N, Pan Y et al (2009) Expression of activated PIK3CA in ovarian surface epithelium results in hyperplasia but not tumor formation. PLoS ONE. doi:10.1371/journal.pone.0004295

Lien EC, Lyssiotis CA, Cantley LC (2016) Metabolic reprogramming by the PI3K-Akt-mTOR pathway in cancer. pp 39–72

Liu P, Cheng H, Santiago S et al (2011) Oncogenic PIK3CA-driven mammary tumors frequently recur via PI3K pathway-dependent and PI3K pathway-independent mechanisms. Nat Med 17:1116–1120. doi:10.1038/nm.2402

LoRusso PM (2016) Inhibition of the PI3K/AKT/mTOR pathway in solid tumors. J Clin Oncol 34:3803–3815. doi:10.1200/JCO.2014.59.0018

Lu ZH, Shvartsman MB, Lee AY et al (2010) Mammalian target of rapamycin activator RHEB is frequently overexpressed in human carcinomas and is critical and sufficient for skin epithelial carcinogenesis. Cancer Res 70:3287–3298. doi:10.1158/0008-5472.CAN-09-3467
Ma XM, Blenis J (2009) Molecular mechanisms of mTOR-mediated translational control. Nat Rev Mol Cell Biol 10:307–318. doi:10.1038/nrm2672
Macara IG, Marinetti GV, Balduzzi PC (1984) Transforming protein of avian sarcoma virus UR2 is associated with phosphatidylinositol kinase activity: possible role in tumorigenesis. Proc Natl Acad Sci U S A 81:2728–2732
Maira SM, Finan P, Garcia-Echeverria C (2010) From the bench to the bed side: PI3K pathway inhibitors in clinical development. Curr Top Microbiol Immunol 347:209–239. doi:10.1007/82-2010-60
Majumder PK, Febbo PG, Bikoff R et al (2004) mTOR inhibition reverses Akt-dependent prostate intraepithelial neoplasia through regulation of apoptotic and HIF-1-dependent pathways. Nat Med 10:594–601. doi:10.1038/nm1052
Majumder PK, Yeh JJ, George DJ et al (2003) Prostate intraepithelial neoplasia induced by prostate restricted Akt activation: the MPAKT model. Proc Natl Acad Sci U S A 100:7841–7846. doi:10.1073/pnas.1232229100
Malanga D, Belmonte S, Colelli F et al (2016) AKT1E17 K is oncogenic in mouse lung and cooperates with chemical carcinogens in inducing lung cancer. PLoS ONE 11:e0147334. doi:10.1371/journal.pone.0147334
Malstrom S, Tili E, Kappes D et al (2001) Tumor induction by an Lck-MyrAkt transgene is delayed by mechanisms controlling the size of the thymus. Proc Natl Acad Sci U S A 98:14967–14972. doi:10.1073/pnas.231467698
Mamane Y, Petroulakis E, Martineau Y et al (2007) Epigenetic activation of a subset of mRNAs by eIF4E explains its effects on cell proliferation. PLoS ONE. doi:10.1371/journal.pone.0000242
Mamane Y, Petroulakis E, Rong L et al (2004) eIF4E–from translation to transformation. Oncogene 23:3172–3179. doi:10.1038/sj.onc.1207549
Mancini ML, Lien EC, Toker A (2016) Oncogenic AKT1(E17K) mutation induces mammary hyperplasia but prevents HER2-driven tumorigenesis. Oncotarget. doi:10.18632/oncotarget.8191
Mandelker D, Gabelli SB, Schmidt-Kittler O et al (2009) A frequent kinase domain mutation that changes the interaction between PI3Kα and the membrane. Proc Natl Acad Sci U S A 106:16996–17001. doi:10.1073/pnas.0908444106
Manning BD, Cantley LC (2007) AKT/PKB signaling: navigating downstream. Cell 129:1261–1274. doi:10.1016/j.cell.2007.06.009
Manning BD, Tee AR, Logsdon MN et al (2002) Identification of the tuberous sclerosis complex-2 tumor suppressor gene product tuberin as a target of the phosphoinositide 3-kinase/akt pathway. Mol Cell 10:151–162
Mavrakis KJ, Zhu H, Silva RLA et al (2008) Tumorigenic activity and therapeutic inhibition of Rheb GTPase. Genes Dev 22:2178–2188. doi:10.1101/gad.1690808
Mayer IA, Arteaga CL (2014) The PI3K/AKT pathway as a target for cancer treatment. Annu Rev Med 67. doi:10.1146/annurev-med-062913-051343
Mellor P, Furber LAA, Nyarko JNKNK, Anderson DHH (2011) Multiple roles for the p85α isoform in the regulation and function of PI3K signalling and receptor trafficking. Biochem J 441:23–37. doi:10.1042/BJ20111164
Mende I, Malstrom S, Tsichlis PN et al (2001) Oncogenic transformation induced by membrane-targeted Akt2 and Akt3. Oncogene 20:4419–4423. doi:10.1038/sj.onc.1204486
Meng L-H, Zheng XS (2015) Toward rapamycin analog (rapalog)-based precision cancer therapy. Acta Pharmacol Sin 36:1163–1169. doi:10.1038/aps.2015.68
Meyer DS, Brinkhaus H, Müller U et al (2011) Luminal expression of PIK3CA mutant H1047R in the mammary gland induces heterogeneous tumors. Cancer Res 71:4344–4351. doi:10.1158/0008-5472.CAN-10-3827

Mirza AM, Kohn AD, Roth RA, McMahon M (2000) Oncogenic transformation of cells by a conditionally active form of the protein kinase Akt/PKB. Cell Growth Differ 11:279–292
Moral M, Segrelles C, Lara MF et al (2009) Akt activation synergizes with Trp53 loss in oral epithelium to produce a novel mouse model for head and neck squamous cell carcinoma. Cancer Res 69:1099–1108. doi:10.1158/0008-5472.CAN-08-3240
Morita M, Gravel SP, Chénard V et al (2013) MTORC1 controls mitochondrial activity and biogenesis through 4E-BP-dependent translational regulation. Cell Metab 18:698–711. doi:10.1016/j.cmet.2013.10.001
Morita M, Gravel SP, Hulea L et al (2015) MTOR coordinates protein synthesis, mitochondrial activity. Cell Cycle 14:473–480. doi:10.4161/15384101.2014.991572
Murugan AK, Alzahrani A, Xing M (2013) Mutations in critical domains confer the human mTOR gene strong tumorigenicity. J Biol Chem 288:6511–6521. doi:10.1074/jbc.M112.399485
Nardella C, Carracedo A, Alimonti A et al (2009) Differential requirement of mTOR in postmitotic tissues and tumorigenesis. Sci Signal 2:ra2. doi:10.1126/scisignal.2000189
Nardella C, Carracedo A, Salmena L, Pandolfi PP (2010) Faithfull modeling of PTEN loss driven diseases in the mouse. Curr Top Microbiol Immunol 347:135–168. doi:10.1007/82-2010-62
Nardella C, Chen Z, Salmena L et al (2008) Aberrant Rheb-mediated mTORC1 activation and Pten haploinsufficiency are cooperative oncogenic events service Aberrant Rheb-mediated mTORC1 activation and Pten haploinsufficiency are cooperative oncogenic events. 2172 2177. doi:10.1101/gad.1699608
Narita M, Young ARJ, Arakawa S et al (2011) Spatial coupling of mTOR and autophagy augments secretory phenotypes. Science 332:966–970. doi:10.1126/science.1205407
Neshat MS, Mellinghoff IK, Tran C et al (2001) Enhanced sensitivity of PTEN-deficient tumors to inhibition of FRAP/mTOR. Proc Natl Acad Sci U S A 98:10314–10319. doi:10.1073/pnas.171076798
Ni J, Liu Q, Xie S et al (2012) Functional characterization of an isoform-selective inhibitor of PI3K-p110β as a potential anticancer agent. Cancer Discov 2:425–433. doi:10.1158/2159-8290.CD-12-0003
Ohne Y, Terunao Takahara T, Maeda T (2009) Evaluation of mTOR function by a gain-of-function approach. Cell Cycle 8:573–579
Onda H, Lueck A, Marks PW et al (1999) Tsc2(±) mice develop tumors in multiple sites that express gelsolin and are influenced by genetic background. J Clin Invest 104:687–695. doi:10.1172/JCI7319
Ooms LM, Binge LC, Davies EM et al (2015) The inositol polyphosphate 5-phosphatase PIPP regulates AKT1-dependent breast cancer growth and metastasis. Cancer Cell 28:155–169. doi:10.1016/j.ccell.2015.07.003
Papa A, Wan L, Bonora M et al (2014) Cancer-associated PTEN mutants act in a dominant-negative manner to suppress PTEN protein function. Cell 157:595–610. doi:10.1016/j.cell.2014.03.027
Parsons DW, Wang T-L, Samuels Y et al (2005) Colorectal cancer: mutations in a signalling pathway. Nature 436:792. doi:10.1038/436792a
Pérez-Mancera PA, Young ARJ, Narita M (2014) Inside and out: the activities of senescence in cancer. Nat Rev Cancer 14:547–558. doi:10.1038/nrc3773
Peterson TR, Sengupta SS, Harris TE et al (2011) MTOR complex 1 regulates lipin 1 localization to control the srebp pathway. Cell 146:408–420. doi:10.1016/j.cell.2011.06.034
Podsypanina K, Ellenson LH, Nemes A et al (1999) Mutation of Pten/Mmac1 in mice causes neoplasia in multiple organ systems. Proc Natl Acad Sci U S A 96:1563–1568. doi:10.1073/pnas.96.4.1563
Podsypanina K, Lee RT, Politis C et al (2001) An inhibitor of mTOR reduces neoplasia and normalizes p70/S6 kinase activity in Pten± mice. Proc Natl Acad Sci U S A 98:10320–10325. doi:10.1073/pnas.171060098
Porstmann T, Santos CR, Griffiths B et al (2008) SREBP activity is regulated by mTORC1 and contributes to Akt-dependent cell growth. Cell Metab 8:224–236. doi:10.1016/j.cmet.2008.07.007

Posor Y, Eichhorn-Gruenig M, Puchkov D et al (2013) Spatiotemporal control of endocytosis by phosphatidylinositol-3,4-bisphosphate. Nature 499:233–237. doi:10.1038/nature12360

Potter CJ, Pedraza LG, Xu T (2002) Akt regulates growth by directly phosphorylating Tsc2. Nat Cell Biol 4:658–665. doi:10.1038/ncb840

Rijal S, Fleming S, Cummings N et al (2015) Inositol polyphosphate 4-phosphatase II (INPP4B) is associated with chemoresistance and poor outcome in AML. Blood 125:2815–2824. doi:10.1182/blood-2014-09-603555

Robinson G, Parker M, Kranenburg TA et al (2012a) Novel mutations target distinct subgroups of medulloblastoma. Nature 488:43–48. doi:10.1038/nature11213.Novel

Robinson G, Parker M, Kranenburg T, Lu C (2012b) Novel mutations target distinct subgroups of medulloblastoma. Nature 488:43–48. doi:10.1038/nature11213.Novel

Robitaille AM, Christen S, Shimobayashi M et al (2013) Quantitative phosphoproteomics reveal mTORC1 activates de novo pyrimidine synthesis. Science 339:1320–1323. doi:10.1126/science.1228771

Ruggero D, Montanaro L, Ma L et al (2004) The translation factor eIF-4E promotes tumor formation and cooperates with c-Myc in lymphomagenesis. Nat Med 10:484–486. doi:10.1038/nm1042

Samuels Y, Waldman T (2010) Oncogenic mutations of PIK3CA in human cancers. Curr Top Microbiol Immunol 347:21–41. doi:10.1007/82-2010-68

Samuels Y, Wang Z, Bardelli A et al (2004) Brevia: high frequency of mutations of the PIK3Ca gene in human cancers. Science 304:554. doi:10.1126/science.1096502

Sancak Y, Thoreen CC, Peterson TR et al (2007) PRAS40 is an insulin-regulated inhibitor of the mTORC1 protein kinase. Mol Cell 25:903–915. doi:10.1016/j.molcel.2007.03.003

Sander S, Calado DP, Srinivasan L et al (2012) Synergy between PI3K signaling and MYC in Burkitt lymphomagenesis. Cancer Cell 22:167–179. doi:10.1016/j.ccr.2012.06.012

Sasaki H, Okuda K, Kawano O et al (2008) AKT1 and AKT2 mutations in lung cancer in a Japanese population. Mol Med Rep. doi:10.3892/mmr-00000009

Sato T, Nakashima A, Guo L et al (2010) Single amino-acid changes that confer constitutive activation of mTOR are discovered in human cancer. Oncogene 29:2746–2752. doi:10.1038/onc.2010.28

Schmit F, Utermark T, Zhang S et al (2014) PI3K isoform dependence of PTEN-deficient tumors can be altered by the genetic context. Proc Natl Acad Sci U S A 111:6395–6400. doi:10.1073/pnas.1323004111

Segrelles C, Lu J, Hammann B et al (2007) Deregulated activity of Akt in epithelial basal cells induces spontaneous tumors and heightened sensitivity to skin carcinogenesis. Cancer Res 67:10879–10888. doi:10.1158/0008-5472.CAN-07-2564

Shaw RJ, Bardeesy N, Manning BD et al (2004) The LKB1 tumor suppressor negatively regulates mTOR signaling. Cancer Cell 6:91–99. doi:10.1016/j.ccr.2004.06.007

Siddiqui N, Sonenberg N (2015) Signalling to eIF4E in cancer. Biochem Soc Trans 43:763–772. doi:10.1042/BST20150126

Simonsen A, Tooze SA (2009) Coordination of membrane events during autophagy by multiple class III PI3-kinase complexes. J Cell Biol 186:773–782. doi:10.1083/jcb.200907014

Skeen JE, Bhaskar PT, Chen CC et al (2006) Akt deficiency impairs normal cell proliferation and suppresses oncogenesis in a p53-independent and mTORC1-dependent manner. Cancer Cell 10:269–280. doi:10.1016/j.ccr.2006.08.022

Song MS, Salmena L, Pandolfi PP (2012) The functions and regulation of the PTEN tumour suppressor. Nat Rev Mol Cell Biol 13:283–296. doi:10.1038/nrm3330

Stambolic V, Tsao MS, Macpherson D et al (2000) High incidence of breast and endometrial neoplasia resembling human Cowden syndrome in pten($\pm$) mice. Cancer Res 60:3605–3611

Stemke-Hale K, Gonzalez-Angulo AM, Lluch A et al (2008) An integrative genomic and proteomic analysis of PIK3CA, PTEN, and AKT mutations in breast cancer. Cancer Res 68:6084–6091. doi:10.1158/0008-5472.CAN-07-6854

Stephens L, Hawkins P (2013) More paths to PI3Kγ PLoS Biol 11:1–3. doi:10.1371/journal.pbio.1001594

Stephens PJ, Tarpey PS, Davies H et al (2012) The landscape of cancer genes and mutational processes in breast cancer. Nature 486:400–404. doi:10.1038/nature11017

Stine ZE, Walton ZE, Altman BJ et al (2015) MYC, metabolism, and cancer. Cancer Discov 5:1024–1039

Sugimoto Y, Whitman M, Cantley LC, Erikson RL (1984) Evidence that the Rous sarcoma virus transforming gene product phosphorylates phosphatidylinositol and diacylglycerol. Proc Natl Acad Sci U S A 81:2117–2121

Sun M, Hillmann P, Hofmann BT et al (2010) Cancer-derived mutations in the regulatory subunit p85α of phosphoinositide 3-kinase function through the catalytic subunit p110α. Proc Natl Acad Sci U S A 107:15547–15552. doi:10.1073/pnas.1009652107

Sun M, Wang G, Paciga JE et al (2001) AKT1/PKBα kinase is frequently elevated in human cancers and its constitutive activation is required for oncogenic transformation in NIH3T3 cells. Am J Pathol 159:431–437. doi:10.1016/S0002-9440(10)61714-2

Thorpe LM, Yuzugullu H, Zhao JJ (2015) PI3K in cancer: divergent roles of isoforms, modes of activation and therapeutic targeting. Nat Rev Cancer 15:7–24. doi:10.1038/nrc3860

Tikoo A, Roh V, Montgomery KG et al (2012) Physiological levels of Pik3caH1047R mutation in the mouse mammary gland results in ductal hyperplasia and formation of ER??-positive tumors. PLoS ONE 7:1–13. doi:10.1371/journal.pone.0036924

Toker A (2012) Achieving specificity in Akt signaling in cancer Alex. Adv Biol Regul 52:78 87. doi:10.1016/j.surg.2006.10.010.Use

Toker A, Marmiroli S (2014) Signaling specificity in the Akt pathway in biology and disease. Adv Biol Regul 55:28–38

Topisirovic I, Ruiz-Gutierrez M, Borden KLB (2004) Phosphorylation of the eukaryotic translation initiation factor eIF4E contributes to its transformation and mRNA transport activities. Cancer Res 64:8639–8642. doi:10.1158/0008-5472.CAN-04-2677

Trejo CL, Green S, Marsh V et al (2013) Mutationally activated PIK3CAH1047R cooperates with BRAF V600E to promote lung cancer progression. Cancer Res 73:6448–6461. doi:10.1158/0008-5472.CAN-13-0681

Truitt MLL, Conn CSS, Shi Z et al (2015) Differential requirements for eIF4E dose in normal development and cancer. Cell 162:59–71. doi:10.1016/j.cell.2015.05.049

Turner KM, Sun Y, Ji P et al (2015) Genomically amplified Akt3 activates DNA repair pathway and promotes glioma progression. Proc Natl Acad Sci U S A 112:3421–3426. doi:10.1073/pnas.1414573112

Utani A (2010) Laminin α3 chain-derived peptide promotes keratinocyte migration and wound closure: clustering of syndecan-4 and integrin β1. Seikagaku 82:327–331. doi:10.1091/mbc.E08

Utermark T, Rao T, Cheng H et al (2012) The p110α and p110β isoforms of PI3K play divergent roles in mammary gland development and tumorigenesis. Genes Dev 26:1573–1586. doi:10.1101/gad.191973.112

Van Keymeulen A, Lee MY, Ousset M et al (2015) Reactivation of multipotency by oncogenic PIK3CA induces breast tumour heterogeneity. Nature 525:119–123. doi:10.1038/nature14665

Vander Haar E, Lee S-I, Bandhakavi S et al (2007) Insulin signalling to mTOR mediated by the Akt/PKB substrate PRAS40. Nat Cell Biol 9:316–323. doi:10.1038/ncb1547

Vander Heiden MG, Cantley LC, Thompson CB (2009) Understanding the Warburg effect: the metabolic requirements of cell proliferation. Science 324:1029–1033. doi:10.1126/science.1160809

Vanhaesebroeck B, Leevers SJ, Panayotou G et al (1997) Phosphoinositide 3-kinases: A conserved family of signal transducers. Trends Biochem Sci 22:267–272

Vasudevan KM, Garraway LA (2010) AKT signaling in physiology and disease. Curr Top Microbiol Immunol 347:105–133. doi:10.1007/82_2010_66

Vivanco I, Sawyers C (2002) The phosphatidylinositol 3-Kinase-AKT pathway in human cancer. Nat Rev - Cancer 2:489–501. doi:10.1038/nrc839

Vogt PK (2012) Retroviral oncogenes: a historical primer. Nat Rev Cancer 12:639–648. doi:10.1038/nrc3320

Vogt PK, Hart JR, Gymnopoulos M et al (2010) Phosphatidylinositol 3-kinase: the oncoprotein. Curr Top Microbiol Immunol 347:79–104. doi:10.1007/82-2010-80
Walker EH, Perisic O, Ried C et al (1999) Structural insights into phosphoinositide 3-kinase catalysis and signalling. Nature 402:313–320. doi:10.1038/46319
Ward PS, Thompson CB (2012) Metabolic reprogramming: a cancer hallmark even warburg did not anticipate. Cancer Cell 21:297–308. doi:10.1016/j.ccr.2012.02.014
Wee S, Wiederschain D, Maria S-M et al (2008) PTEN-deficient cancers depend on PIK3CB. Proc Natl Acad Sci U S A 105:13057–13062. doi:10.1073/pnas.0802655105
Wendel H-G, De Stanchina E, Fridman JS et al (2004) Survival signalling by Akt and eIF4E in oncogenesis and cancer therapy. Nature 428:332–337. doi:10.1038/nature02369
Wendel HG, Silva RLA, Malina A et al (2007) Dissecting eIF4E action in tumorigenesis. Genes Dev 21:3232–3237. doi:10.1101/gad.1604407
Whitman M, Downes CP, Keeler M et al (1988) Type I phosphatidylinositol kinase makes a novel inositol phospholipid, phosphatidylinositol-3-phosphate. Nature 332:644–646
Wu CYC, Carpenter ES, Takeuchi KK et al (2014a) PI3K regulation of RAC1 is required for KRAS-induced pancreatic tumorigenesis in mice. Gastroenterology 147:1405–1416.e7. doi:10.1053/j.gastro.2014.08.032
Wu H, Shekar SC, Flinn RJ et al (2009) Regulation of class IA PI 3-kinases: C2 domain-iSH2 domain contacts inhibit p85/p110α and are disrupted in oncogenic p85 mutants. Proc Natl Acad Sci U S A 106:20258–20263. doi:10.1073/pnas.0902369106
Wu R, Baker SJ, Hu TC et al (2013) Type i to type II ovarian carcinoma progression: mutant Trp53 or Pik3ca confers a more aggressive tumor phenotype in a mouse model of ovarian cancer. Am J Pathol 182:1391–1399. doi:10.1016/j.ajpath.2012.12.031
Wu R, Hu TC, Rehemtulla A et al (2011) Preclinical testing of PI3K/AKT/mTOR signaling inhibitors in a mouse model of ovarian endometrioid adenocarcinoma. Clin Cancer Res 17:7359–7372. doi:10.1158/1078-0432.CCR-11-1388
Wu Y, Kim J, Elshimali Y et al (2014b) Activation of Akt1 accelerates carcinogen-induced tumorigenesis in mammary gland of virgin and post-lactating transgenic mice. BMC Cancer 14:266. doi:10.1186/1471-2407-14-266
Xu P-Z, Chen M-L, Jeon S-M et al (2012) The effect Akt2 deletion on tumor development in Pten (±) mice. Oncogene 31:518–526. doi:10.1038/onc.2011.243
Yamaguchi H, Kawazu M, Yasuda T et al (2015) Transforming somatic mutations of mammalian target of rapamycin kinase in human cancer. Cancer Sci 106:1687–1692. doi:10.1111/cas.12828
Yang J, Kalogerou M, Samsel PA et al (2014) Renal tumours in a Tsc2(±) mouse model do not show feedback inhibition of Akt and are effectively prevented by rapamycin. Oncogene: 1–10. doi:10.1038/onc.2014.17
Yang Q, Modi P, Newcomb T et al (2015) Idelalisib: first-in-class PI3K δ inhibitor for the treatment of chronic lymphocytic leukemia, small lymphocytic leukemia, and follicular lymphoma. Clin Cancer Res 21:1537–1542. doi:10.1158/1078-0432.CCR-14-2034
Yap TA, Bjerke L, Clarke PA, Workman P (2015) Drugging PI3K in cancer: refining targets and therapeutic strategies. Curr Opin Pharmacol 23:98–107. doi:10.1016/j.coph.2015.05.016
Young CD, Nolte EC, Lewis A et al (2008) Activated Akt1 accelerates MMTV-c-ErbB2 mammary tumourigenesis in mice without activation of ErbB3. Breast Cancer Res 10:R70. doi:10.1186/bcr2132
Yuan W, Stawiski E, Janakiraman V et al (2013) Conditional activation of Pik3ca(H1047R) in a knock-in mouse model promotes mammary tumorigenesis and emergence of mutations. Oncogene 32:318–326. doi:10.1038/onc.2012.53
Yuzugullu H, Baitsch L, Von T et al (2015) A PI3K p110β-Rac signalling loop mediates Pten-loss-induced perturbation of haematopoiesis and leukaemogenesis. Nat Commun 6:8501. doi:10.1038/ncomms9501
Zaytseva YY, Valentino JD, Gulhati P, Mark Evers B (2012) MTOR inhibitors in cancer therapy. Cancer Lett 319:1–7. doi:10.1016/j.canlet.2012.01.005

Zhai Y, Kuick R, Tipton C et al (2016) Arid1a inactivation in an Apc- and Pten-defective mouse ovarian cancer model enhances epithelial differentiation and prolongs survival. J Pathol 238:21–30. doi:10.1002/path.4599

Zhao JJ, Liu Z, Wang L et al (2005) The oncogenic properties of mutant p110α and p110β phosphatidylinositol 3-kinases in human mammary epithelial cells. Proc Natl Acad Sci U S A 102:18443–18448. doi:10.1073/pnas.0508988102

Zhao L, Vogt PK (2008a) Class I PI3K in oncogenic cellular transformation. Oncogene 27:5486–5496. doi:10.1038/onc.2008.244

Zhao L, Vogt PK (2008b) Helical domain and kinase domain mutations in p110 of phosphatidylinositol 3-kinase induce gain of function by different mechanisms. Proc Natl Acad Sci 105:2652–2657. doi:10.1073/pnas.0712169105

Zoncu R, Efeyan A, Sabatini DM (2011) mTOR: from growth signal integration to cancer, diabetes and ageing. Nat Rev Mol Cell Biol 12:21–35. doi:10.1038/nrm3025

MicroRNA Dysregulation to Identify Novel Therapeutic Targets

Carlo M. Croce

Abstract This paper describes how we discovered the juxtaposition of the MYC gene to the human immunoglobulin loci and how that finding was extended to characterize molecularly the t(14;18) chromosome translocation of follicular lymphoma and to clone the BCL2 gene. BCL2 is also overexpressed in CLL, the most common human leukemia. We discovered that most of human CLLs have a deletion of two microRNAs residing in the same polycistronic RNA, miR-15a and miR-16-1, and that these two microRNAs are negative regulators of BCL2. Thus, loss of miR-15/16 leads to overexpression of BCL2 that can be targeted by the new drug, venetoclax, that was recently approved by the FDA for the treatment of aggressive CLLs.

Contents

C.M. Croce (✉)
Department of Cancer Biology and Genetics, The Ohio State University Wexner Medical Center, Columbus, OH, USA
e-mail: carlo.croce@osumc.edu

Current Topics in Microbiology and Immunology (2017) 407:191–204
DOI 10.1007/82_2017_34

Published Online: 27 June 2017

1 Introduction

1.1 A Long Journey

A long time ago my laboratory was interested in human gene mapping by taking advantage of somatic cell hybrids between mouse and human cells that segregate human chromosomes, allowing the establishment of a correlation between the presence or absence of a specific human trait and the presence or absence of a specific human chromosome. Although, at that time, probes for the human immunoglobulin loci were not available, we decided to map the human immunoglobulin loci by taking advantage of somatic cell hybrids between mouse myeloma cells and a variety of human B-cell lines (Croce et al. 1979). Since these hybrids express human immunoglobulins, we thought that we could map the chromosomal location of the immunoglobulin loci by looking at the segregation of the expression of the different human immunoglobulin chains in the hybrids (Croce et al. 1979). A complication of this approach was the phenomenon of allelic exclusion, meaning that only one of the loci of heavy and light chain is expressed while the other is silenced (Croce et al. 1979).

In 1979, we examined the expression of human heavy chains (μ, γ and α) in the hybrids between mouse myeloma and human B-cell lines and observed that the expression of the human immunoglobulin heavy chains in the hybrids required the presence of human chromosome 14 (Croce et al. 1979). This was the first mapping of genes that require genetic rearrangement to be expressed, like immunoglobulin and T-cell receptor genes (Croce et al. 1979).

2 Burkitt Lymphoma

After this mapping, we decided to map the chromosomal location of the human immunoglobulin light chain genes. In 1981, in a paper in Nature, we showed that the immunoglobulin λ chain locus is on human chromosome 22 (Erikson et al. 1981). The fact that the human heavy chain locus is on human chromosome 14 and the lambda light chain locus was on human chromosome 22, made us speculate that in Burkitt lymphoma, that carries translocation between human chromosome 8 and either human chromosome 14, 2, or 22, the rearrangement could involve the human immunoglobulin loci, the enhancers of which could dysregulate a cellular proto oncogene (Erikson et al. 1981). This hypothesis we made in 1981 (Erikson et al. 1981) received further support when the laboratory of Dr. Philip Leder, then at NIH, mapped the K locus of human immunoglobulin light chain genes to human chromosome 2 (Kirsch et al. 1982). Some of the hybrids were produced at that time were between different mouse cell lines and human Burkitt lymphoma cells carrying the t(8;14) translocation (Erikson et al. 1982). Some of the hybrids retained human chromosomes $14q^{+}$ or $8q^{-}$ that resulted from the reciprocal chromosome translocation t(8;14), (Fig. 1) the hallmark of approximately 80% of Brukitt lymphomas. The others carry the variant translocations (Fig. 1). Since (in 1981) we obtained probes for the constant and variable chain of human immunoglobulin heavy chain, we were able to ask the question of whether the human immunoglobulin heavy chain locus (that we mapped to human chromosome 14) was directly involved in the chromosomal translocation observed in Burkitt lymphomas with the t(8;14) chromosome translocation (Fig. 1). By having hybrids with

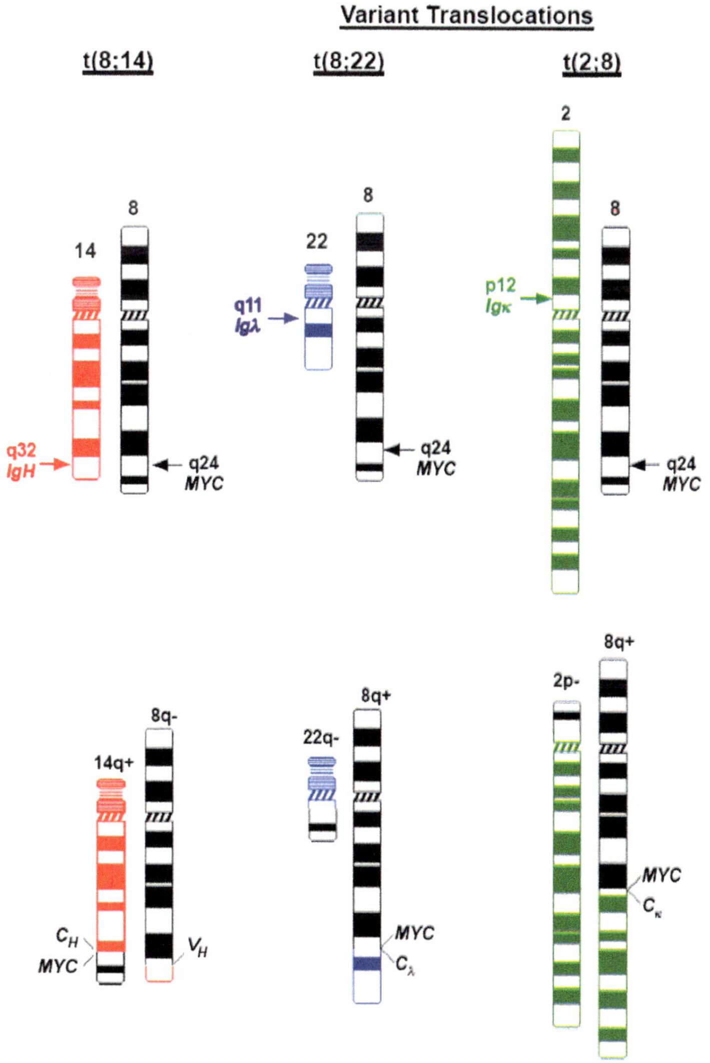

Fig. 1 The three different translocations in Burkitt lymphoma. We first demonstrated that the t (8;14) translocation involved the immunoglobulin heavy chain locus (Erikson et al. 1982). Then that the t(8;14) translocation juxtaposes the MYC gene to the immunoglobulin heavy chain locus. In 1983 we resolved the molecular genetics of the variant translocations

the $14q^+$ and the $8q^-$ from Burkitt lymphoma cells, it became easy to show that while the genes for the constant regions of the immunoglobulin heavy chain locus remained on the $14q^+$ chromosome, the genes for the variable genes were translocated to the $8q^-$ chromosome, demonstrating that the human immunoglobulin heavy chain locus was directly involved in the translocation t(8;14) in Burkitt lymphoma (Fig. 1) (Erikson et al. 1982). Thus, this result proved that Burkitt lymphoma was caused by the juxtaposition of a human immunoglobulin locus to a presumable oncogene on human chromosome 8 (Erikson et al. 1981).

3 Follicular Lymphoma

At the same time, our laboratory was involved in mapping the human homologues of retroviral oncogenes in collaboration with Dr. Riccardo Dalla-Farvera at NCI (Dalla-Favera et al. 1982), a probe Dr. Dalla-Favera obtained was for the human homologue of the MYC oncogene present in the avian myelocytomatosis virus that can induce B-cell lymphomas in chicken (Dalla-Favera et al. 1982). In 1982, we observed that the human homologue of this retroviral oncogene was present in the hybrids with the $14q^+$ chromosome only and in those with normal chromosome 8, but not in those with the $8q^-$, demonstrating that the MYC gene of Burkitt lymphoma was juxtaposed to the human immunoglobulin locus in Burkitt lymphomas with the t(8;14) chromosome translocation through a head to head recombination event (Dalla-Favera et al. 1982, 1983; Marcu et al. 1983) resulting in the dysregulation of the expression of the MYC gene involved in the translocation (ar-Rushdi et al. 1983; Nishikura et al. 1983) while the normal MYC allele on human chromosome 8 remained transcriptionally silent (ar-Rushdi et al. 1983; Nishikura et al. 1983) (Fig. 1).

This was a stunning discovery because it showed that a specific human cancer could be caused by the activation of a specific oncogene, MYC. In 1983, our laboratory investigated also the two variant chromosomal translocations in Burkitt lymphomas and we found that those translocations juxtapose the human MYC oncogene to the human immunoglobulin light chain loci (Croce et al. 1983; Erikson et al. 1983) (Fig. 1). Since 100% of Burkitt lymphomas carry translocations that activate the MYC oncogene and 100% of the cells of each Burkitt lymphoma carry the same translocation, we could infer that MYC must be the driver of this malignancy and its activation must be the earliest, or one of the earliest, genetic event leading to Burkitt pathogenesis (Dalla-Favera et al. 1982, 1983; ar-Rushdi et al. 1983; Croce et al. 1983; Erikson et al. 1983; Marcu et al. 1983; Nishikura et al. 1983). The concept of cancer driver and of targeted therapy are essentially based on this fundamental discovery (Dalla-Favera et al. 1982, 1983; ar-Rushdi et al. 1983; Croce et al. 1983; Erikson et al. 1983; Marcu et al. 1983; Nishikura et al. 1983).

Our studies of Burkitt lymphomas suggested to us that it should have been possible to clone and characterize unknown oncogenes by taking advantage of chromosome translocations involving the human immunoglobulin loci.

At first, we studied the t(11;14) translocation characterization of mantle cell lymphoma and we were able to clone the t(11;14) chromosomal break points by walking away from the immunoglobulin heavy chain locus into the chromosome 11 and cloned the BCL1 locus (Tsujimoto et al. 1984b), which turned out to be the Cyclin D1 gene. Interestingly, sequencing of the breakpoints involved in the translocations indicated that the t(11;14) chromosome translocation must have been caused by an error during the process of V-D-J joining, where the recombinase made an error by joining a segment of human chromosome 11 to a segment of human chromosome 14 or the basis of signal sequences for V-D-J joining (Tsujimoto et al. 1985c).

After the characterization of the t(11;14) chromosome translocation, we focused on the t(14;18) chromosome translocation characteristic of follicular lymphoma, the second most common lymphoma in man (Tsujimoto et al. 1984a, 1985a). This translocation is also observed in 20–30% of diffuse large cell lymphomas (DLCLs), the most common lymphoma in man. Thus, the t(14;18) translocation is the most common cytogenetic alteration in human lymphomas.

Through the work of a fantastic postdoctoral fellow in the lab, Yoshihide Tsujimoto, in 1984 we were able to clone and characterize the t(14;18) chromosome translocation and in 1985 to characterize the gene we called BCL2 (Tsujimoto et al. 1984a, 1985a) (Fig. 2).

In 1986, we reported the entire sequence of the gene and of its protein (Tsujimoto and Croce 1986). We also found that the t(14;18) translocation occurred through mistakes in V-D-J joining and characterized the variant chromosome translocations that juxtaposed the BCL2 gene to the light chain loci of immunoglobulins (Tsujimoto et al. 1985b; Adachi et al. 1989). Analysis of the

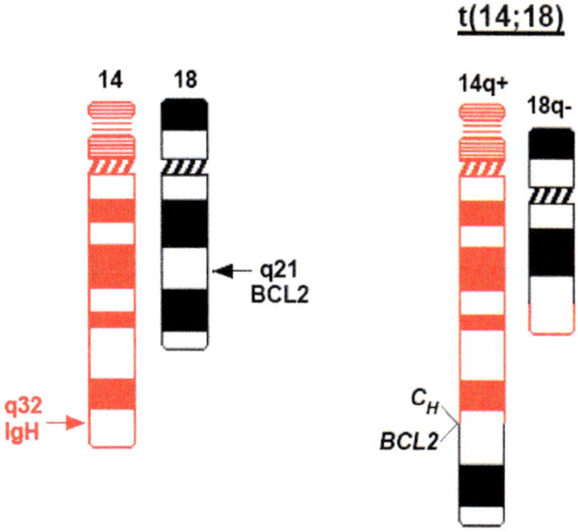

Fig. 2 This is the t(14;18) translocation in hallmark of follicular lymphoma. In this translocation the BCL2 is juxtaposed to the immunoglobulin heavy chain locus

joints between the BCL2 locus and of the heavy chain locus, clearly indicated that the t(14;18) translocations occurred during the physiological process of V-D-J joining (Tsujimoto et al. 1985b).

Thus, this study identified a gene at 18q21 that is involved in essentially all follicular lymphomas and in a significant fraction (20–30%) of diffuse large cell lymphomas.

At the time immediately after the discovery of BCL2, the mechanism of action of the Bcl2 protein was not clear, until in 1988 when the Australian group of Vaux, Adams, and Cori found that Bcl2 inhibits the process of programmed cell death or apoptosis (Vaux et al. 1988). Thus, oncogenes cannot only affect the process of cell proliferation but also that of cellular survival by inhibiting apoptosis.

4 Chronic Lymphocytic Leukemia

Following these discoveries, it was found that the Bcl2 protein is located in mitochondria and dimerizes with the Bax protein. It was also found that is over-expressed in chronic lymphocytic leukemia (CLL). At this point, it became obvious that Bcl2 could become an important target for treatment, but nobody before was able to inhibit protein–protein (Bcl2-Bax) interactions with small molecules. After it was found that the BH3 domain of Bcl2 mediates dimerization, Steve Fesig at Abbott determined the Bcl-XL + BH3 3D structure and then the Bcl2 3D structure (Fig. 5, shows the interaction between the Bcl2 pocket and venetoclax).

While we were continuing to clone and characterize oncogenes and tumor suppressor genes on the basis of consistent chromosomal alterations in human tumors, such as ALL1 (MLL1) (Gu et al. 1992), FHIT (Ohta et al. 1996; Sozzi et al. 1996), TCL1 (Russo et al. 1989), etc., we also attempted to figure out the molecular genetic events that results in the pathogenesis of the most common human leukemia, chronic lymphocytic leukemia (CLL).

Figure 3 shows the four most common cytogenetic alterations observed in CLL. For the most part, CLL is an indolent B-cell malignancy, like follicular lymphoma, although a significant fraction of indolent CLLs can progress to aggressive CLL because of additional genomic alterations. The 13q14 deletion that is observed by cytogenetics in approximately 55% of CLLs occurs, for the most part, in indolent CLLs, or in indolent CLLs that have progressed to aggressive CLL. The deletion of 11q23 is observed in approximately 12% of CLLs and is characteristic of aggressive CLLs. Trisomy of chromosome 12 is observed in indolent CLLs or in indolent CLLs that have progressed to aggressive CLL because of activating mutations of the Notch 1 gene (Balatti et al. 2012), while the deletion of 17p, where many year ago we mapped the gene for p53, occurs in only 7–10% of CLL, predominantly aggressive CLLs (Fig. 3).

Since the loss at 13q14 is the most common alteration in CLL, we decided to characterize the genomic region involved in the loss with the scope of identifying genes, the loss of which may be critical for the development of CLL. By carrying

Fig. 3 Shows the most common chromosomal rearrangements in CLL

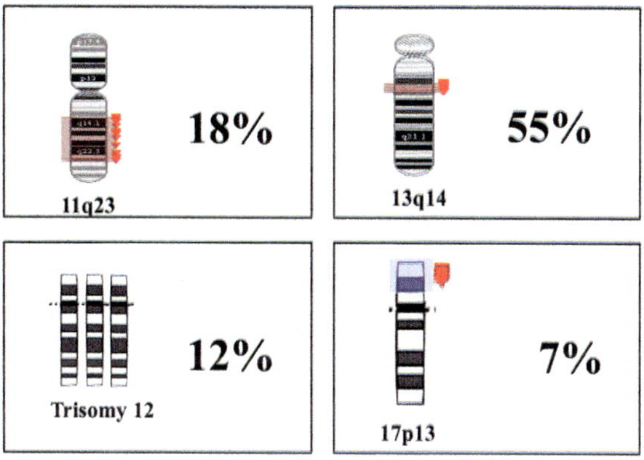

out loss of heterozygosity study (LOH), we identified a region of approximately 700 kB that is the epicenter of loss and the genes present in the region (Fig. 4a). Then we asked the question of whether any gene present in the region was specifically mutated or lost in CLL. After a research that lasted approximately 7 years, we were unable to find any gene in the critical region that was specifically mutated or lost in CLL. Thus, after the LOH study that indicated the loss of a specific region of 13q14 in CLL, we were unable to identify a specific gene that may be involved in the pathogenesis of CLL.

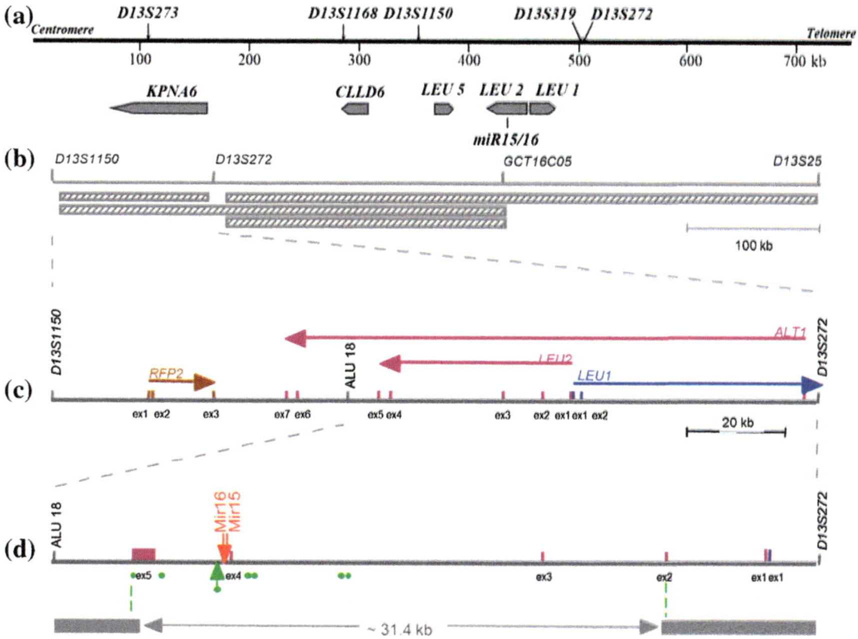

Fig. 4 In **a** and **b** we show the epicenter of loss, a region of approximately 700 kB, in CLL. In **a** we described the region involved in the t(2;13) translocation (*green arrow*) and the region of 31.4 kB involved in the small deletion (**d**)

At this point we decided to try to find rare chromosomal translocations in CLL that may interrupt the 13q14 chromosomal region, providing us with the precise location of the CLL gene. We found that Michael Keating at MD Anderson Cancer Center had a patient with a CLL carrying a t(2;13) chromosome translocation with a break at 13q14. We got the cells from Dr. Keating and made somatic cell hybrids to immortalize the genome of the CLL cells and analyzed the hybrid clones for the presence of the chromosome involved in the translocation. A postdoctoral fellow in the lab, George Calin, was involved in the study and was able to map and clone the breakpoint (Fig. 4). Figure 4 shows the mapping of the critical region at 13q14 involved in the LOH (upper horizontal bar) and the t(2;13) breakpoint in this patient's CLL (green arrow). When we looked at whether the break involved a gene, presumably the one involved in CLL, we found none. The breakpoint, however, occurred in the precise region defined by the LOH study of CLLs. At this point we were still in the dark. The break came from a discussion with Kanti Rai, the father figure of CLL research. He told us he had a patient with familial retinoblastoma and CLL. Since the gene for retinoblastoma (Rb) is also on 13q14, but way on the left of the epicenter of loss, we thought that CLL might have had a loss of both the Rb and of the CLL gene. Thus, we tried to determine whether that was the case. We obtained the CLL cells from this patient from Dr. Rai, made somatic cell hybrids between the CLL cells and mouse cells, and analyzed the hybrids for the segregation of the two chromosomes 13 in the hybrids. We found that one of the two chromosomes 13 of this patient had a very small deletion approximately 31.4 kB, which occurred precisely in the same region of the translocation breakpoint observed in the CLL of the previous patient (Fig. 4) (Calin et al. 2002). Thus, the CLL gene had to be there, but no protein coding gene was present in the involved region.

5 miR-15a and miR-16-1 Are Deleted in CLL

At this point, we started thinking that the gene involved in CLL might not be a protein coding gene. In 1993, Victor Ambros found that mutations of a gene of *Caenorhabditis elegans*, Lyn 4, affected the development of *C. elegans*, but it encoded a short non coding RNA, a microRNA. After 1993, nobody seemed to be interested in microRNAs, until 1998 when small interfering RNAs (siRNAs) were discovered. Since microRNAs seemed to act quite similarly to siRNAs, this discovery boosted further studies of microRNAs, and in the year 2000 it was found that Drosophila melanogaster (fruit fly) had in its genome microRNA genes and in year 2001 it was found that mice, rats, and human beings carried microRNA genes. In 2001 we discovered where the CLL gene had to be (Fig. 4 upper bar) and asked the question of whether microRNA genes mapped precisely in the small region defined by the translocation breakpoint and by the small deletion (Fig. 4 lower bar). At this point, we found that two microRNA genes, miR-15a and miR-16-1, mapped precisely to that region, indicating that loss of function of miR-15/16 was the cause of the development of CLL, and providing the first evidence that alterations in non coding genes could be the cause of disease (cancer) (Calin et al. 2002, 2005).

We know that less than 2% of the human genome is protein coding, while the remaining 98% was considered junk (garbage), although most of it is transcribed. Our results indicated that the non-coding genome could be involved in cancer causation. Immediately after this finding, we examined many CLLs for loss of miR-15/16 and found that these two microRNA genes are lost in approximately 70% of CLLs providing evidence that such alterations must be occurring very early during the pathogenesis of indolent CLLs. At this point we sequenced many microRNAs in a large panel of 75 CLLs and 160 normal controls, and found rare cases of mutations in CLLs (Calin et al. 2005). Among the mutated microRNA genes we found miR-15/16 were mutated in two cases out of 75 (Calin et al. 2005). Interestingly, the mutations occurred 7 nucleotides 3′ of miR-16 and affected the processing of the precursor by the Drosha complex (Calin et al. 2005). One of the two CLL patients with the mutation was a female who previously developed breast cancer, her mother died of CLL (therefore, this was a familial CLL) and her sister died of breast cancer (Calin et al. 2005). Thus, loss of function of miR-15/16 can also cause familial CLL (Calin et al. 2005). Interestingly, a specific mouse strain, the NZB mouse, develops an indolent form of CLL late in life, like in humans. Elizabeth Raveche mapped the disease in the mouse to mouse chromosome 14 in a region that is homologous to region 13q14 in humans (Raveche et al. 2007). Thus, she sequenced the locus in the mouse and found a mutation 6 nucleotides 3′ to miR-16 that also affected the processing of the precursor; such mutation was present only in the NZB mouse strain. Therefore, loss of function of miR-15/16 causes the indolent form of CLL both in humans and the mouse.

At this point, we needed to define the critical target(s) of miR-15/16 that may be responsible for the development of CLL. We were stunned to find that at the top of the list of predicted targets of miR-15/16, was the gene we identified in 1984, BCL2. This prediction made a lot of sense because in humans, there are two major indolent B-cell malignancies, one is follicular lymphoma that is caused by a t (14;18) chromosome translocation that activates the BCL2 gene and the other is CLL that is caused, for the most part, by the loss of miR-15/16. We discovered that BCL2 is a target of miR-15/16 and that the level of expression BCL2 is inversely correlated to the loss of miR-15/16. Hence, the overexpression of BCL2 in CLL, which was reported extensively, is caused by the loss of specific microRNAs that are negative regulators of BCL2 expression (Cimmino et al. 2005).

In summary, the overexpression of BCL2, the driver in CLL is caused by the genetic loss of two negative regulators of this gene, miR-15/16. These two microRNAs map to the same polycistronic RNA.

We also found that MCL1, another gene of the BCL2 family of anti-apoptotic genes is also targeted by miR-15/16 (Fabbri et al. 2011).

6 Interaction of Venetoclax with BCL2

Following our discovery and sequencing of BCL2 (Tsujimoto et al. 1984a, 1985a; Tsujimoto and Croce 1986), a pharmaceutical company, Abbott, decided to develop small molecule inhibitors of Bcl2. This was a very difficult task because at that time people thought that was impossible to inhibit protein–protein interactions (Bcl2-Bax) with small molecules. Fortunately, under the scientific leadership of Steve Fesig, Abbott determined at first the 3D structure of the BH3 domain of BCLXL and then of BCL2 and were able to develop a series of inhibitors (Shuker et al. 1996; Chang et al. 1997). The last of which was specific for Bcl2, ABT199 or venetoclax (Fig. 5) (Souers et al. 2013). This drug inhibits Bcl2 at nanomolar concentrations, is capable of inducing leukemic cell lysis in CLL patients, and complete remission even without detectable minimal residual disease. This drug has been approved for the treatment of aggressive CLL, by the FDA last April. Thus, it took 32 years from the discovery of BCL2 to the FDA approval of a drug that targets the driver of malignant transformation in CLL. The history of this development, "*Finally, An Apoptosis-Targeting Therapeutic for Cancer*", has been recently published in Cancer Research (Croce and Reed 2016).

Of course it is likely that CLL patients we treat with venetoclax may develop resistance to the drug. CLL patients carry billions of leukemic cells, among which a very few may not respond to the drug because of different mechanisms. We

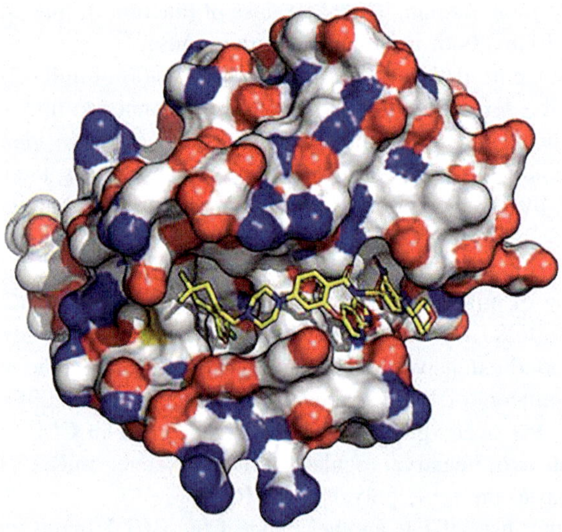

Fig. 5 The interaction between the BH3 domain of the human Bcl-2 protein and venetoclax is represented based on the crystal structure of (4(4-{[4-(4-chlorophenyl)-5,6-dihydro-2H-pyran-3yl] methyl}piperazin-1-yl)-*N*-{[3-nitro-4-(tetrahydro-2H-pyran-4-ylamino)phenyl]sulfonyl}benzamide), a close analog. The Connolly surface of the protein is colored by mapped atom type (carbon *white*; nitrogen *blue*; oxygen *red*; sulfur *yellow*) (Croce and Reed 2016)

reasoned that since miR-15/16 target a number of genes, it may be possible to use an additional drug to kill CLL cells by avoiding the possibility of resistance. Since we found that MCL1 is also overexpressed by CLL cells because of the loss of miR-15/16, a drug specifically targeting Mcl1 may be a possibility, but at the moment it is not available.

Interestingly, very recently we have discovered that ROR1, an orphan receptor, is also regulated by miR-15/16 and found coordinated overexpression of BCL2 and ROR1 in CLLs. Since antibodies against ROR1 are available, we are proposing a combination therapy of CLL by using venetoclax and monoclonal antibodies against ROR1 for the cure of CLL.

References

Adachi M, Cossman J, Longo D, Croce CM, Tsujimoto Y (1989) Variant translocation of the bcl-2 gene to immunoglobulin lambda light chain gene in chronic lymphocytic leukemia. Proc Natl Acad Sci U S A 86(8):2771–2774

ar-Rushdi A, Nishikura K, Erikson J, Watt R, Rovera G, Croce CM (1983) Differential expression of the translocated and the untranslocated c-myc oncogene in Burkitt lymphoma. Science 222 (4622):390–393

Balatti V, Bottoni A, Palamarchuk A, Alder H, Rassenti LZ, Kipps TJ, Pekarsky Y, Croce CM (2012) NOTCH1 mutations in CLL associated with trisomy 12. Blood 119(2):329–331

Calin GA, Dumitru CD, Shimizu M, Bichi R, Zupo S, Noch E, Aldler H, Rattan S, Keating M, Rai K, Rassenti L, Kipps T, Negrini M, Bullrich F, Croce CM (2002) Frequent deletions and down-regulation of micro-RNA genes miR15 and miR16 at 13q14 in chronic lymphocytic leukemia. Proc Natl Acad Sci U S A 99(24):15524–15529

Calin GA, Ferracin M, Cimmino A, Di Leva G, Shimizu M, Wojcik SE, Iorio MV, Visone R, Sever NI, Fabbri M, Iuliano R, Palumbo T, Pichiorri F, Roldo C, Garzon R, Sevignani C, Rassenti L, Alder H, Volinia S, Liu CG, Kipps TJ, Negrini M, Croce CM (2005) A MicroRNA signature associated with prognosis and progression in chronic lymphocytic leukemia. N Engl J Med 353(17):1793–1801

Chang BS, Minn AJ, Muchmore SW, Fesik SW, Thompson CB (1997) Identification of a novel regulatory domain in Bcl-X(L) and Bcl-2. EMBO J 16(5):968–977

Cimmino A, Calin GA, Fabbri M, Iorio MV, Ferracin M, Shimizu M, Wojcik SE, Aqeilan RI, Zupo S, Dono M, Rassenti L, Alder H, Volinia S, Liu CG, Kipps TJ, Negrini M, Croce CM (2005) miR-15 and miR-16 induce apoptosis by targeting BCL2. Proc Natl Acad Sci U S A 102(39):13944–13949

Croce CM, Reed JC (2016) Finally, an apoptosis-targeting therapeutic for cancer. Cancer Res 76 (20):5914–5920

Croce CM, Shander M, Martinis J, Cicurel L, D'Ancona GG, Dolby TW, Koprowski H (1979) Chromosomal location of the genes for human immunoglobulin heavy chains. Proc Natl Acad Sci U S A 76(7):3416–3419

Croce CM, Thierfelder W, Erikson J, Nishikura K, Finan J, Lenoir GM, Nowell PC (1983) Transcriptional activation of an unrearranged and untranslocated c-myc oncogene by translocation of a C lambda locus in Burkitt. Proc Natl Acad Sci U S A 80(22):6922–6926

Dalla-Favera R, Bregni M, Erikson J, Patterson D, Gallo RC, Croce CM (1982) Human c-myc onc gene is located on the region of chromosome 8 that is translocated in Burkitt lymphoma cells. Proc Natl Acad Sci U S A 79(24):7824–7827

Dalla-Favera R, Martinotti S, Gallo RC, Erikson J, Croce CM (1983) Translocation and rearrangements of the c-myc oncogene locus in human undifferentiated B-cell lymphomas. Science 219(4587):963–967

Erikson J, Martinis J, Croce CM (1981) Assignment of the genes for human lambda immunoglobulin chains to chromosome 22. Nature 294(5837):173–175

Erikson J, Finan J, Nowell PC, Croce CM (1982) Translocation of immunoglobulin VH genes in Burkitt lymphoma. Proc Natl Acad Sci U S A 79(18):5611–5615

Erikson J, ar-Rushdi A, Drwinga HL, Nowell PC, Croce CM (1983) Transcriptional activation of the translocated c-myc oncogene in burkitt lymphoma. Proc Natl Acad Sci U S A 80(3):820–824

Fabbri M, Bottoni A, Shimizu M, Spizzo R, Nicoloso MS, Rossi S, Barbarotto E, Cimmino A, Adair B, Wojcik SE, Valeri N, Calore F, Sampath D, Fanini F, Vannini I, Musuraca G, Dell'Aquila M, Alder H, Davuluri RV, Rassenti LZ, Negrini M, Nakamura T, Amadori D, Kay NE, Rai KR, Keating MJ, Kipps TJ, Calin GA, Croce CM (2011) Association of a microRNA/TP53 feedback circuitry with pathogenesis and outcome of B-cell chronic lymphocytic leukemia. JAMA 305(1):59–67

Gu Y, Nakamura T, Alder H, Prasad R, Canaani O, Cimino G, Croce CM, Canaani E (1992) The t (4;11) chromosome translocation of human acute leukemias fuses the ALL-1 gene, related to Drosophila trithorax, to the AF-4 gene. Cell 71(4):701–708

Kirsch IR, Morton CC, Nakahara K, Leder P (1982) Human immunoglobulin heavy chain genes map to a region of translocations in malignant B lymphocytes. Science 216(4543):301–303

Marcu KB, Harris LJ, Stanton LW, Erikson J, Watt R, Croce CM (1983) Transcriptionally active c-myc oncogene is contained within NIARD, a DNA sequence associated with chromosome translocations in B-cell neoplasia. Proc Natl Acad Sci U S A 80(2):519–523

Nishikura K, ar-Rushdi A, Erikson J, Watt R, Rovera G, Croce CM (1983) Differential expression of the normal and of the translocated human c-myc oncogenes in B cells. Proc Natl Acad Sci U S A 80(15):4822–4826

Ohta M, Inoue H, Cotticelli MG, Kastury K, Baffa R, Palazzo J, Siprashvili Z, Mori M, McCue P, Druck T, Croce CM, Huebner K (1996) The FHIT gene, spanning the chromosome 3p14.2 fragile site and renal carcinoma-associated t(3;8) breakpoint, is abnormal in digestive tract cancers. Cell 84(4):587–597

Raveche ES, Salerno E, Scaglione BJ, Manohar V, Abbasi F, Lin YC, Fredrickson T, Landgraf P, Ramachandra S, Huppi K, Toro JR, Zenger VE, Metcalf RA, Marti GE (2007) Abnormal microRNA-16 locus with synteny to human 13q14 linked to CLL in NZB mice. Blood 109 (12):5079–5086

Russo G, Isobe M, Gatti R, Finan J, Batuman O, Huebner K, Nowell PC, Croce CM (1989) Molecular analysis of a t(14;14) translocation in leukemic T-cells of an ataxia telangiectasia patient. Proc Natl Acad Sci U S A 86(2):602–606

Shuker SB, Hajduk PJ, Meadows RP, Fesik SW (1996) Discovering high-affinity ligands for proteins: SAR by NMR. Science 274(5292):1531–1534

Souers AJ, Leverson JD, Boghaert ER, Ackler SL, Catron ND, Chen J, Dayton BD, Ding H, Enschede SH, Fairbrother WJ, Huang DC, Hymowitz SG, Jin S, Khaw SL, Kovar PJ, Lam LT, Lee J, Maecker HL, Marsh KC, Mason KD, Mitten MJ, Nimmer PM, Oleksijew A, Park CH, Park CM, Phillips DC, Roberts AW, Sampath D, Seymour JF, Smith ML, Sullivan GM, Tahir SK, Tse C, Wendt MD, Xiao Y, Xue JC, Zhang H, Humerickhouse RA, Rosenberg SH, Elmore SW (2013) ABT-199, a potent and selective BCL-2 inhibitor, achieves antitumor activity while sparing platelets. Nat Med 19(2):202–208

Sozzi G, Veronese ML, Negrini M, Baffa R, Cotticelli MG, Inoue H, Tornielli S, Pilotti S, De Gregorio L, Pastorino U, Pierotti MA, Ohta M, Huebner K, Croce CM (1996) The FHIT gene 3p14.2 is abnormal in lung cancer. Cell 85(1):17–26

Tsujimoto Y, Croce CM (1986) Analysis of the structure, transcripts, and protein products of bcl-2, the gene involved in human follicular lymphoma. Proc Natl Acad Sci U S A 83 (14):5214–5218

Tsujimoto Y, Finger LR, Yunis J, Nowell PC, Croce CM (1984a) Cloning of the chromosome breakpoint of neoplastic B cells with the t(14;18) chromosome translocation. Science 226 (4678):1097–1099

Tsujimoto Y, Yunis J, Onorato-Showe L, Erikson J, Nowell PC, Croce CM (1984b) Molecular cloning of the chromosomal breakpoint of B-cell lymphomas and leukemias with the t(11;14) chromosome translocation. Science 224(4656):1403–1406

Tsujimoto Y, Cossman J, Jaffe E, Croce CM (1985a) Involvement of the bcl-2 gene in human follicular lymphoma. Science 228(4706):1440–1443

Tsujimoto Y, Gorham J, Cossman J, Jaffe E, Croce CM (1985b) The t(14;18) chromosome translocations involved in B-cell neoplasms result from mistakes in VDJ joining. Science 229 (4720):1390–1393

Tsujimoto Y, Jaffe E, Cossman J, Gorham J, Nowell PC, Croce CM (1985c) Clustering of breakpoints on chromosome 11 in human B-cell neoplasms with the t(11;14) chromosome translocation. Nature 315(6017):340–343

Vaux DL, Cory S, Adams JM (1988) Bcl-2 gene promotes haemopoietic cell survival and cooperates with c-myc to immortalize pre-B cells. Nature 335(6189):440–442

The Evolution of Tumor Formation in Humans and Mice with Inherited Mutations in the p53 Gene

Arnold J. Levine

Abstract While tumors are very heterogeneous in their origins, mutations in the p53 gene and inactivation of p53 gene functions are the most common feature that predispose to the formation of cancers in humans. Inherited p53 mutations lead to different tumor types at very different frequencies and at very different ages than somatic p53 mutations. The reasons for this are explored. When the first mutation arises in a stem cell (a gatekeeper mutation) it selects for a specific subset of second mutations which in turn select for mutations in a third subset of genes. The nature of the first mutation in a tumor determines, by selection, the functional types of subsequent mutations. Inherited mutations occur at different developmental times and in different orders of mutational sequences than somatic mutations. The excess risk of developing a cancer with an inherited p53 mutation is two- to three-fold in endodermal derived tissues compared with 100- to 1000-fold for ectodermal and mesenchymal derived tissues. By contrast, endodermal derived tumors with somatic p53 mutations occur at very high frequencies (70–100%). These evolutionary restrictions upon the mutational path that tumor development may take could open up new avenues for therapy and prevention.

Contents

A.J. Levine (✉)
Institute for Advanced Study, Princeton, NJ 08540, USA
e-mail: alevine@ias.edu

Current Topics in Microbiology and Immunology (2017) 407:205–222
DOI 10.1007/82_2017_5

Published Online: 28 March 2017

1 Background

In 1979, four research groups described a cellular tumor antigen, which was detected in viral, chemical and spontaneously transformed cells and came to be called the p53 protein (Linzer and Levine 1979; Lane and Crawford 1979; DeLeo et al. 1979; Kress et al. 1979). The p53 protein was shown to form a protein complex with the viral SV40 oncogene product, the large T-antigen (Linzer and Levine 1979; Lane and Crawford 1979; Kress et al. 1979) but was also shown to be found at high concentrations in chemically transformed cells (DeLeo et al. 1979) and in cell lines from testicular teratocarcinomas (Linzer and Levine 1979). Animals bearing tumors inoculated with these cell lines all produced tumors that produced antibodies directed against the p53 protein, classifying it as a tumor antigen. Subsequently, the p53 protein was found in a protein complex with the Adenovirus oncoprotein, the E1B 55Kd tumor antigen (Sarnow et al. 1982) and the Human Papilloma virus 12 and 18 E6 oncoprotein (Werness et al. 1990) demonstrating that several unrelated and diverse small DNA tumor viruses target this same p53 cellular protein by forming protein complexes. The Papilloma virus E6 protein was shown to aid in the polyubiquitination and ultimate degradation of the p53 protein (Scheffner et al. 1990) suggesting that the p53 protein was targeted by these viruses for inactivation or degradation. Oren and Levine (Oren and Levine 1981) first demonstrated that the levels of the p53 protein in non-transformed cells were regulated by a post-translational process producing a protein with a short half-life (degraded in 6–20 min).

Several groups were able to clone p53 c-DNAs from a diverse set of cell lines in culture. Two of these clones were shown to transform normal cells in culture (Eliyahu et al. 1984) while one of these clones not only failed to transform cells in culture but also prevented cells from being transformed by oncogene products (Finlay et al. 1989). The p53 c-DNA clones that transformed cells in culture were shown to arise from mutations in the p53 gene producing a protein that acted in a dominant negative fashion to transform cells (Hinds et al. 1989). The fact that the p53 gene functioned as a tumor suppressor gene preventing cancers was confirmed in cell lines (Wolf and Rotter 1984) and in human colorectal carcinomas where both alleles of this gene were found to harbor the same mutation (Nigro et al. 1989). Human families with germ line heterozygous mutations in the p53 gene were shown to develop tumors with a penetrance of about 93% (Hainaut and Pfeifer 2016) and knock out mice with a p53 gene deletion develop tumors with almost 100% penetrance (Donehower 1996) demonstrating that the p53 gene and its protein help to prevent tumors in mice and humans. The DNA sequencing projects of human tumors carried out over the past few years have demonstrated that p53 mutations are the single most common mutations found in human cancers suggesting a special role for the p53 protein in preventing cancers. There is a great diversity in the frequencies of p53 mutations in human cancers (serous ovarian cancers about 100% and testicular teratocarcinomas about 2%) suggesting that the cell type of origin and tissue type play an important role in p53 functions.

What are the functions of the p53 protein? It rapidly became clear that the p53 protein acted as a transcription factor binding to a specific set of DNA sequences to regulate gene transcription (Beckerman and Prives 2010; Zambetti et al. 1992; El-Deiry et al. 1992). Among the first set of genes shown to be regulated by the p53 protein were the p21 gene which blocks cell cycle progression in the G-1 phase of the cell cycle (El-Deiry et al. 1992) and the MDM-2 gene (Momand et al. 1992) which is an E-3 ubiquitin ligase that polyubiquitinates the p53 protein and leads to its degradation. The crystal structure of the p53–MDM-2 complex was elucidated (Kussie et al. 1996) and this led to the development of a class of drugs, called the nutlins (Vu et al. 2013) that disrupt the p53–MDM-2 protein complex and this activates the wild type p53 protein which causes a reduction in tumor growth by killing the cancer cells. The fact that the p53 protein promotes the transcription of the MDM-2 gene while MDM-2 leads to the degradation of the p53 protein, creates an auto-regulatory feedback loop (Wu et al. 1993) and the oscillation of these two proteins 180 degrees out of phase in the cell (Bar-Or et al. 2000; Lahav et al. 2004). As the number of diverse genes regulated by the p53 transcription factor became clear (Riley et al. 2008) several transcriptional programs that enhance tumor suppression were uncovered. The activation of the p53 protein, so that it functions as a transcription factor, could lead to apoptosis or cellular senescence killing the cell before it can become cancerous. p53 was shown to mediate G-1 and G-2 cell cycle arrest. p53 regulated the number of centrosomes produced during the cell cycle. It had a large impact upon the metabolic profiles used by cells regulating the insulin-like growth factor pathway, mitochondrial functions, glutathione production and shifts between the Warburg effect and normal oxidative phosphorylation. p53 regulated cell mobility, invasiveness and wound healing. The p53 protein was shown to regulate many genes involved in DNA repair processes and epigenetic changes in cells. Each of these diverse transcriptional programs is regulated by an activated p53 protein that is modified by a variety of enzyme activities that are also employed to modify histones forming the chromatin template for transcription. These p53 protein modifications (phosphorylation, methylation, acetylation, ubiquitination, etc.), arise in a cell undergoing one or more types of stress.

A wide variety of DNA damaging agents promote p53 activation, and an increase in p53 levels. The high levels and protein modifications result in p53 dependent transcriptional programs resulting in the transcription of genes whose products are involved in DNA repair and/or cell death (Maltzman and Czyzyk 1984; Kastan et al. 1991). These two publications by Maltzman in 1984 (Maltzman and Czyzyk 1984) and Kastan in 1991 (Kastan et al. 1991) demonstrate a common issue in scientific research; an important discovery made seven years before and forgotten by the field is rediscovered and becomes a central core in our understanding of how cancers are prevented by the p53 protein. Maltzman was trained at Stanford in a laboratory that worked on DNA repair processes. He carried out his postdoctoral work at Princeton working on the p53 protein in the Levine laboratory and then took a position at Rutgers. There he met Evelyn Witkin who had discovered the rec A gene central in controlling cell division in E. coli and responding to DNA strand breaks. Witkin was reading a paper about the p53 protein and its fluctuations in the cell cycle (Reich and Levine 1984) and saw real similarities between Rec A functions and p53 functions and she told Maltzman about that. Maltzman (Maltzman and Czyzyk 1984) (in 1984) then carried out the experiments that demonstrated that the p53 protein responds to DNA damage by ultraviolet exposure increasing its concentration (no longer being regulated by MDM-2 which would be found eight years later), acquiring post-translational modifications, and transcribing a set of genes to kill the cell (which would be uncovered ten years later). No one followed up these experiments until Kastan rediscovered that p53 responds to DNA damage in 1991. At that point, the p53 field was ready and realized that the p53 protein functioned to respond to stresses like DNA damage.

Today, a large number of cellular stresses, in addition to DNA damage, are known to activate the p53 protein and initiate a transcriptional program that responds to stress. Hypoxia, nutrient deprivation, telomere shortening, oncogene activation by Ras, myc or viral oncogenes, epigenetic reprograming of cells and virus infections, cytokine exposures and failures in ribosomal biogenesis are all activators of a vigorous p53 response. Just which transcriptional program is then carried out by p53 to repair or kill the cell depends upon a large number of variables: the cell and tissue type, whether a cell is transformed by an oncogene or not, the chromosomal ploidy of a cell, the stage of differentiation of a cell (stem cell, progenitor cell, differentiated cell) and the lineage of a cells' differentiation program. The p53 transcriptional program can synthesize cytokines or interferons and call in the immune system to clear out dead cells and present neo-antigens to the immune response. A role for p53 in the central nervous system where it can sense DNA damage and respond with cell death and neurodegeneration is a viable hypothesis that needs to be tested in more detail. A role for p53 in sensing and responding to the microbiome is also suggested by some observations and now requires a set of follow-up experiments. The p53 protein is known to monitor ribosomal biogenesis making sure cells will produce enough ribosomes for cell division to proceed. Inherited mutations in some ribosomal protein genes give rise to defects in an optimal rate of ribosomal biogenesis and activation of p53, which can result in Diamond Blackfan Anemia in humans do to inhibition of cell cycle progression and death of reticulocytes during red blood cell production in the bone marrow. Clearly, the p53 tumor suppressor is a double edge sword both protecting against errors during cell division and preventing cancers from arising but also responding to stresses by killing essential cells that can then compromise normal physiology.

2 The Li-Fraumeni Syndrome

The Li-Fraumeni Syndrome (Oliveri et al. 2010) is an autosomal dominant disorder that results in human families that develop a characteristic subset of tumors commonly at a young age. About 70% of these families harbor p53 mutations in the heterozygous state, which reduce to homozygosity at a high rate and begin a

Table 1 The impact of the germ line cell types upon the excess risk of developing a tumor with inherited p53 mutations or the frequency of somatic p53 mutations in tumors that arise over a lifetime

Germ line contribution	Inherited excess risk	Somatic mutation frequency (%)
Ectoderm	10–1000×	2–35
Mesoderm	10–100×	50–60
Endoderm	2–3×	70–100

process, which results in one or more cancers appearing over a lifetime with a penetrance of 93%. The International Association of Cancer Research (IARC, Lyon, France) collects and stores a great deal of information about these families and the analysis described herein was carried out by Pierre Hainaut and is based upon the R16 version of the IARC database (November, 2012). These data sets describe 634 families with p53 germ line mutations producing 2483 cancer cases with a 30,000 person-year follow-up of these individuals. One of the surprising observations made with these families was that individuals who inherit a p53 mutation develop cancers during their lifetime that derive from an ectodermal and mesenchymal lineage with a ten- to one-thousand-fold excess risk than are observed in wild type p53 populations, while the excess risk of developing a tumor derived from the endodermal lineage is only two- to three-fold higher than the wild type population (Table 1).

This is in contrast to tumors with spontaneous somatic p53 mutations acquired over a lifetime in endodermal tumors of the colon (75% mutations), lung (75% p53 mutations), pancreas (80% p53 mutations), and serous ovarian tumors (100% p53 mutations). Clearly, there is a remarkable difference in the frequencies of p53 mutations that cause endodermal tumor types produced by inherited p53 mutations and spontaneous p53 somatic mutations and the reasons for this remain unclear. Note, however, that inherited p53 mutations position the p53 mutation to be the first mutation in a series of mutations that cause the cancer. Thus, the p53 mutation can be classified as the gatekeeper mutation and this modifies the cell so that additional mutations in other genes can be selected for that contribute to cell growth and division. Spontaneous p53 mutations that occur in somatic tissues over a lifetime are not commonly the gatekeeper or first mutation in a series, so the selection for the next set of mutations in progressing to a tumor alters the evolution of that tumor. This explanation suggests that the order of mutations in the formation of a tumor determines the subsequent genes selected for based upon the nature of the gate-keeper mutation. There are a growing number of studies that support this hypothesis.

In mice that have inherited p53 mutations in both alleles, the next mutation is always a PTEN mutation, followed by a cdk6 and cyclin D amplification and finally a notch pathway defect (Fig. 1) (Dudgeon et al. 2014).

Cells carrying a p53 mutation are more susceptible to promoting genomic instability. If the rare cell that has p53 mediated genomic instability acquires a

P53	→	PTEN	→	Cycle D Cdk-6	→	NOTCH Pathway
-/-		-/-		Amplified		Splicing Alterations

Fig. 1 The order of selection of mutations in thymic lymphomas from ten different $p53^{-/-}$ mice, the p53 mutation, –/–, was inherited, the PTEN deletions arose before the rearrangement of the T-cell receptors in the progenitor T cells. Gene amplifications were then observed in cyclin D1, 2 or 3 genes and the cdk6 gene followed by mutations in the notch pathway. The p53 mutations give rise to genomic instability. The PTEN mutations give rise to the high utilization of glucose and metabolic alterations permitting rapid cell division. The cyclin D-cdk6 mutations accelerate the cell cycle times and the notch mutations block T cell development and differentiation keeping the tumor stem cells progenitor cells

homozygous PTEN mutation, it provides higher glucose levels for the cell to alter its metabolic program so as to support enhanced cell division, increasing the clone size. This is followed by enhancing the rate limiting steps in the G-1 phase of the cell cycle, cyclin D-cdk4/cdk6 phosphorylation of Rb. In $p53^{-/-}$ cells, this is accomplished by gene amplifications of any one of the three cyclin D-1, 2 or 3 genes and the gene for cdk6. By shortening the G-1 phase of the cell cycle and constantly bring cells back into cell cycle while the need for high levels of glucose is maintained, one continues the clonal expansion of the precursor to a thymic lymphoma (Dudgeon et al. 2014). Finally, mutations in the notch pathway restrict the T-cells from fully differentiating keeping a progenitor cell in division (the tumor is a CD-4+/CD-8+ double positive T-cell lymphoma). Each of these four steps in the development of the tumor is ordered and contribute to different cellular processes: 1. Genomic instability and loss of apoptosis, 2. An increased ability to take up glucose and to employ Warburg aerobic glycolysis to supply both energy and substrates for growth, 3. Enhanced rates of cell division by reducing the rate limiting aspects of the G-1 phase of the cell cycle and 4. Eliminating the final steps in differentiation of T cells giving rise to a thymic lymphoma.

In colon cancers, the most common gatekeeper mutation is in the Wnt pathway (APC mutation or beta catenin mutation), which is followed by a Ras mutation, then a TGF-beta mutation and finally a p53 mutation (Fearon and Vogelstein 1990). That the order of these mutations was critical to the development of colon cancers was demonstrated in mice (Takeda et al. 2015). Insertional mutagenesis was employed to promote colon cancers in mice that inherited (the gatekeeper mutation) an APC mutation, a Ras mutation, a TGF-beta mutation or a p53 mutation. Tumors arose fastest in time in the genetic background that carried the APC mutation. They arose second in time in the RAS-mutated background, third in time in the TGF-beta background and it took the longest time to observe tumors of the colon in mice with a mutation in the p53 gene. Since these mutations likely occur spontaneously in populations that do not start with a gatekeeper mutation and therefore occur stochastically over time and cell divisions, the simplest explanation for these observations is that the order of mutations results from the selection of the next mutation in cells that divide and survive so as to expand the lineage ultimately leading to a rapidly growing and dividing cancer. In other words, a Wnt pathway mutation next selects for a Ras mutation to get the cells into cycle, followed by the loss of the negative regulator of growth and division, TGF-beta. All of these mutations may happen in the stem cell of the colon keeping the benign tumor or polyp in situ or in the crypt. A final p53 mutation enhances cell migration, invasion and penetration of the colon wall, giving rise to a colon or colorectal carcinoma. This line of reasoning does not eliminate the possibility that the first or previous mutation results in the enhancement of the mutation rate in a cell, resulting in an optimal order of mutations that give rise to a tumor rapidly. These ideas do help to explain why one can find tumor suppressor gene mutations and oncogene cancer driver mutations in cells in the body that have not given rise to a cancer and must be awaiting a mutation in a gene that expands the clone and results in a malignant cancer. The fact that many stem cells (colon, skin, pancreas, breast, etc.) rely upon stem cells that are driven by the Wnt pathway predicts that Wnt pathway mutations might well be a common gatekeeper mutation.

A second interesting observation that derives from the Li-Fraumeni data set is that the age range at which different tumor types occur and are diagnosed in a Li-Fraumeni patient are limited to specific times during a life span and are reproducible in many different patients. For example, adrenalcorticocarcinomas (ADC) occur between the ages of 6 months to 4 years; choroid plexus papillomas (CPT) between 6 months and 3 years, medullary blastomas (MED) between 2 and 9 years and Rhabdomyosarcomas (RMS) between 1 and 4 years of age, all at 50- to 1000-fold higher excess risk than observed in the wild type population (Table 2).

Liposarcomas (LPS) and osteogenic sarcomas (OST) commonly occur from 1 to 15 to 20 years of life with a ten-fold excess risk compared to the wild type population. Breast cancers (BC) occur in 50% of the Li-Fraumeni females between

Table 2 The ages of diagnosis and excess risk of different tissue types of inherited p53 mutations ADC, Adrenalcortococarcinomas; CPT, Choroid Plexus Tumors; MED, Medulloblastomas; RMS, Rhabdomytosarcomas; LPS, Liposarcomas; OST, Osteogenic sarcomas; BC, Breast Cancers, LMS, Leiomyosarcomas

Tumor type	Ages	Excess risk
ADC	6 months–4 years	100–1000×
CPT	6 months–3 years	100–1000×
MED	2–9 years	100–1000×
RMS	1–4 years	100–1000×
LPS, OST	1–20 years	10–1000×
BC	18–45 years	50% of females
LMS	20–55 years	10–100×

18 and 45 years of age and leiomyosarcomas (LMS) and liposarcomas (LPS) occur about 20–55 years of age (Table 2). In other words, in patients with Li-Fraumeni syndrome and heterozygous for a p53 mutation, tumors of specific tissue types occur at specified times or ages during a life span. This suggests that p53 loss and functions act differently in different tissues. Alternatively, p53 functions could be most important in actively growing tissues or stem or progenitor cells which are determined by developmental programs active at different times during the life span of a person.

Interestingly, the excess risk of developing a first cancer in an individual with a heterozygous p53 mutation falls below 1.0, or lower than the wild type population, when the individual with a p53 mutation is over the age of 65–70 years. This of course is a time when the wild type population is increasingly at risk for developing cancers. So why would a person at age 75 or 80 years of life with a p53 mutation that failed to promote a detectable cancer over a lifetime actually have a lower risk of developing a cancer than a person born with two wild type alleles in the p53 gene? This observation may be ascribed to good luck, remaining in a good environment with a low mutation rate over a lifetime or the presence of genetic suppressors in an individual that are able to compensate for the presence of only one copy of a wild type p53 gene over a lifetime. The fact that enzymatic activities, like histone acetyltransferases or methylases, that modify the p53 protein can impact its activity and specificity as a transcription factor, suggest one class of p53 modifiers, activators or suppressor polymorphisms that may increase the levels or activity of a p53 protein that is made from a single wild type copy of DNA so that it would be equal or better than two copies of wild type p53 in a person. Mutations or polymorphisms in enzymatic activities that alter the epigenetic program might also result in greater longevity, a healthier life at older ages and resistance to the development of cancers. This provides a hypothesis that is testable in the 7% of individuals that have p53 mutations, with Li-Fraumeni syndrome, but fail to develop cancers over their lifetime. It may also be testable in centenarians that are in good health and have never developed a cancer.

Individuals that have inherited p53 mutations and develop cancers throughout their lifetimes are at risk for different tumor types at different ages of developmental time (Table 2). An examination of the tissue types susceptible to developing cancers in Li-Fraumeni patients with a $p53^{\pm}$ genotype occurs during the time of life with very active cell divisions and the activities of tissue-specific stem cells in the mesenchymal and ectodermal lineages, but not the endodermal lineage. These data link p53 functions to selected stem cell populations derived from ectodermal and mesenchymal lineages during their most active rates of cell growth and division. Here, p53 functions as a tumor suppressor by eliminating actively dividing cells that undergo stresses that create high error rates in DNA sequences, processes of cell division, faithful segregation of chromosomes or even undergo abnormal epigenetic alterations.

3 Li-Fraumeni Syndrome in Mice

The observations made in humans with a $p53^{\pm}$ genotype provide interesting associations and correlations, but the diverse genetic backgrounds of humans and different environmental exposures over a lifetime introduce quite a few variables that can influence the results and interpretations of these data sets. Even the rare occurrence of identical twins with $p53^{\pm}$ genotypes that are brought up in the same family and environment provide limited examples and environmental differences that are difficult to control. For example, a Li-Fraumeni family in Malaysia had a pair of identical twins who were confirmed to both have the same p53 mutation and almost identical DNA sequences (there are about sixty differences in nucleotide sequences mostly in non-coding regions) and were brought up in similar environments in the same family. One twin developed independent tumors at 4, 9 and 21 years of age while the other twin has not developed any tumors as of today. This difference could be an example of an environmental triggering event (exposure to X-ray DNA damage, etc.) in only one of the twins but this is not recorded in the medical records. At the minimum, we can conclude that random or stochastic events could be a determining factor in tumor initiation.

One solution to the limited number of identical twins with Li-Fraumeni Syndrome in humans is to create mice with $p53^{-/-}$ genotypes and $\pm$ genotypes from an inbred strain and who share the same cages in an animal colony over a lifetime. We can then record the frequencies of tumors, tumor types that develop and the ages of onset of the tumors. This type of study produces better statistics, permits one to vary the genetic background and even the environment such as the exposure to radiation. $p53^{-/-}$ mice of many inbred stains develop thymic lymphomas as a tumor early in life and if they survive they can develop tumors of the mesenchymal and ectodermal lineages later in life. Thymic lymphomas are never or rarely observed in human Li-Fraumeni Syndrome patients. p53 heterozygous, $p53^{\pm}$ mice develop tumors later in life with more mesenchymal and ectodermal tumors observed and fewer thymic lymphomas. If these mice are irradiated with X-rays, then tumors develop earlier in life and both leukemias, lymphomas and solid tumors arise.

A p53 male mouse homozygous for the p53 172 mutation in the C57 Bl/6 genetic background was mated to $p53^{+/+}$ females from eight different genetic backgrounds, and the type of tumors that were formed in the F-1 hybrid offspring mice were determined at autopsy by histological study analyzed by a pathologist. This experiment was carried out to determine whether different inbred stains of mice carried dominant (in the F-1 offspring) suppressors or modifiers of the tissue types that would form tumors in a $p53^{\pm}$ mouse. Does the genetic background of the host help to determine or modify the probability of tumor formation due to loss of the p53 gene function? Table 3 presents the results of this experiment. Based upon this analysis, it is quite clear that the genetic background of the host can influence the frequencies and types of tumors that arise in cells with a p53 deficiency.

Table 3 Tumor tissue types in p53$^{\pm}$ F-1 mice from eight different genetic backgrounds p53 wild type mice from the genetic background indicated in the strain column were crossed with C57Bl/6 mice of the p53$^{-/-}$ phenotype producing p53$^{\pm}$ F-1 mice. The frequency of tissue types of tumors that arose over one and a half years are indicated

Train	Lymphoid Hyperplasia Lymphoma	Osteosarcoma	Adenocarcinoma	Spindle Cell Sarcoma	Angioma/ Angiosarcoma	Squamous Carcinoma	Hemangioma	Lipoma Lipomatous Tumor/ Liposarcoma	Totals
129sl	5	3	1		1	1	1	1	13
A/J		5	1	2	1			5	14
BalbC	2	5	6	2	7	3	1		26
C3H	6	2	1	1					10
C57	8	1	4	2	1		2		18
DBA	6	1	2		1		1	1	12
NOD	18	12	2	6	1	1	2		42
SWR	21	14	8	3	3	6	1		56
Total Cases	66	43	25	16	15	11	8	7	191

The great majority of tumors have lost the wild type copy of the p53 gene in the tumor but occasionally the tumors are heterozygous for the p53 gene. There are some genetic backgrounds (C3H, C57Bl/6, DBA) where 50–60% of the mice develop thymic lymphomas, some (NOD, SWR) where 40–43% of the mice develop thymic lymphomas and others like AJ mice had no thymic lymphomas (out of 14 tumors produced) or BALB/c mice had only 4% thymic lymphomas (2/26 tumors produced). 5/7 liposarcomas produced in 191 tumors observed in these mice from diverse genetic backgrounds were produced in the AJ mice. 7/15 angiosarcomas were made in the BALB/c F-1 mice. Mice with diverse genetic backgrounds have different predispositions to develop different tumor types, and the loss of the p53 tumor suppressor gene appears to exaggerate these predispositions. In all of these mice, 66/191 tumors were thymic lymphomas (34% of the tumors formed), and this type of tumor is not observed in humans with a $p53^{\pm}$ genotype. Clearly, the species under study introduces differences we need to understand if we are to use these animals as models for the human disease.

A number of these tumors have had their DNA sequenced so that one can ask whether all thymic lymphomas have the same genes mutated when p53 mutations act as the gatekeeper independently of the genetic background or the age of the animal when the tumor developed? In $p53^{-/-}$ mice in the C57Bl/6 genetic background, all the thymic lymphomas harbored a common set of mutations (Dudgeon et al. 2014). The p53 mutation is inherited. The next mutation selected for is a homozygous loss of the PTEN gene. That is followed by gene amplifications in the cyclin D genes and the cdk6 gene or alternatively the deletion of Rb (retinoblastoma gene), which is the substrate for the cyclin D-cdk6 kinase. Finally, mutations in the notch pathway are found that compromise the differentiation of the T cells that form the lymphoma (Table 2). In the SWRxC57Bl/6 F-1 mice with a $p53^{\pm}$ genotype, thymic lymphomas can be detected as early as 4.7 months after birth or as late as 18.5 months after birth. We are sequencing these early and late thymic lymphomas to determine if the differences in the time it took for the formation of a tumor result in differences in the genes that contribute to the formation of that tumor. Similarly, we observe osteogenic sarcomas arising in SWRxC57Bl/6 F-1 mice at 11 months of age and at 19.6 months of age. We are sequencing these two tumors to examine the variables of age of tumor formation, mutation frequencies, and accumulation over a life span. These experiments permit us to study the nature of the driver mutations in tumors with diverse incubation periods in mice that are genetically identical and environmentally constant and when the tumor tissue type producing a tumor is identical.

4 Quantitation of the Clonal Evolution of Thymic Lymphomas in Mice with Mutant p53 Proteins

The thymic lymphomas produced in $p53^{-/-}$ mice in the C57Bl/6 genetic background permit us to ask a specific set of questions about the cellular clonal evolution of these tumors. During fetal life, a precursor to the T-cell lineage migrates to the thymus where the T cells will produce unique receptors by both recombination of the V-D-J genes and the insertion of nucleotides by terminal transferase to produce four different chains of the T-cell receptor dimer. Each T cell that is produced has a distinct T-cell receptor amino acid sequence in the hyper-variable region of the gene and protein. This region at the DNA level spans about 200 nucleotides in the DNA copy of a T-cell receptor and it has been possible to set PCR primers on either side of this DNA region, amplify the sequences from millions of T-cell receptors produced in a thymus in a quantitative and reproducible fashion and sequence these receptors (Dudgeon et al. 2014). This permits us to ask a number of questions: Are the thymic lymphomas that are formed several weeks after the birth of a mouse clonal, that is derived from a cell with a single T-cell receptor sequence, or are they oligoclonal? If they are oligoclonal tumors what is the frequency of transformation in the thymus? How many new cancer cells arise per day or week? Do different clones that arise then compete so that an oligoclonal tumor early in life resolves itself into a dominant clone later in life, which then kills the host? What are the nature of the additional mutations that permit some malignant cell clones to be the most fit for survival? By sequencing the DNA from heterogeneous or homogeneous tumors can we determine a frequency of point mutations in the genome? Can we determine the frequency of copy number variations or gene amplifications, deletions, inversions, insertions, aneuploidy or other types of mutations that are mediated by the absence of the p53 gene and protein? By carrying out these experiments (Dudgeon et al. 2014), it has been possible to answer some of these questions. First, mice that are born with p53 mutations in both alleles do not have evidence of a thymic lymphoma. p53 is necessary but not sufficient for the production of these tumors. By nine weeks of life, the first indication of thymic lymphomas is observed, so that additional mutations are required to produce a tumor when p53 mutations are the gatekeeper. At nine weeks, the thymic lymphomas are oligoclonal, with 2–10 different T-cell receptors being detected in a thymus. The clones continue to be produced with new receptors (de novo) and some existing clones continue to expand becoming dominant clones in the thymus and other lymphoid organs (up to 70–90% of the tumor). Between nine weeks of life as malignant clones arise and twenty weeks of life as some mice die of their disease, new clones continue to arise in the thymus at an average frequency of 0.13–0.80 clones per day (from nine weeks to twenty weeks) with that much variation between different mice of the same genetic background and the same gatekeeper mutation (Table 4).

Sequencing the DNA from these thymic lymphomas at 20 weeks (70–90% homogeneous tumors) demonstrates a point mutation frequency of single nucleotides of one per megabase. This is not extraordinarily high, being a similar frequency for human breast or pancreatic cancers. However, aneuploidy is very common; with 4–8 chromosomes per tumor with one or more aneuploidy chromosomes (chromosome 5) reproducible between tumors from different mice. There were also a large number of copy number variations (276–422 per tumor) producing deletions, amplifications, inversions, recombinations in every tumor. Up to 175 different genes or regions of chromosomes were commonly amplified in most tumors (Table 4). Chromothripsis, where a chromosome is broken into many parts

Table 4 The clonal evolution of thymic lymphomas in $p53^{-/-}$ mice, the DNA exons from three independent thymic lymphomas that arose in three different $p53^{-/-}$ mice were sequenced giving a frequency of point mutations per megabase and the number of copy number variations (CNV) per genome in the tumors. The T-cell receptors from these tumors were sequenced and the number of abnormally high levels of clonal T-cell receptors (tumor cells) detected in the tumor indicated the number of independently transformed cells that produce a tumor over the period of 9–20 weeks expressed as the transformation frequency per day. These data are from 20 different tumors

1. Frequency of point mutations	1/megabase
2. Number of CNV	276–422/genome
3. Number of transformed clones (9–20 weeks)	0.13–0.80/day

and reassembled in a more random order, was fairly common in p53 mutant cells of these tumors. It appears that the absence of p53 as a gatekeeper mutation permits a much higher frequency of copy number variations in the genome and most of the losses of tumor suppression genes arise from deletions and most of the oncogene activations occur from gene amplifications and aneuploidy. In this case, p53 is correctly named "the guardian of the genome."

5 Conclusions

Evidence is presented that cancers arise from the presence of a first or gatekeeper mutation in a stem cell or progenitor cell. The nature and functions of a gatekeeper mutation then determine the order of subsequent mutations, which are selected for by a clonal expansion. Each subsequent mutation only functions to contribute to the formation of a cancer when it is selected for by the functions of the previous mutations. Mutations may well arise in a random fashion but are only selected for by the properties of a cell that will enhance its fitness from that mutation. This helps to explain why inherited mutations in tumor suppressor genes (where they are gatekeepers) produce tumors of different tissue types, then when the same mutation in the same gene acts as a somatic mutation later in life (not a gatekeeper mutation). The order of a set of mutations plays a critical role in the tissue type, outcome and evolution of a cancer. This also helps to explain the ever more common observation that mutations in cancer driver genes are detected (by sequencing or single cell sequencing) in stem cells or progenitor cells of people that do not have a cancer. Such mutations are predispositions in the sense they are waiting to be paired with a gatekeeper mutation that permits them to be selected for in the population.

This hypothesis would seem to predict that all tumors of a particular tissue tumor type would follow a single order of selected mutations and produce homogeneous tumors, when in fact we are finding great heterogeneity between tumors of the same tissue types. This can largely be understood by the hundreds of additional and unique mutations that arise in these tumors (genomic instability, epigenomic instability). They may contribute to invasiveness, metastasis, metabolic differences, altered mutation rate, altered DNA repair, drug resistance, etc. These mutations may arise during the competition for fitness of clones that are produced by the tumor cells. But the take home message for successful treatment is our focus upon inhibiting or activating those gatekeepers and subsequently selected mutant genes that are fundamental drivers of the tumor. BRAF mutations alone produce a nevus or benign tumor. In a melanoma, BRAF mutations can be inhibited and contribute to remission of the tumor, but resistance arises and the tumor proceeds. If it is correct that most stem cells utilize the Wnt pathway for driving cell growth and creating a pool of stem cells to produce progenitor cells and stem cells are the place where gatekeeper mutations become established, then inhibitors of the Wnt pathway might be the best drugs to treat cancers. However, we need to understand how to block a Wnt pathway gene without stopping normal stem cell replication. We rely upon these stem cells for regenerating ourselves throughout life. Perhaps we can develop drugs that block cell division or promote cell death in a cell that has two mutations like p53 and PTEN (thymic lymphomas) or Wnt and RAS (colon cancers). We clearly need to better understand the fundamentals of the origins and evolution of cancers, and we need real novelty in planning the path to treatments and prevention of cancers.

As part of this challenge to the field, we have a real clue that must be followed. The single most common mutation in human cancers occurs in the p53 gene. These mutations are most often missense mutations largely localized to the DNA binding domain of this transcription factor. There are more than 220 different mutant alleles but each when measured has a similar defect, the protein only binds poorly to the DNA sequence that the wild type p53 protein binds to and each allele tested is a poorer transcription factor for a set of p53 regulated genes. These mutations are therefore loss of function mutations. Can we restore their functions or find synthetic lethal gene functions that kill cells with a p53 mutation? Many of these missense mutant p53 proteins appear to bind to other cellular proteins, and there is increasingly convincing evidence that these protein–protein interactions contribute to several gain of function phenotypes that promote growth, alter metabolism, add to drug resistance, enhance tumorigenic potential, enhance invasiveness of cancer cells and may alter the tumor types observed with different alleles of p53 mutations. If this is correct, will it be possible to develop drugs that eliminate the gain of function phenotype? Will this be a useful therapeutic approach? One would think that such a drug may have only limited or no side effects if it indeed acted to inhibit a mutant gain of function, what normal cellular function would it act upon?

Finally, if we had an efficient way to introduce the wild type p53 gene or c-DNA into all of the cancer cells in a patient, we could then kill these cells. To accomplish this, we require two new technologies we do not have. We require a delivery system (viral gene therapy, DNA or RNA, reagent or chemical) that targets cancer cells and not normal cells, and we require a targeting therapy that is efficient enough to kill all the cancer cells in the body. While to date this has not been a very productive path, there are some new ideas or approaches that should be explored. Can we activate and organize the immune system to work on these questions that require this type of specificity and selectivity and efficiency, three interesting properties of the immune system. Can we modify immune cells to recognize cancer cells but not normal cells? Could immune cells or products of immune cells be used to enhance specificity and the efficiency of killing tumor cells? Alternatively, is there a role for the microbiome to communicate with the immune system leading to tumor-specific cell killing, activation of p53 in a cancer cell or introducing wild type p53 c-DNAs? These kinds of ideas are likely to be funded and explored by private foundations that are taking a leading role in the exploration of ideas that might lead us to new technologies that in turn will move the field in new directions. This is basic science and this is needed.

Acknowledgements This work was supported by a grant from the National Cancer Institute, 5PO1 CA087497-15. The author would like to thank Drs. P. Hainaut and C. Chan for their help in the bio-informatics that was the basis for many of the ideas and conclusions expressed in this manuscript.

References

Bar-Or RL, Ruth Maya R, Segel LA, Alon U, Levine AJ, Oren M (2000) Generation of oscillations by the p53-Mdm2 feedback loop: a theoretical and experimental study. Proc Natl Acad Sci USA 97:11250–11255

Beckerman R, Prives C (2010) Transcriptional Regulation by p53. In: Levine AJ, Lane DP (eds) The p53 family. Cold Spring Harbor Laboratory Press, Cold Spring Harbor, NY, pp 63–80

DeLeo AB, Jay G, Appella E, Dubois GC, Law LW, Old LJ (1979) Detection of a transformation-related antigen in chemically induced sarcomas and other transformed cells of the mouse. Proc Natl Acad Sci USA 76:2420–2424

Donehower LA (1996) The p53-deficient mouse: a model for basic and applied cancer studies. Semin Cancer Biol 7:269–278

Dudgeon C, Chan C, Kang W, Sun Y, Emerson R, Robins H, Levine AJ (2014) The evolution of thymic lymphomas in p53 knockout mice. Genes Dev 28:2613–2620

El-Deiry WS, Kern SE, Pietenpol JA, Kinzler KW, Vogelstein B (1992) Definition of a consensus binding site for p53. Nat Genet 1:45–49

Eliyahu D, Raz A, Gross P, Givol D, Oren M (1984) Participation of p53 cellular tumour antigen in transformation of normal embryonic cells. Nature 312:646–649

Fearon ER, Vogelstein B (1990) A genetic model for colorectal tumorigenesis. Cell 61:759–767

Finlay CA, Hinds PW, Levine AJ (1989) The p53 proto–oncogene can act as a suppressor of transformation. Cell 57:1083–1093

Hainaut P, Pfeifer GP (2016) Somatic TP 53 mutations in the era of genome sequencing. In: Lozano G, Levine AJ (eds) The p53 protein from cell regulation to cancer. Cold Spring Harbor Laboratory Press, Cold Spring Harbor, NY, pp 279–300

Hinds P, Finlay C, Levine AJ (1989) Mutation is required to activate the p53 gene for cooperation with the ras oncogene and transformation. J Virol 63:739–746

Kastan MB, Onyekwere O, Sidransky D, Vogelstein B, Craig RW (1991) Participation of p53 protein in the cellular response to DNA damage. Cancer Res 51:6304–6311

Kress M, May E, Cassingena R, May P (1979) Simian virus 40-transformed cells express new species of proteins precipitable by anti-simian virus 40 tumor serum. J Virol 31:472–483

Kussie PH, Gorina S, Marechal V, Elenbaas B, Moreau J, Levine AJ, Pavletich NP (1996) Structure of MDM2 oncoprotein bound to the p53 tumor suppressor transactivation domain. Science 274:948–953

Lahav G, Rosenfeld N, Sigal A, Geva-Zatorsky N, Levine AJ, Elowitz MB, Alon U (2004) Dynamics of the p53-Mdm2 feedback loop in individual cells. Nat Genet 36:147–150

Lane DP, Crawford LV (1979) T antigen is bound to a host protein in SV 40-transformed cells. Nature 278:261–263

Linzer DIH, Levine AJ (1979) Characterization of a 54,000 MW cellular SV40 tumor antigen present in SV40–transformed cells and uninfected embryonal carcinoma cells. Cell 17:43–52

Maltzman W, Czyzyk L (1984) UV irradiation stimulates levels of p53 cellular tumor antigen in nontransformed mouse cells. Mol Cell Biol 4:1689–1694

Momand J, Zambetti GP, Olson DC, George D, Levine AJ (1992) The mdm–2 oncogene product forms a complex with the p53 protein and inhibits p53 mediated transactivation. Cell 69:1237–1245

Nigro JM, Baker SJ, Preisinger AC, Jessup JM, Hostetter R, Cleary K, Bigner SH, Davidson N, Baylin S, Devilee P et al (1989) Mutations in the p53 gene occur in diverse human tumour types. Nature 342:705–708

Oliveri M, Hollstein M, Hainaut P (2010) TP 53 in human cancers: origins, consequences, and clinical use. In: Levine AJ, Lane DP (eds) The p53 family. Cold Spring Harbor Laboratory Press, Cold Spring Harbor, NY, pp 63–80

Oren M, Levine AJ (1981) Immunoselection of Simian Virus 40 large T antigen messenger RNAs from transformed cells. Virology 113:790–793

Reich NC, Levine AJ (1984) Growth regulation of a cellular tumor antigen, p53, in nontransformed cells. Nature 308:199–201

Riley T, Sontag E, Chen P, Levine AJ (2008) Transcriptional control of human p53-regulated genes. Nat Rev Molec Cell Biol 9:402–412

Sarnow P, Ho YS, Williams J, Levine AJ (1982) Adenovirus E1b-58kd tumor antigen and SV40 large tumor antigen are physically associated with the same 54 kd cellular protein in transformed cells. Cell 28:387–394

Scheffner M, Werness BA, Huibregtse JM, Levine AJ, Howley PM (1990) The E6 oncoprotein encoded by human papillomavirus 16 or 18 promotes the degradation of p53. Cell 63: 1129–1136

Takeda H, Wei Z, Koso H, Rust AG, Yew CC, Mann MB, Ward JM, Adams DJ, Copeland NG, Jenkins NA (2015) Transposon mutagenesis identifies genes and evolutionary forces driving gastrointestinal tract tumor progression. Nat Genet 47:142–150

Vu B, Wovkulich P, Pizzolato G, Lovey A, Ding Q, Jiang N, Liu JJ, Zhao C, Glenn K, Wen Y, Tovar C, Packman K, Vassilev L, Graves B (2013) Discovery of RG7112: a small-molecule MDM2 inhibitor in clinical development. ACS Med Chem Lett 4:466–469

Werness BA, Levine AJ, Howley PM (1990) Association of human papillomavirus types 16 and 18 E6 proteins with p53. Science 248:76–79

Wolf D, Rotter V (1984) Inactivation of p53 gene expression by an insertion of Maloney murine leukemia virus-like DNA sequences. Mol Cell Biol 4:1402–1410

Wu XW, Bayle HJ, Olson D, Levine AJ (1993) The p53–mdm–2 autoregulatory feedback loop. Genes Develop 7:1126–1132

Zambetti GP, Labow M, Levine AJ (1992) A mutant p53 protein is required for the maintenance of the transformed cell phenotype in p53 plus *ras* transformed cells. Proc Natl Acad Sci USA 89:3952–3956

A Personal Note As an undergraduate at The State University of New York at Binghamton, Harpur College, I took a course in microbiology and became fascinated with viruses. Viruses were too small to observe in a light microscope, were obligate intracellular parasites and yet reproduced themselves in an infected cell. How much genetic information did it take to reproduce oneself? Are viruses the smallest and simplest form of life? If so, would the study of viruses tell us what it was about life, life processes and reproduction of self that would lead us to the fundamental principles of biology. In the 1960s, I was a graduate student at The University of Pennsylvania and a postdoctoral fellow at The California Institute of Technology and worked with viruses (adenoviruses, SV40 and bacteriophages). That is when I learned it might well take a lifetime to answer some of these questions. I also was exposed to many scientists in the literature, at meetings and in courses taught at Cold Spring Harbor, Aspen Colorado, Woods Hole Massachusetts, and Bar Harbor Maine. The 1960s were a time for virus isolation, classification, uncovering the styles of virus replication and the role of the host cell. Among the leaders in describing these events in the viruses of chickens was Peter Vogt. His clarity, organization, attention to detail and insights set the standard for the field. By the 1970s, he laid the groundwork and actively contributed to the major events and publications that changed the field, providing the evidence that RNA tumor viruses, the retroviruses, contained oncogenes that were the mutant forms of cellular genes. By contrast, the DNA tumor viruses that I worked with contained their own oncogenes. Why? While the paths of the RNA and DNA virologists began to separate, Peter continually provided direction to virology and virologists that pushed both fields to explore common questions and new ideas. Peter always saw, discussed and explored those common themes that led us to answer those fundamental processes of what it took to reproduce oneself. At the end of the 1970s, we used those DNA tumor viruses as a tool to uncover a cellular protein, p53, which was an anti-oncogene or a tumor suppressor gene. It is clear that the tumor viruses played an important role in uncovering our present day understanding of human cancers.

In addition to his scientific contributions, Peter contributed to the development of the field of virology so that it had outlets to communicate, so that it had standards to aspire to, so that it moved toward original directions guided by each generation of leaders of the field. His service to the community of virologists assured productivity, a style of leadership that was inclusive and soft, uncovering new questions and directions that guided the field, and always holding to personal and professional standards everyone admired and aspired to fulfill. This issue of Current Topics in Microbiology and Immunology is a fitting tribute to Peter Vogt and his continuing and enormous contribution to our field. Thank you Peter.